D1173800

THE
MICHELSON ERA
IN AMERICAN
SCIENCE
1870–1930

AIP
CONFERENCE
PROCEEDINGS 179

RITA G. LERNER
SERIES EDITOR

THE
MICHELSON ERA
IN AMERICAN
SCIENCE
1870–1930

CLEVELAND, OH 1987

EDITORS:

STANLEY GOLDBERG
UNIVERSITY OF MARYLAND

ROGER H. STUEWER
UNIVERSITY OF MINNESOTA

AMERICAN INSTITUTE OF PHYSICS   NEW YORK 1988

L.C. Catalog Card No. 88-83369
ISBN 0-88318-379-X
DOE CONF 8710389

Printed in the United States of America.

CONTENTS

PART IV. AMERICAN SCIENCE IN THE AGE OF
MICHELSON AND BEYOND

PREFACE

The articles in this volume relate, in one way or another, to a single experiment: the Michelson-Morley ether-drift experiment of 1887. About one-half of the articles in the collection are based on papers delivered at a two-day symposium, "The Michelson Era in American Science: 1870–1930," held at Case Western Reserve University in Cleveland, Ohio, 28–29 October 1987, commemorating the centennial of the experiment.

The Michelson-Morley experiment was designed to detect the motion of the earth through the luminiferous ether. At the time, physicists did not question the existence of the ether. The experiment, like many physical probes of nature, was simple in conception, yet uncompromisingly demanding of the art of contemporary craftsmanship. The state of the mechanical arts and its relationship to the pursuit of physical science in late nineteenth-century America is well documented in Section I, which also addresses the more general question of the link between state-of-the-art shop practice and engineering theory.

In spite of the loving care that Michelson and Morley devoted to building their apparatus, and in spite of the thought and concern that went into collecting their data, the upshot was that no evidence of the motion of the earth relative to the supposedly ubiquitous medium was found. Immediate interpretations of this null result did not question the existence of the ether. The ether was as necessary to late nineteenth-century conceptions of physical reality as quarks have become for those of the late twentieth century.

At first, natural philosophers concentrated on the question of how matter and ether must interact to account for the null results of the Michelson-Morley and other, subsequent, second-order experiments. Only later, after the introduction of the theory of relativity, was the experiment widely cited as evidence showing that the ether was an extraneous concept in physics. This did not mean that proponents of the ether suddenly disappeared; it meant that the climate of opinion among physicists regarding the basic constructs underlying physical reality had shifted.

The extent to which the Michelson-Morley experiment as such contributed to this shift is still unclear. Many laymen and physicists believe that the Michelson-Morley experiment was a crucial experiment in the classical sense—an experiment that forced a shift away from ether theories and one that had an immediate and direct influence on Albert Einstein. This belief is not shared by many historians and philosophers of science. In fact, the role of the Michelson-Morley experiment in the rejection of ether theories and in the creation of the special theory of relativity has been actively debated for at least the last 30 years. This is one of the issues addressed by the articles in Section II, where questions of the historical and philosophical significance of the experiment are considered. While readers will not find a consensus among the authors, they will find expressions of the range of opinions that have controlled and directed the course of this ongoing debate.

There have been many instances of so-called crucial experiments in the history of science, but none have enjoyed the continuing renown of the Michelson-Morley experiment. In part this is because of its close association—conceptually, not historically—with the special theory of relativity and its creator, Albert Einstein, who has come to represent the archetype of the creative scientist. In part, however, it is also because of the reputation and stature of Albert Abraham Michelson. Michelson's six-decade career in experimental physics began in the 1870s and spanned the period in which American science, including American physics, approached a position of parity with European physics. Aspects of Michelson's work and of the changing character of the problems dominating physics on both sides of the Atlantic during the period in which he was active are considered in Section III.

Since World War II, many Americans have won the Nobel Prize in physics. Albert Michelson was the first American to be so honored, and his achievement, in 1907, not only enhanced the reputation of American science, it also assured Michelson a place as one of America's foremost physicists. The articles in Section IV examine the changing character of American science during Michelson's scientific lifetime and in subsequent decades. Among the topics considered are the internal functioning of American social institutions in science and the deliberations that resulted in nominating Americans for Nobel Prizes in science.

***

With regard to the symposium that formed the basis for this volume, special thanks go to Philip L. Taylor, Professor of Physics at Case Western Reserve University, who served as Executive Director of the Michelson-Morley Centennial Celebration and to Alan J. Rocke, Associate Professor of the History of Science and Technology at Case Western Reserve, who served as chairman of the organizing committee for the symposium. Other members of the organizing committee, besides the editors of this volume, were Professor Martin J. Klein (Yale University) and Professor Robert Rynasiewicz (Johns Hopkins University). The symposium and this volume have been made possible entirely through the generous support of Mr. and Mrs. James P. Storer and the General Dynamics Corporation.

When the participants in the symposium gathered in Cleveland at the end of October 1987, the city's celebration of the centenary of the Michelson-Morley experiment already had been in full swing for six months. It had elicited the participation of thirteen of Cleveland's educational and cultural institutions in an extraordinary cooperative enterprise, and a total of 131 speakers, including 17 Nobel laureates, participated. Among the celebratory events were symposia on the legacy of Michelson and Morley to modern physics, on the work of Edward W. Morley, and on science, arts, and the humanities, as well as historical exhibits on the experiment, student competitions, exhibits and planetarium programs exploring the physics of light and color, major art exhibits at The Cleveland Museum of Art, The Cleveland Institute of Art, The Cleveland Center for Contemporary Art, and The Cleveland Museum of Natural History. Musical events included a concert by The Cleveland Orchestra that featured a specially commissioned piece by Philip Glass, and a concert by the Cleveland Institute of Music Chamber Orchestra featuring commissioned pieces by Karel Husa and by Donald Erb. This list is by no means exhaustive.

Visitors to Cleveland's University Circle during 1987 were perhaps astonished and bemused to find the area festooned with brightly colored banners proclaiming the theme of the celebration: "Light, Space and Time." Local vendors sold tee shirts and other memorabilia depicting, in bright pink, the well-known schematic of the Michelson-Morley interferometer. It was a celebration worthy of the occasion being celebrated.

Such an imaginative and wide-ranging program required time, energy, and money. But perhaps more than anything, it required organization. One can only imagine the labyrinth of committees responsible for overseeing each of the individual events. Such a grand plan required a planner—a person who had the vision to see the celebration in its totality and who could inspire and command supporters and participants. That person was Dorothy Humel Hovorka, Case Western Reserve University Trustee. Mrs. Hovorka oversaw the creation of the Michelson-Morley Centennial Corporation and served as the Corporation's Chairman. Those of us who had the privilege of working with her know how much the scope and depth of the celebration depended upon Mrs. Hovorka's vision, her seemingly unfailing store of energy, and her ability to infuse others with her enthusiasm and commitment to the project. She is a wonderful whirlwind who leaves order and purpose in her wake.

Stanley Goldberg
University of Maryland,
Baltimore County

Roger H. Stuewer
University of Minnesota

September 1988

I

The Crafting of Great Experiments

PRECISION ENGINEERING AND EXPERIMENTAL PHYSICS: WILLIAM A. ROGERS, THE FIRST ACADEMIC MECHANICIAN IN THE U.S.

Chris J. Evans and Deborah Jean Warner

PRECISION ENGINEERS AND ACADEMIC MECHANICIANS

Albert A. Michelson thought of himself as a physicist, and that is the context within which he is usually considered. Michelson might also be viewed as an academic mechanician. We coin this hybrid term to describe those men who applied science to engineering and engineering to science. Unlike skilled machinists who used intuition and craft skill to produce good designs, academic mechanicians had an explicit understanding of the physical principles behind good machine design. Like the so-called scientific mechanicians, who were able to apply their skills to a wide variety of industrial challenges,[1] academic mechanicians applied their understanding to a variety of scientific and industrial problems requiring high-precision machine design. The academic mechanician was a particularly American phenomenon. He bridged the gap between science and engineering, and had often been trained in both disciplines. Academic mechanicians were familiar with one another's work, and they often built on one another's achievements. They also maintained close connections with engineering institutions, often relying on practical mechanics to work out the details of, build, and operate their innovative machines.

There has been a tendency to treat Michelson as a heroic experimenter working on his own.[2] It is clear, however, that the major areas in which Michelson worked embroiled him in problems that today we recognize as those of precision engineering.[3] While Michelson may have invented the interferometer, his interferometric measurement techniques were based in large part on Fizeau's 1864 interferometric dilatometer and Jamin's refractometers. Similarly, in developing engines for ruling diffraction gratings, Michelson was well aware of the extraordinary attention that had to be paid to such nontrivial mechanical problems as the elimination of vibration, compensation for self-loaded deformations and thermal expansion, and generation of exactly repeatable small increments of movement.

The first important community of precision engineers had emerged in London around the turn of the 19th century. Jesse Ramsden and his successors were joined by Troughton, Simms, Donkin, Maudslay, Allan, and Ross. Although constrained to earn their living as instrument makers and machine tool builders, these men (and Barton, who worked at the Mint) established a culture of fine mechanics and precision measurement. A similar community emerged in Paris around mid-century, with men such as Froment, Perreaux, and Gambey. De la Rive and Thury founded the Societé Genevoise d'Instrumente de Physique in the early 1860s, and Zeiss and Abbe began working together at Jena soon thereafter.

By the 1840s the U.S. boasted several precision engineers. Some, such as Joseph Brown and Samuel Darling, came out of an indigenous craft tradition; and some, such as Joseph Saxton and William Wurdemann, had been trained at least in part in Europe. A community of academic mechanicians emerged in the period after the Civil War. Working at the interface of physics and precision engineering, they combined an appreciation of the questions of science with an understanding of problems and possibilities of precision machine design. The three outstanding academic mechanicians -- William A. Rogers, Henry A. Rowland, and Albert A. Michelson -- were also the most notable experimental physicists working in the U.S. at that time.[4] All three developed engines for ruling diffraction gratings, the first precision instrument "designed for scientific research and made in America that was truly superior to the European product."[5] In this paper the work and influence of the lesser known Rogers is explored.

Figure 1
William A. Rogers

WILLIAM A. ROGERS (1832-98)

Rogers was an active and respected scientist.[6] A member of the American Association for the Advancement of Science, he regularly gave three or four talks at each meeting between 1880 and 1896. He served as president, vice-president and, when Michelson was president, on the committee for Section B. He was elected a fellow of the American Academy of Arts and Sciences in 1873 and a member of the National Academy of Sciences in 1885. He was also president of the American Microscopical Society and an honorary fellow of the Royal Microscopical Society.

Rogers was also an engineer whose designs and publications demonstrated an excellent fundamental understanding of fine mechanism and precision instrumentation. He built comparators, ruling engines, and linear and circular dividing engines. He recognized the limitations posed by the elasticity of materials and seems to have understood concepts and constraints such as long-term dimensional instability and thermal resonance that many modern machine designers still ignore. His intercomparisons between the Imperial Yard and the meter fostered interest in basic measurement in both science and industry. The practical length standards he developed led to the claim that:

> Pratt & Whitney established the inch. The standard was accurate to millionths of an inch.[7] Fortunately for Pratt & Whitney, there was more truth in their gauges than their advertising copy, and for that, as we will show, they had to thank Rogers.

Rogers was born at Waterford, Connecticut, in 1832, and brought up in New York state. He attended Alfred Academy, obtained an M.A. from Brown University in 1857, and then returned to Alfred as a teacher in the academy. He was appointed professor of mathematics and astronomy at Alfred University in 1859 and remained there until 1870. This period at Alfred, which has been described as "years of activity and preparation rather than of scientific productivity," encompassed the Civil War, when Rogers spent fourteen months in the Navy. Rogers established an observatory at Alfred[8] and spent two long stints at the Harvard College Observatory. He also spent nearly a year studying industrial mechanics at Yale's Sheffield Scientific School before becoming professor of industrial mechanics at Alfred. Rowland received a practical education at Rensselaer Polytechnic Institute, but Michelson at the Naval Academy apparently did not.

Rogers's most productive years began in 1870 when he resigned from Alfred[9] to

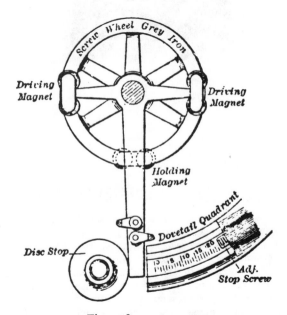

Figure 2
W. A. Rogers's electromagnetic leadscrew advance mechanism incorporating a profiled cam to compensate for periodic errors.

become an Assistant at the Harvard College Observatory in charge of the newly erected 8-inch meridional circle. Rogers soon ran into difficulties obtaining adequate spider lines for the eyepiece micrometer. That turned his attention to methods of producing graticules and fine lines, a theme that underlies a significant part of his subsequent work.

For Rogers it was but a small step from ruling graticules to ruling diffraction gratings, and over the next several years he developed a series of ruling engines that were substantially better than previous ones. Rogers established the basic concepts but did not actually make these engines. In one instance he relied on Charles Van der Woerd, a superintendent at the Waltham Watch Co., for the "detailed arrangement" and supervision of manufacture. In another the work was done by the machinist George F. Ballou.

The ruling engines of Rogers, and later those of Rowland and Michelson, employed a high-precision leadscrew to move the cutter from one groove to the next. Traditional methods of producing leadscrews often introduced a number of types of errors, the most objectionable of which were periodic. To reduce these errors Rogers developed an improved method of lapping screws and introduced the use of the "free nut." Both of these techniques were taken up and further improved by Rowland, and contributed to the success of his ruling engine.

Rogers also tackled the ratchet and pawl on the screw head commonly used to advance the ruling carriage from one groove to the next. He recognized that the errors of the gear cutter that produced the ratchet transfer one-for-one as periodic errors in the grating or scale. Rather than attempt to correct these errors, he developed an electromagnetic advance mechanism that alternately "locked" and "unlocked" a lever arm that moved through an adjustable arc (Fig. 2). Besides avoiding the periodic errors arising from the ratchet, this device allowed Rogers to rule lines at, essentially, any spacing and to incorporate a cam as one stop. This cam can be profiled to compensate for centering errors and for periodic errors of the screw itself. Rogers later used this elegant advance mechanism in his circular dividing engines and microtomes.

Rogers was also involved in the introduction into the U.S. of the leadscrew corrector bar--a device which can compensate for the full gamut of repeatable machine errors. The general form of this important mechanism had first been used by Donkin, in 1826, who had adapted an idea of Maudslay's for correcting linear screw errors. Donkin used a series of small steps to vary the correction along the length of the leadscrew. The idea was taken one step further by Thury in his 1865 linear dividing engine at Societe Genevoise.[10] By the turn of the century all SIP dividing engines (which appear to have been fairly common in American colleges) incorporated such a corrector. Rogers stated his intention to use such a corrector in his second ruling engine, but since this does not still exist, we cannot know for certain if he did. It is certain, however, that he used a leadscrew corrector bar on a screw cutting lathe he built in the early 1890s, and described in several publications. Michelson used a similar device on one of his diffraction grating ruling engines.

In parallel with the development of his ruling engines, Rogers undertook a study of interactions between the diamond cutter and the glass surfaces being ruled. (In the early stages they were glass surfaces; later he ruled speculum metal.) He found the variation of hardness with crystal orientation, tried tools made simply by cleaving, developed his own diamond polishing expertise, and then experimented with a range of different tool types. In other work, he encountered materials problems -- arising both from lack of homogeneity and from residual stresses in the glass -- and related the various types of chip formed to the characteristics of the resulting line. Besides ruling glass directly, Rogers developed his own "resist," which he ruled prior to etching with hydrofluoric acid. He also worked extensively with the microscopy community, developing resolution targets and a "standard" centimeter traceable to international standards.

Rogers's ruling engines were very good. Moreover, although details are still lacking, it is clear that they influenced the design of the still better engines of Rowland and Michelson. We know, for instance, that Rowland was able to study a Rogers engine, that Rogers sold Rowland diamond ruling tools, that he calibrated length standards for him, and that he identified a screw error in Rowland's engine from measurements of his rulings.[11] Rogers's influence on Michelson is inferred from similarities in the design of their ruling engines. Certainly the two men could talk together about ruling engines while serving together on the same AAAS committee.

Rogers's ruling engines ushered in the "classical period" of diffraction gratings.[12] From that point on, Americans dominated the ruling art, at least until the end of the 1960s, and arguably still do. The tradition of leadership by academic mechanicians -- the Babcocks at Mount Wilson, Anderson and Strong at Johns Hopkins, and Harrison and Stroke at MIT -- has also remained.

By the mid-1880s Rogers had turned his full attention, and his dividing and ruling expertise, to the problems of standards of length. He had become aware of inadequacies in length standards in the course of his work on standard scales for microscopy.[13] Specifically, he perceived that there was no well-defined relationship between yard and meter standards then in use in the U.S.

With the encouragement of the Rumford Committee of the American Academy of Arts and Sciences, Rogers was able to obtain an authorized copy of the then U.S. standard, Bronze 11, and a centimeter scale from Brunner Freres.[14] Although Brunner Freres was one of the premier instrument makers in Paris, Rogers believed that the only "true" standards were Bronze 1, the British national standard yard, and the Metre des Archives in Paris. Thus in late 1879, with an appealing directness, he travelled to London, where the Warden of the Standards made a "transfer" from Bronze 1. In modern terminology, Rogers had his standard calibrated. From London, Rogers went on to Paris. Circumstances precluded his obtaining a "transfer" direct from the Metre des Archives, but Tresca, at the Conservatoire Nationale des Arts et Metiers, agreed to calibrate Rogers's standard by comparison with another metre, No. 19, the relation of which to the Metre des Archives had previously been well characterized.

Armed with "authorized" standards, Rogers set about intercomparing these and other bars. His first comparator was built with the imperfect facilities at Harvard. Although "of the most simple form" -- a pair of micrometer microscopes on a moving carriage -- it gave Rogers valuable experience of the effects of moving loads on the deformation of structures and of angular motion in systems that do not comply with the Abbe principle.[15] Money for a better comparator was to come not from science, but from industry.

By the 1870s the lack of screw thread standardization had begun to cause severe problems in railway repair shops. Nominally standard nuts produced in, say, Philadelphia, would not fit bolts made in New York. By the end of the decade matters had become so bad that the Master Car Builders' Association engaged Pratt & Whitney, a prominent tool making firm, to furnish standard United States, or 'Franklin Institute' thread screw gauges.[16] Pratt & Whitney in turn sought out William A. Rogers. Pratt & Whitney wanted "absolute" traceability to the Imperial Yard. Rogers wanted a new comparator. The immediate result was what came to be known as the Rogers-Bond comparator. (Fig. 3).

Rogers developed the conceptual design for the comparator, but the detailed design was worked out by George Meade Bond (1852-1935), an experienced machinist then in his final year of a mechanical engineering course at the Stevens Institute of Technology.[17] Following graduation Bond was hired by Pratt & Whitney. As manager of their standards and gauge department, a post he held until 1902, Bond built a highly respected line of instruments, examples of which are still to be found in gauge laboratories.

Rogers's requirements for the comparator -- presumably influenced by his studies of comparators in England and France, and at the U.S. Coast Survey -- included "the complete separation of the standards to be compared from the frame-work to which the microscopes are attached."[18] This requirement was met by mounting the microscopes on a frame which was mounted on granite-capped brick piers which are totally separated from the carriage that supports the standards. Other requirements included the "easy comparison of line and end measures," an "invariable reference plane," and "mechanism that will allow easy comparison of the sub-divisions of any unit."

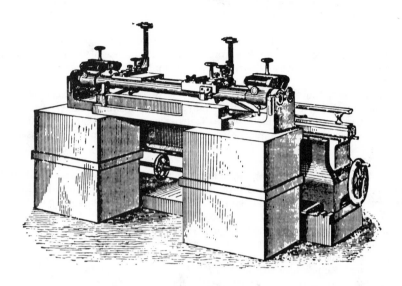

Figure 3a
Original Rogers-Bond comparator

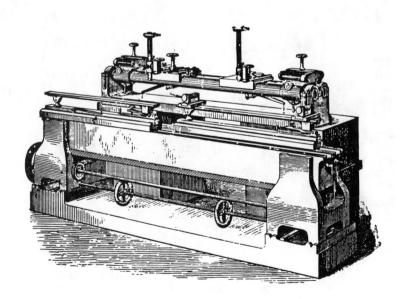

Figure 3b
Modified Rogers-Bond comparator

The Rogers-Bond comparator incorporated many noteworthy features. One was the reversal designed to check for horizontal straightness errors in the comparator. Using a tracelet carriage riding on the ways, Rogers traced a line on the top polished surface of a properly supported bar; the bar was then reversed and aligned so that the tracelet made a mark at precisely the same (small) distance from each end of the original line. Another line was ruled and any variation in the gap between the lines indicated straightness errors.[19]

For those measurements in which the microscope carriages must be moved, the comparator is provided with a counterweight system to eliminate the effects of deformation of the rails with varying loads. Electro-magnetic stops are provided to ensure repeatable movements of the microscope carriages. For levelling, Rogers used a shallow pan of mercury, making adjustments so that the surface was perfectly in focus at any point along the ways.

By May 1881 two comparators had been built. One was installed at Pratt & Whitney, which by this time had hired Bond, and the other was installed in a specially constructed "comparing room" at Harvard. Rogers and Bond obtained a patent (#332,110) in 1885 on a simplified form of the comparator which has far less flexibility in operation. One of these comparators, used for many years by Pratt & Whitney, is now in the collections of the National Museum of American History (Fig. 4). Rogers also marketed ruling engines and dividing engines, and he worked with the machinist George Ballou, who marketed a comparator similar to the Rogers-Bond design.

In 1886, enticed by money sufficient to build a new comparing room said to be "the most complete and convenient which at that time had been designed," Rogers accepted

Figure 4
Simplified, patent model Rogers-Bond comparator
(Smithsonian photo # 73-6437)

the chair of physics at Colby College in Waterville, Maine. Here he built a 100-inch comparator, which he used to supply standards to many universities and other laboratories, and to test scales produced by others. He worked on improving the accuracy of thermometry for standards work. He measured a number of Rowland's gratings and investigated the thermal characteristics of the speculum Rowland ruled.

Rogers's interest in thermal effects in metrology led to further collaboration with Edward Morley. In 1879 the two men had worked together on metrological applications of microscopy. In the 1890s they worked on the application of interferometry to length measurement. While Michelson in Paris was using interferometry to measure the absolute length of the standard meter bar, Rogers and Morley developed Morley's original idea of using an interferometer in a comparator.

Rejecting the use of dilatometers such as Fizeau's that were limited to small samples, Rogers and Morley aimed to accommodate entire meter bars. Although they were never explicit about their reasons, they may well have recognized that the coefficient of linear expansion obtained from a small sample in a dilatometer might not give a true picture of the thermal behavior of the total bar. Their work in vacuum suggests that they recognized the difficulties posed by air path temperature and pressure fluctuations.

Their basic approach was to spring load mirrors against both ends of both bars, using a Michelson interferometer to measure the relative movement of the "near" and "far" ends of the bars. Rogers built the apparatus at Colby. Morley spent his summers there. Five years of experimental misery followed. Experiments in air, with one bar packed in ice and the other heated, were impossible due to condensation on the optics. They then decided to work in vacuum. In their new apparatus the bars to be compared were housed in rolled brass boxes, but they were unable to hold the vacuum. Even today design and construction of a comparator of this sort would not have been undertaken lightly. For their third system (Fig. 5):

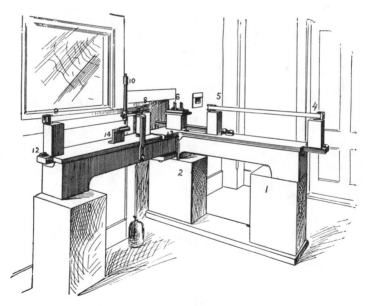

Figure 5
The Rogers-Morley Interferometric
Comparator and Dilatometer

> The two brass boxes and the box for receiving the diagonal mirrors were cast as one piece, which weighed about 500 lbs. Every part of this casting was covered on the outside with solder to the depth of one-eighth of an inch...with an ordinary soldering iron...the boxes, with their connections are nearly 60 inches in length and 6 x 6 inches in cross-section. One hundred and fifty pounds of solder were required.[20]

By summer of 1895, the apparatus was ready for the "final" experiments. Further experimental problems followed, however, and Rogers's health was deteriorating rapidly. Except for some data on Jessop's steel, no results from this research were published.

In the fall of 1897 Rogers tendered his resignation, effective on 1 April 1898, intending to return to Alfred University, where a new physics laboratory had been built to his specifications, and equipped with $10,000 worth of apparatus which he had accumulated during his career. Rogers's departure was not prompted by any desire to leave Colby, but by the knowledge that he was dying. He was an ardent, life-long member of the Seventh Day Baptists, and wished to spend his final years among "fellow believers."[21] He died, however, on 1 March 1898.

TECHNOLOGICAL CONVERGENCE AND INVENTOR DIVERGENCE

The phrase "technological convergence" was coined by Nathan Rosenberg to explain the rapid growth of the American machine tool industry in the nineteenth century.[22] Dissatisfied with accepted economic models, Rosenberg argued that, in the early period, there was a class of mechanicians who would build just about anything. Thus developments and new ideas in one field could be applied, almost immediately, in another. What was true in industry was equally true in the realm of precise scientific instrumentation. Dividing machines, lathes, telescopes and comparators relied on the same essential elements. They all had leadscrews, bearings, and slideways. They were technologically convergent.

William A. Rogers, who applied a fundamental understanding of physics, materials science, and design to a variety of problems, well exemplifies technological convergence. He also exemplifies what has been termed "inventor divergence."[23] He developed great competence in the arts of fine line production that he applied to such diverse products as diffraction gratings, resolution targets for microscopy, eyepiece micrometers, and scales. He also applied widely his expertise in high precision measurement, developing and calibrating micrometric standards for microscopy, supplying line standards to numerous laboratories, and providing the scientific basis for the Pratt & Whitney gauges and measuring machines.

NOTES

1. B. Sinclair, *Philadelphia's Philosopher Mechanics: A History of the Franklin Institute, 1824-1865* (Baltimore: 1974).

2. L. S. Swenson, "Michelson and Measurement," *Physics Today*, May 1987:24-30. D. M. Livingston, *The Master of Light* (New York: 1979).

3. C. Evans, "Precision Engineering -- An Evolutionary Perspective" (M.Sc. thesis, Cranfield Institute of Technology, UK, 1987); R. V. Jones, "Precision Engineering: From Supertankers to Integrated Circuits (Comment)" *Precision Engineering* 1 (Jan. 1979).

4. L. M. Rutherfurd does not fit conveniently into either category. See, for example, D. J. Warner, "Lewis M. Rutherfurd: Pioneer Astronomical Photographer and Spectroscopist," *Technology and Culture*, 1971, *12*:190-216.

5. D. J. Warner, "Rowland's Gratings: Contemporary Technology," *Vistas in Astronomy*, 1986, *29*:125-30.

6. E. W. Morley, "Memoir of William Augustus Rogers, 1832-1898," *Biographical Memoirs of the National Academy of Sciences*, Vol. 4 (1902), pp. 187-199, and A. Searle, "Biographical Memoir of William A. Rogers. Part II. Astronomical Work," ibid., Vol. 6, pp. 109-117. See also *Science*, 1 April 1898, *7*:447-50; *Physical Review*, 1898, *4*:315-19; *Dictionary of American Biography*, 1935, pp. 114-15; *The National Cyclopaedia of American Biography*, vol. 9 (1967), pp. 530-31; *Appleton's Cyclopaedia of American Biography*, Vol. 5, pp. 310-11 (1894); *Transactions of the American Society of Mechanical Engineers*, 1898, *19*:968-972.

7. *Accuracy for 70 years, 1860-1930* (Hartford, Conn.: Pratt & Whitney Co., 1930).

8. Equipped with Fitz and Young telescopes, a Bond chronometer, and appropriate micrometers. See J. L. Stull, "A Moment in Time: History of the Alfred University Observatory," in *Alfred University Sequecentennial History* (Alfred, N.Y.: 1986).

9. C. F. Randolph, "President Allen and William A. Rogers," 1943, ms. in the Alfred University Archive. This suggests that strained relations with Allen, president of the University, helped push Rogers toward the post at Harvard.

10. See Evans, "Precision Engineering," for a detailed discussion of corrector mechanisms.

11. The Rowland papers (Special Collections, Milton S. Eisenhower Library, Johns Hopkins University) contain some twenty letters from Rogers written between 1883 and 1893. See also W. A. Rogers, "On the Determination of the Absolute Length of Eight Rowland Gratings," *Proceedings of the American Society of Microscopists*, 1885, *7*:151-177.

12. C. Evans, "Precision Engineering"; G. R. Harrison, "The Production of Diffraction Gratings, I. Development of the Ruling Art," *Journal of the Optical Society of America*, June 1949, *39*:413-26.

13. R. H. Ward, "History of the National Committee on Micrometry," *Proceedings of the American Society of Microscopists*, 1883, *6*; W. A. Rogers, "A Practical Method of Securing Copies of the Standard Centimeter Designated 'Scale A'," *Proceedings of the American Society of Microscopists* (1889); R. H. Ward, "Annual Address of President," *Proceedings of the National Microscopical Congress* (1878) and of the *American Society of Microscopists* (1879).

14. W. A. Rogers, "On the Present State of the Question of Standards of Length," *Proceedings of the American Academy of Arts and Sciences*, 1880, *15*:273-312. W. A. Rogers, "Studies in Metrology," *Proceedings of the American Academy of Arts and Sciences*, 1883, *18*:287-398.

15. Rogers, "Studies in Metrology," p. 299. Rogers would not have called it the Abbe principle, since Abbe did not publish the idea until 1892.

16. G. M. Bond, "Standard Measurements," *Trans. American Society of Mechanical Engineers*, 1881, *2*:81-92.

17. C. C. Tyler in *Transactions of the American Society of Civil Engineers*, 1935, *100*; and B. H. Blood in *Trans. American Society of Mechanical Engineers*, 1936, *58*.

18. Rogers, "Studies in Metrology."

19. There is a class of theoretically possible error (with a symmetric wave form and a wavelength exactly equal to the tracing length) which this test will not detect; repeating the test with one different end point will, however, indicate those errors also, although there is no evidence that Rogers adopted that procedure.

20. E. W. Morley and W. A. Rogers, "On the Measurement of the Expansion of Metals by the Interferential Method," *Physical Review*, 1896, *4*:1-22, 106-127.

21. W. A. Rogers, letter of resignation, 1898, Special Collections, Colby College.

22. N. Rosenberg, "Technological Change in the Machine Tool Industry, 1840-1910," *Journal of Economic History*, 1963, 414-46.

23. Defined by Carolyn Cooper in "The Role of Thomas Blanchard's Woodworking Inventions in 19th Century American Manufacturing Technology" (Ph.D. thesis, Yale Univ., 1985).

THE CONTEXT OF SCIENCE: THE COMMUNITY OF INDUSTRY AND HIGHER EDUCATION IN CLEVELAND IN THE 1880s[*]

Darwin H. Stapleton

Historians have approached the Michelson-Morley experiment primarily as a scientific event. I propose that it can be approached equally well as a technological event. The late Derek Price reminded us that the great achievements of modern science have been indissolubly linked to advances in instrumentation.[1] It was never so true as with the Michelson-Morley experiment. In this paper I discuss some aspects of its technological history, focusing particularly on the interrelationship of the experimenters and their instrument-makers, Worcester R. Warner and Ambrose Swasey.

Cleveland was in the later nineteenth century one of the leading manufacturing centers of the United States. One normally associates Cleveland's leadership with the manufacture of iron and steel, the city's foremost industry during the Michelson-Morley era, but closely related was the machine-tool and foundry sector, with the third largest industrial production in the city. Growing rapidly were the electro-chemical industries.[2]

Industrial research was beginning to play a role in each of these industries, and many of the new researchers had academic preparation. For every figure with minimal formal schooling in the electrical industry, like Thomas A. Edison, there was a Charles F. Brush, Cleveland's pioneer electrical industrialist, who had higher technical education. And there was another dimension to the interrelationship of industry and higher education. By the 1880s it was not unusual for American industrialists to call on academics for advice and assistance, nor was it unknown for them to act as patrons of individual scientists.[3]

Although we lack a standard history of industrial research in this era it certainly appears that Cleveland's industrial leadership was in the vanguard of both industrial research and the development of a close connection with academic researchers and institutions.[4] Two events of 1880 symbolize that connection: the founding of Case School of Applied Science and the establishment of the Brush Electric Company.

Case School was provided for in the will of Leonard Case, Jr., a wealthy Clevelander deeply involved with the industrial-commercial community of the city, and a man of strong scientific interests. The executors of his estate and the trustees of the school, fellow businessmen and industrialists, remained faithful to Case's wishes and quickly established an engineering course with a strong science orientation. The trustees hired a faculty with inclinations toward research and publication -- perhaps because the first president of the institution, John N. Stockwell, a personal friend of Leonard Case, was an astronomer and mathematician deeply into research. Hiring Albert A. Michelson and granting him an immediate leave to complete his researches in Germany fit the character of the new engineering school.[5]

Charles Brush sought higher technical education before the Case School was founded, and received a degree in mining engineering from the University of Michigan in 1869. He returned to Cleveland to set up shop as an analytical chemist. But his strongest interest was electricity, and his experiments led to the invention of an outdoor arc-lighting system first successfully demonstrated in Cleveland in 1879. The next year the Brush Electric Company was formed to market the system, and a laboratory was con-

---

[*]Some of this paper is based on a study of the history of University Circle in Cleveland, which I am carrying out under the auspices of University Circle, Incorporated, supported by a grant from The Cleveland Foundation.

structed especially for Charles Brush. There in the 1880s he refined his electrical system and did crucial work on storage batteries.[6]

Electro-chemical technology strongly interested academic scientists, and Brush cultivated several by donating equipment to them; in return he received their advice and opinions. His equipment was used at Western Reserve College as early as 1877, and Reserve professor Charles F. Mabery presumably worked at the Brush shops on an electrical furnace for aluminum alloys. Brush gave Michelson an arc-lighting set soon after Michelson's arrival in Cleveland. But Brush's most enduring collaboration was with Edward W. Morley.[7]

Morley's first recorded encounter with Brush's scientific benevolence is included in a letter to Morley's father in 1883:

> Week before last on Friday I went to the Brush Electric works to buy a Dynamo-Electric Machine; but came away with one as a gift, and directions to go to the superintendent and tell him what I wanted made for us. It is to be done this week, and sent to us by the company. It would [have] cost us three hundred dollars. So I shall have so much more to buy other apparatus. [8]

Brush and Morley had a continuing relationship: Morley acknowledged Brush's support in the preface to his definitive work on oxygen and hydrogen published in 1895, and the two men coauthored two brief papers in later years.[9]

Morley and Brush shared an interest not only in science but in industrial research. Morley became a chemical consultant to several Cleveland area firms soon after his appointment to Western Reserve College in 1869 and continued through the 1880s. He did mineralogical and metallurgical analyses for the iron industry, studies of refining problems for Standard Oil, and various commissions for the city gas works, a linseed oil firm, and others.[10]

Morley also kept in close contact with the technical community of Cleveland by his active membership in the Civil Engineers' Club of Cleveland, a group with broad interests in science and technology. At the meetings Morley regularly met the machine-tool makers Warner and Swasey.[11]

Skilled researchers other than Morley and Brush were employed by Cleveland firms. For example, the Sherwin-Williams Company (paint) hired its first professionally trained chemist in 1884. The National Carbon Company (one of the ancestors of Union Carbide) was founded in 1886 to manufacture carbon electrodes; it had a full-time researcher on hand beginning in 1889. Recent scholarship has shown that the 1880s were watershed years for the implementation of industrial research programs by firms at the frontiers of American technology.[12]

The most prominent industrial researcher in Cleveland was Herman Frasch, a German immigrant with pharmaceutical training. In 1877 he came to Cleveland to develop refining technology for a Cleveland petroleum company associated with John D. Rockefeller. In 1886 he was hired by Rockefeller's Standard Oil to focus on desulfurizing oil from the new fields of western Ohio and Indiana. Frasch set up a laboratory and experimental refinery in Cleveland and a year later applied for patents on a new process that soon was worth millions of dollars. While celebrating 1987 as the centennial of one great scientific and technological event in Cleveland, we should perhaps be reminded that it was also the centennial of Frasch's and Standard Oil's early successes in industrial research, which revolutionized the American petroleum industry.[13]

It is clear that many Cleveland businessmen regarded research as worthy of support, whether in the form of industrial research, which had the potential of direct economic benefits, or of science, in which benefits were normally indirect. The Cleveland machine-tool firm of Warner and Swasey was possessed of both views. It dealt regularly with academics, sought orders for huge astronomical instruments that offered the possibility of substantial profits, yet simultaneously believed in offering substantial discounts on smaller instruments and in carrying out special jobs for scientists.[14]

Warner and Swasey probably did not view their relationship with Edward Morley and Albert Michelson as unique. They had consciously developed an astronomical instrument business, which involved mostly telescope-making, but included optical and precision-instrument orders. In doing so, they worked closely with academic scientists and apparently conversed with them as colleagues.

Worcester Warner's interest in astronomy had begun as an avocation during his years in the Pratt and Whitney shops, and he had acquired a personal knowledge of astronomical instrumentation by visiting observatories in the United States and Europe in the late 1870s. He made the acquaintance of some leading astronomers in America. With Swasey, who adapted his machine-tool design abilities to telescopy, he envisioned telescopes as a high-profile adjunct to their new machine-tool business.

Their first telescope was sold to Beloit College in 1880, and of the sale price of $2,000, one fourth was returned to the college to support astronomical research. In 1881 they got a contract to erect the revolving dome for the University of Virginia's new observatory, and that same year began negotiations with the James Lick Trust of California to build parts of what was to be the most powerful telescope in the world. Eventually the fledgling machine-tool company in Cleveland received the contract to build the entire 36-inch refractor, completing it in 1887 (*annus mirabilis!*) after extensive negotiations, and after completion of smaller contracts for Lick had proved their superior skills in scientific instrumentation.

In the years from 1884 to 1891, Warner and Swasey built a total of thirty-five telescopes and a number of other instruments, many for academic institutions. The Warner and Swasey records for those years note orders from the University of Michigan, Wesleyan University, Vanderbilt University, Swarthmore College, John Hopkins University, Vassar College, Smith College, and others. In a manner reminiscent of Brush's attitude toward scientific customers, and consonant with their gesture of patronage of refunding part of the price of the Beloit telescope, Warner and Swasey sometimes lowered their fees for scientific work. When they repaired a "telescope clock" for Samuel P. Langley of the Allegheny Observatory in Pennsylvania, they reduced their standard charge by 25 percent, "because," the bookkeeper noted, "it looks so large."[15]

Judging by the number of entries in the Warner and Swasey shop order books and notebooks, Morley and Michelson were their most frequent customers for scientific instruments in the 1880s. I have counted twenty-seven separate orders for Michelson, Morley or both (not including several orders for Case and Western Reserve generally). The earliest is Michelson's for brass tubes and a tuning fork (for speed-of-light determination) in November of 1882, in his first semester at Case; the next is his order for a planometer, possibly to check the surfaces of the mirrors that were so critical to Michelson's light experiments. We have Michelson's letter noting that the device did not meet his expectations, though "not due to any fault in construction -- which was all that could be desired."[16]

I believe that this kind of critical feedback, whereby the scientist reports his admiration of the device, but guides the instrument-maker closer to his experimental needs, governed the dialogue between the two sets of partners, although I have located only a few fragments of it in surviving correspondence. Certainly Michelson had this sort of dialogue with Elmer Sperry in later years when he called upon Sperry to construct the critical apparatus for his speed of light measurement on Mount Wilson.[17]

Such a dialogue was possible in Cleveland in the 1880s because it could take place both on the scientific level, where Warner and Swasey were competent amateurs, and on the technical level, where Morley and Michelson had significant skills. As I have pointed out earlier, Morley was deeply involved in industrial research; he surely had learned much about the technical side of the businesses for which he consulted. Moreover, we know that he delighted in creating or repairing technical devices, such as clocks, a personal telegraph system, and a eudiometer and manometer for his experiments.[18]

Michelson's technical skills are less obvious than Morley's, but form an undercurrent to his science. His first paper published after his arrival in Cleveland described his invention of an electrical device for determining the frequency of a tuning fork.[19] A lifetime spent in the perfection and use of instruments such as the interferometer and the diffraction grate ruling engine suggests that his imagination tended toward the mechanical. One must also take into account his Annapolis education and naval training, both of which would have forced him to become acquainted with a variety of devices.

One such device was the sextant, which any midshipman had to learn to use both for determining position at sea and for surveying. It is intriguing to consider that a sextant of the 1870s is superficially much like an interferometer of the 1880s. The sextant's combination of a telescopic viewer (for his 1882 measuring device Michelson specified a "sextant-telescope"), several reflecting mirrors, and a half-silvered mirror to match up two light paths is very suggestive of Michelson's interferometer design. I can imagine Michelson envisioning his interferometer as a precision optical device utilized by an observer passing through the lapping ethereal sea, much as a sextant, perhaps the most sophisticated optical instrument on board a sailing ship, was used by a navigator on the rolling ocean.[20]

The record of the collaboration of the machine-tool makers and the scientists is clearest in the years 1885–88, when Michelson and Morley carried out their most important collaborative experiments. The first series of orders was from 28 February 1885 through 25 March 1886. Warner and Swasey then filled seven separate orders for equipment, including one on 30 November 1885 that lists them jointly as "Profs. Morley and Michelson" and one of 19 January 1886 that specifically notes that it was for "philosophical apparatus."[21]

All this equipment probably was destined for use in the repetition of the Fizeau experiment, and included the tubing, valves and pumps for controlling the water medium, and a blower for the rotating mirror. The Warner and Swasey shop book entries are tantalizingly brief, but provide a few nuggets. One entry notes that a job was performed at a 40 percent discount; another includes a charge for working on a set screw, presumably for Michelson's interferometer -- the first evidence that the scientists were entrusting precision work to Warner and Swasey.

And finally, it is interesting that the last entry in this series, the one involving the manufacture of the blower, is dated only one day before Michelson wrote to Sir William Thomson to say that both the air and water medium had been tested to confirm the results of Fizeau. One wonders whether Warner and Swasey merely repaired an old blower, or replaced a damaged one, or whether upon receiving the blower Morley and Michelson rushed through their experiment in one day. Given their later performance of the epochal ether-drift determination in a few hours over three days, a single-day experiment seems possible.[22]

After the Fizeau apparatus orders, there appear in the Warner and Swasey sales book three orders from Michelson for what was termed a "dynamometer," but which apparently was not to be one of the standard definition, i.e., a device for measuring power. Rather, from the mentions of discs in each order and the course of Michelson's research, I believe that Michelson probably was pursuing some aspect of light research using the phenomena of phosphorescence. In the 1880s Boisbaudran and Crookes were demonstrating the value of studying elemental spectra through phosphorescent light, probably through devices similar to Becquerel's phosphoroscope. It utilized pierced, rapidly rotating discs to transmit the brief bursts of phosphorescent light coming from an illuminated object while cutting off the prior reflected light.[23]

It is not clear whether Michelson had used spectroscopy in his research before the mid-1880s, but Morley was surely familiar with the field. In 1876 he had given two papers on diffraction gratings at the AAAS. That same year a Cleveland iron and steel company donated a spectroscope to Western Reserve, and Morley presumably used it for mineralogic and forensic analyses. Moreover, Michelson's colleague Charles Mabery (appointed to Case in 1883) was a leading spectroscopist. His spectroscope was twice repaired by Warner and Swasey in 1887.[24] In any case, the preparations that Michelson made for an experiment using the disc device were probably ruined by the great fire of October 1886, which destroyed the Case Main Building and reportedly most of Michelson's equipment.[25]

About a year after their Fizeau apparatus was completed, Michelson and Morley again ordered instruments from Warner and Swasey for a light-wave experiment, this time for an ether-drift measurement. In May 1887 they were billed for eighty-five hours spent making "apparatus for carrying [a] mirror" and on 30 June Warner and Swasey's sales book records two invoices. One was for

> 50 Hrs. making 4 plates with projections[;] this includes time on special Jig Taps & Die[;] 58 lbs. cast iron [@] .03[;] 3 Hrs. Forging on Shifting Device[;] 2 Hrs. M[ach]ine work on same.

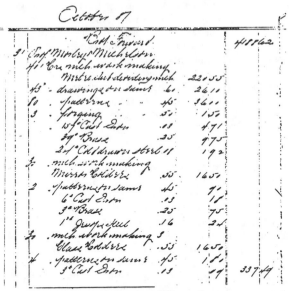

Figure 1

This 1887 entry in a Warner and Swasey shop book records the initial 593 hours spent making a "metre sub-dividing" machine for Morley and Michelson, including drawings, patterns, forging, and machine work. This device was the comparer for the sodium wave-length measurements of 1888. Warner and Swasey sales book, 31 October 1887, box 38, Warner and Swasey Collection, Special Collections, Freiberger Library, Case Western University, Cleveland, Ohio.

And the second for "1 Hr. pattern work[;] 7 lbs. cast iron[;] 1 Hr. pattern work."[26] The dates of these orders accord with Michelson and Morley's plans, especially with Morley's statement of 17 April 1887 that "We have not got the apparatus done yet, and shall not be likely to get it done for a month or two."[27] Moreover, the recently discovered mirror holders for the 1887 interferometer are indeed made of cast iron, as the experimenters themselves described them.[28]

Of course the information from the sales book does not reveal much about the instrument itself and the process of making it. We still must rely on indirect evidence. For example, Loyd Swenson has pointed out that while the crucial part of the interferometer was its micrometer, the record is silent on its maker. Yet we know that Warner and Swasey were manufacturing micrometers, and in July 1886 supplied for a telescope a micrometer-adjusted cross-hair eyepiece that must have been very similar to the micrometer for the 1887 interferometer. I think there can be little doubt that Morley and Michelson would have turned to Warner and Swasey for this gem of fine machinery.[29]

We have a more authoritative record of Warner and Swasey's work on the instrument for Michelson and Morley's final collaboration, the "comparer" for the sodium-wave determination. There are drawings of it in the shop notebook and two or perhaps three entries in the sales book. It was a more time-consuming and detailed job than the ether-drift interferometer, but their part appears to have been completed in late winter of 1888.[30]

They apparently sent it to John Brashear in Pittsburgh for the installation of the optical parts that he had made; and the device was returned to Cleveland early in May 1888. Morley had ordered the comparer by August of the previous year and had expected it as soon as September. He complained bitterly about the delay, blaming it on Warner and Swasey and threatening privately not to take any more business to them.[31]

It should be noted that the comparer was in effect a combined interferometer and spectroscope. It can be viewed as a culmination of Michelson and Morley's experimental work over the previous three years, and as the result of Warner and Swasey's development of instruments to meet their needs during that time.

*　*　*

This review of the Cleveland context of the Michelson-Morley experiments suggests that two aspects of the experiment have been undervalued in prior accounts.

First, although many writers have praised the interferometer and have recognized that instrumentation was an important aspect of their experiments, none have seriously considered the role of the instrument makers, Warner and Swasey. By my account, the instrument emerged out of an auspicious context. Warner and Swasey were deeply involved in the making of precision optical instruments and regularly designed instruments to suit the needs of academics. At least one other Cleveland industrialist, Charles Brush, also supplied laboratory apparatus to academics, and Cleveland's leadership in industrial research made industrial-academic connections common.

Michelson and Morley found Warner and Swasey ready to respond to their particular needs. Perhaps the machine-tool makers should be termed eager for their business, since they went so far as to give Michelson and Morley special discounts at times. Michelson and Morley entered into a continuing dialogue with Warner and Swasey, giving them a series of orders from 1882 through 1888 that required increasing degrees of difficulty and precision. It appears to have been one of the most fruitful cases of the development of scientific instrumentation in modern physics.

Second, my account of the Michelson-Morley experiment indicates that Morley's role was considerably more important than most writers have suggested. While it may have been true that a major reason for the collaboration was, as Loyd Swenson has said, that Morley's laboratory was better equipped than was Michelson's, there was far more to bind the two men together and to make them a team: namely, spectroscopy and access to instrument-makers.[32]

Morley was experienced in spectroscopy, clearly more so than Michelson when they began their collaboration in 1885. Since spectroscopy played no role in Michelson's research before this point, and an increasing role thereafter, I think that Morley probably played a major part in introducing Michelson to the experimental aspects of the subject.

Morley had connections to the Cleveland industrial community on a much deeper level than Michelson, because of his industrial research activities, and probably was the main channel of communication between the experimenters and Warner and Swasey. I find it indicative that joint orders from the two experimenters are always entered in the Warner and Swasey sales book as from "Morley and Michelson," the reverse of the order in which they have come down to us in historical texts.

On balance, then, I would like to suggest that the most famous of the Michelson-Morley experiments could only have happened in Cleveland on the Case School-Western Reserve campus in 1887. It was not an inevitable part of the research program of Albert Michelson, although it was of course directly derived from his earlier work. For Michelson, Cleveland was not a scientific backwater, but a location that nurtured and enlarged his research program. There he took a giant step toward the Nobel Prize.[33]

NOTES

1. William J. Broad, "Is Technology the Hero of [the] Scientific Revolution?" *The Plain Dealer*, 21 August 1984.

2. Department of the Interior, Census Office, *Compendium of the Tenth Census (June 1, 1880)* (Washington: Government Printing Office, 1883), 2: 1054-57; Department of the Interior, Census Office, *Compendium of the Eleventh Census: 1890* (Washington: Government Printing Office, 1894), 2: 782-85.

3. A useful survey that covers a variety of aspects of the industrial-academic intersection is John Rae, "The Application of Science to Industry," in *The Organization of Knowledge in Modern America, 1860-1920*, ed. Alexandra Oleson and John Voss (Baltimore and London: The Johns Hopkins Univ. Press, 1979), pp. 249-68.

4. S.v., "Science," and "Technology and Industrial Research," in *The Encyclopedia of Cleveland History*, ed. David D. Van Tassel and John Grabowski (Bloomington, Ind.: Indiana Univ. Press, 1987).

5. Clarence H. Cramer, *Case Institute of Technology: A Centennial History, 1880-1980* (Cleveland: Case Western Reserve Univ., 1980), pp. 1-40.

6. Darwin H. Stapleton, "The Rise of Industrial Research in Cleveland, 1870-1930," prepared for a volume of essays edited by Elizabeth Garber.

7. C. F. Brush to J. W. Langley, 23 December 1876, C. F. Brush to G. F. Barker, 9 February 1877, C. F. Brush to C. F. Brackett, 1 October 1877, Box 4, Brush Papers, Special Collections, Freiberger Library, Case Western Reserve University, Cleveland, Ohio; statements of shipments to S. P. Langley (2 November 1882), T. Sterling (13 November 1882), and Smithsonian Institution (31 March 1884), box 14, Brush Papers; Alfred Cowles, *The True Story of Aluminum* (Chicago: Henry Regnery Company, 1958), pp. 32-34, 40; Harold C. Passer, *The Electrical Manufacturers, 1875-1900* (Cambridge, Mass.: Harvard Univ. Press, 1953), pp. 225-26; Cramer, *Case Institute of Technology*, p. 26; Nathan Reingold, ed., *Science in Nineteenth-Century America: A Documentary History* (New York: Octagon Books, 1979), p. 305.

8. Edward W. Morley to S. B. Morley, 27 April 1883, Morley letters, Case Western Reserve University Archives, Cleveland, Ohio. Three years later Morley went to the Brush works and asked for a new battery as if such a gift were routine: Edward W. Morley to S. B. Morley, 8 April 1886, Morley letters.

9. Howard R. Williams, *Edward Williams Morley: His Influence on Science in America* (Easton, Pa.: Chemical Education Publishing Co., 1957), p. 278; Edward W. Morley, *On the Densities of Oxygen and Hydrogen and on the Ratio of Their Atomic Weights* (Washington: Smithsonian Institution, 1895), p. vi. The latter citation reads "Mr. C. F. Wason [*sic*] presented me with an electric motor which has been of great service," certainly an error for "C. F. Brush."

10. Williams, *Edward Williams Morley*, pp. 168-69; Edward W. Morley to S. B. Morley, 28 January 1877, 16 December 1883, 29 March 1884, 27 December 1885, 13 March 1887, 17 April 1887, 15 December 1887, 29 March 1889, 14 April 1889, 25 April 1889, Morley letters.

11. See, for example: *Journal of the Association of Engineering Societies* 1888, 7: 110-111, 146, 157, 226; *The Golden Anniversary Book of the Cleveland Engineering Society* (Cleveland: n.p., 1930), p. 117.

12. *The Story of Sherwin-Williams* (Cleveland: Sherwin-Williams, 1954), n.p.; *The Industries of Cleveland: A Resume of the Mercantile and Manufacturing Progress of the Forest City* (Cleveland: Elstner, 1888), p. 160; Circuit Court of the United States, Northern District of Ohio, Eastern Division, *Clarence M. Barber vs. The National Carbon Company, et al.*, [c. 1895], pp. 17, 24-26; Leonard S. Reich, *The Making of American Industrial Research: Science and Business at GE and Bell, 1876-1926* (Cambridge: Cambridge Univ. Press, 1985); Lillian Hoddeson, "The Emergence of Basic Research in the Bell Telephone System, 1875-1915," *Technology and Culture*, 1981, *22*: 512-44.

13. William R. Sutton, "Herman Frasch" (Ph.D. diss., Louisiana State Univ., 1984), chaps. 1-4; Allan Nevins, *Study in Power: John D. Rockefeller, Industrialist and Philanthropist* (New York and London: Charles Scribner's Sons, 1953), Vol. 2: pp. 101-2.

14. Unless otherwise noted, my history of Warner and Swasey relies upon the fine study of Edward J. Pershey, "The Early Telescope Work of Warner and Swasey" (Ph.D. diss., Case Western Reserve Univ., 1982).

15. Sales Book, 1880-87, box 38, Warner and Swasey Collection, Special Collections, Freiberger Library, Case Western Reserve University, Cleveland, Ohio (hereafter Sales Book, W&S).

16. Sales Book, W&S; A. A. Michelson to Warner and Swasey, 16 April 1883, box 18, Warner and Swasey Collection; and Reingold, ed., *Science in Nineteenth-Century America*, pp. 281, 302, have Michelson's mentions of tuning forks for his apparatuses of 1878 and 1882.

17. Thomas Parke Hughes, *Science and the Instrument-maker: Michelson, Sperry and the Speed of Light. Smithsonian Studies in History and Technology, no. 37* (Washington: Smithsonian Institution Press, 1976).

18. Williams, *Edward Williams Morley*, pp. 45, 55-56, 164-68; Loyd S. Swenson, Jr., *The Ethereal Aether: A History of the Michelson-Morley-Miller Aether-Drift Experiments, 1880-1930* (Austin and London: Univ. Texas Press, 1972), p. 46.

19. Albert A. Michelson, "A Method for Determining the Rate of Tuning-forks," *American Journal of Science* 3rd ser., January 1883, *25*: 61-64. In 1885, Michelson directed a student's electrical experiments and helped him obtain carbon plates from a Cleveland factory: William L. Meriam diary, 23 March 1885, 8 April 1885, 13 April 1885, 20 April 1885, 2 May 1885, 4 May 1885, Case Western Reserve University Archives.

20. My initial acquaintance with the properties of the sextant came from writing a note on Hadley's quadrant in *The Virginia Journals of Benjamin Henry Latrobe, 1795-1798*, ed. Edward C. Carter II (New Haven: Yale Univ. Press, 1977), Vol 1: pp. 213-20. In August 1987 I examined a superb exhibit of sextants at Mystic Seaport, Mystic, Connecticut. Michelson's specification for a "sextant-telescope" is in Reingold, ed., *Science in Nineteenth-Century America*, p. 303. Note the statement that the beam-splitter for the 1885 experiment was "a half silvered surface": A. A. Michelson and E. W. Morley, "Influence of Motion of the Medium on the Velocity of Light," *Am. J. Sci.*, 3rd ser., 1886, *31*: 380.

21. 28 February 1885, 31 March 1885, 27 May 1885, 30 November 1885, 21 December 1885, 19 January 1886, 28 January 1886, 26 March 1886, Sales Book, W&S.

22. 30 November 1885, 19 January 1886, 26 March 1886, Sales Book, W&S; Dorothy Michelson Livingston, *The Master of Light*, pp. 116-17; Swenson, Jr., p. 93; and Reingold, ed., *Science in Nineteenth-Century America*, pp. 281, 284, have Michelson's references to a blower for his revolving mirror for speed-of-light experiments, and Morley's comment in Reingold, on p. 312, indicates that the Fizeau experiment was yet to be performed at the end of January 1886 and was at least a week away.

23. 31 March 1886, 17 May 1886, 30 September 1886, Sales Book, W&S; Charles C. Gillespie, ed., *Dictionary of Scientific Biography* (New York: Scribner's, 1970-1980), Vol. 1: 556, Vol. 2: 254, Vol. 3: 478. *The Dictionary of Scientific Biography*'s index has a chronological listing of references to spectroscopy that provides useful entry into the state of the field in the mid- and later nineteenth century.

24. 27 June 1876, Minutes of the Western Reserve College Trustees, CWRU Archives; Williams, *Edward Williams Morley*, p. 274; Cramer, *Case Institute of Technology*, pp. 25-27; Yakov M. Rabkin, "Technological Innovation in Science: The Adoption of Infrared Spectroscopy by Chemists," *Isis*, March 1987, *78*: 37; 24 February 1887, 26 May 1887,

Sales Book, W&S. Morley's involvement in industrial research is reported above. His biographer notes Morley's frequent involvement in medical analyses for legal purposes. Morley was a professor at the Medical College of Western Reserve from 1873 to 1888. Williams, *Edward Williams Morley*, pp. 158-63, 169-73.

25. The literature assumes that the equipment in Michelson's laboratory in October 1886 included the ether-drift interferometer. From the entries in the Warner and Swasey shop book, cited above and below, it appears that the interferometer was yet to be constructed, and the recently completed spectroscope-phosphoroscope was the instrument then at the center of Michelson's work. Cramer, *Case Institute of Technology*, pp. 30-31, 51; Livingston, *The Master of Light*, pp. 121-23.

26. 26 May 1887, 30 June 1887, Sales Book, W&S.

27. Reingold, ed., *Science in Nineteenth-Century America*, pp. 312-13.

28. Albert A. Michelson and Edward W. Morley, "On a Method for Making the Wave Length of Sodium Light the Actual and Practical Standard of Length," *J. Assoc. Eng. Soc.*, May 1888, 7 :153-56. This has a slightly different text than the article by the same title in the *Am. J. Sci.*

29. "Micrometer for 6" telescope by Fauth & Co.," 13 July 1886, Notebook, 1886-1888, box 18, W&S; 18 September 1888, Sales Book, 1887-1892, W&S; *A Few Astronomical Instruments from the Works of Warner & Swasey* (Cleveland, Ohio: Warner & Swasey, 1900), plates V, XII, XXXV. The calibrated adjustment heads on the micrometers in *A Few Astronomical Instruments*, which were manufactured in the 1880s and 1890s, are identical to the one sketched in the sales book for the Fauth telescope. The astronomical position micrometer in plate V is probably contemporary with the Michelson-Morley micrometer.

30. "Sketch of Experimental apparatus for the determining of the length of a wave of Sodium light, Profs. Morley & Michelson," (2 pp.), 23 January 1888, "Experimental Measuring Machine for Prof. Michelson," 2 February 1888, "Measuring Machine for Prof. Michelson," 20 February 1888, Notebook, 1886-1888, W&S; 31 October 1887, 5 November 1887, 30 December 1887, Sales Book, W&S.

31. Livingston, *The Master of Light*, pp. 138-39; Edward W. Morley to Warner & Swasey, 18 August 1887, box 18, W&S; Edward W. Morley to S. B. Morley, 10 May 1888, Morley letters.

32. Swenson, Jr. *The Ethereal Aether*, pp. 46-47.

33. Michelson was awarded the Nobel Prize for "his optical precision instruments and the spectroscopic and metrological investigations carried out with their aid." Elisabeth Crawford, *The Beginnings of the Nobel Institution: The Science Prizes, 1901-1915* (Cambridge: Cambridge Univ. Press, 1984), pp. 173-74.

THE DIMENSIONAL REVOLUTION:
THE NEW RELATIONS BETWEEN THEORY AND EXPERIMENT IN ENGINEERING IN THE AGE OF MICHELSON

Edwin T. Layton, Jr.

The Michelson-Morley experiment of 1887 was one of a number of steps that led to the modern revolution in physics. A revolution took place in engineering at about the same time as that in physics. The revolution in physics is known to all; that in engineering, while not precisely unknown, is taken for granted and remains little understood.[1]

The revolution in physics involved quantum theory, relativity, and other developments. Similarly, the revolution in engineering represented the convergence of several independent lines of historical evolution. The topic is a large one, and not well-developed historically. The present paper is a preliminary survey of one particular part of the revolution in engineering. The rise of dimensional analysis, broadly conceived, produced a revolution in engineering -- we can call it "the dimensional revolution." It had two closely interrelated parts: fruitful scale-model experiments, and more sophisticated, less-idealized engineering theories.

The theoretical and experimental advances in engineering were intimately related; they constituted two sides of the same coin. Scale-model experiments required knowledge of the laws of similarity, which linked experiments performed at small scale to the behavior of full-sized engineering artifacts. Precise, scale-independent experimental data required better theories so that observation and theory could be matched with one another. In practice this meant less idealization and increased complexity to take into account messy real-world phenomena, such as friction and viscosity. More powerful theories depended upon the discovery of key dimensionless parameters and their incorporation into engineering theories. Indeed, the statement of engineering theory in terms of dimensionless parameters was one of the foremost characteristics of the new, scientific engineering.[2]

The revolution in physics produced startling new predictions, such as the conversion of mass into energy. Dimensional analysis itself predicted no dramatic new phenomena to catch the public eye. Neither did the new theories, which were focused on complex, but generally unspectacular, properties of matter and energy. Nevertheless, the dimensional revolution in engineering had a profound impact on virtually all technology. More fruitful and precise experiments, along with better theories, made engineering more scientific, causing an acceleration of technological development in many directions. The effect was unobvious and indirect, but without model experimentation and powerful engineering theories in fluid mechanics and aerodynamics, it is hard to imagine the spectacular growth of aeronautics since the Wright Brothers. Without the modern science of heat transfer it is hard to imagine the maturing of jet propulsion. The further to obscure things, many modern developments depend upon the interaction of several new sciences or theories. Such, for example, is the case with modern rocketry and space exploration, in which heat transfer, combustion theory, fluid mechanics and other sciences or theories combined to make possible startling new potentialities for technology.

Dimensional analysis and improved theories ended a long-standing gap between theory and experiment in engineering. Though the gap was not total it was significant; the methods developed by John Smeaton and other scientifically minded engineers helped, but they did not overcome the problem. One can date the beginning of the liberation of engineering from the tyranny of scale effects (perhaps a bit arbitrarily) to 1883 and the publication, only four years prior to the historical Michelson-Morley experiment, of a paper by Osborne Reynolds. In this paper Reynolds developed laws of similitude and demonstrated the existence of a dimensionless number (later named after him) that measured the critical transition in a fluid from laminar flow to turbulent flow.[3]

Though Reynold's 1883 paper on fluid motion was particularly noteworthy, several of his other papers played a vital role in the dimensional revolution in engineering. In dates they bracket the Michelson–Morley experiment. They included papers by Reynolds on developing scale models of hydraulic systems such as the Mersey River system.[4] Another fundamental paper by Reynolds set forth a famous and fruitful analogy between the turbulent exchange of momentum and of heat, a critical insight linking heat transfer with mass transfer.[5] Yet another revolutionary paper by Reynolds was his theory of lubrication, published in 1886. An engineer, Beauchamp Tower, carried out and published extensive experiments on lubrication; this work was then developed theoretically by Reynolds in a paper that lay the foundations for the modern science of friction and lubrication.[6]

What made Reynolds's papers such a historical turning point in the history of engineering? There are two answers to this question, one technical and the other social and institutional. On the technical front, Reynolds was addressing the idealization of physical theory. Messy phenomena such as turbulence and friction had been idealized out of physics. The reasons were good; these phenomena did not lend themselves to theoretical treatment. The price of progress in natural philosophy had been that these phenomena be ignored. But turbulence and friction were central problems in engineering that could not be ignored. Engineers used various empirical expedients; but the result was a gap between physical theory and engineering practice. It was Reynolds's significance that he played a critical role in the process whereby certain important gaps between theory and experiment in engineering were largely closed. To do this he developed scale-model experiments and a dimensionless parameter that was to prove of critical importance for bringing the archetype of chaotic phenomenon under the rule of law.

Reynolds's paper was also the product of an important social and institutional change. Reynolds had assumed a new engineering chair, the second of its kind in England, at Owens College, Manchester, in 1868. Here he pursued experimental and theoretical investigations in very much the spirit of contemporary physics. And, in a deeper sense, this fact was more important than any specific result. Reynolds's research can be seen as part of a broad process whereby engineering research came increasingly to be conducted in university laboratories, published in research journals, and associated with the activities of scientifically oriented professional societies. There was a converse of this; chemists and physicists, beginning at about the same time, came increasingly to be employed in industrial research laboratories.[7] Research in engineering also came increasingly to share many of the values associated with basic research, though engineers and research laboratory directors gave them a technological twist by changing the rank-order, placing practical benefits ahead of theoretical advances. Engineering research, like scientific research in industry, had come to share key elements of the institutional framework and many of the intangible values developed by the communities of basic science. By borrowing the institutional structure and many of the values of the basic sciences, engineering became something of a "mirror image twin" of them.[8]

Osborne Reynolds was a pioneer shaping the scientific social role for engineers. Reynolds's predecessor in this task, W. J. M. Rankine, had developed a theory of the engineering sciences as something intermediate between pure science and engineering practice.[9] Reynolds did much to establish and legitimate the hybrid role of the engineering scientist. In the classic paper of 1883 on the transition from laminar to turbulent flow (to use the modern words for the phenomena), Reynolds held that his results had "both a practical and a philosophical aspect."[10] But Reynolds noted that his results, "as viewed in their philosophical aspect, were the primary objects of the investigation."[11] While there are individual differences, theoreticians in engineering, like Reynolds, go beyond puzzle solving. They get psychic rewards from the elegance, generality, and harmony of their theories. In this respect theorists in engineering and the basic sciences are alike.

However, Reynolds and subsequent engineering scientists have differed from practitioners in the basic sciences in other attitudes and orientations. Above all, Reynolds identified himself with engineering, and manifested an enduring commitment to its problems and needs.[12] In practice this meant that Reynolds used sophisticated methods akin to those of physics to address problems of vital importance to engineering. Implicitly this loyalty to engineering meant that Reynolds, however he might love theory and science, placed a higher value on "doing" than "knowing," and this commitment shaped the problems he chose to study and the nature of the results he sought to get from them.[13]

The dimensional revolution in engineering inaugurated a much closer relationship between engineering and physics. Engineering borrowed ideas, institutions, personnel, instrumentation, and much more from physics. These extensive borrowings by engineers can be seen as part of a long cycle in science-technology interactions. During the Renaissance it was engineering that had been in the lead; it provided critical stimulus to the Scientific Revolution of the seventeenth century. Engineers had been attempting to study machines and structures scientifically since classical antiquity. The Hellenistic engineers derived their famous catapult rule by experimental methods akin to modern "parameter variation."[14] The idea that there existed law-like regularities in technology received a renewed impetus in the Renaissance. The efforts of Renaissance engineers to mathematicize machines and engineering and to develop appropriate experimental methodologies provided important stimuli to the later development of experimental methods and mathematical theory in natural philosophy from Galileo onward. Alexander Keller,[15] Ladislao Reti,[16] Thomas Settle,[17] and others,[18] have shown that engineers developed vigorous scientific traditions that influenced the nascent physics of Galileo and his successors.[19]

The precocious development of science by Renaissance engineers encountered critical difficulties. Engineers were severely hampered by scale effects. As Vitruvius had noted much earlier:

> In some machines the principles are of equal effect on a large and on a small scale; others cannot be judged of models. Some there are whose effects in models seem to approach the truth, but vanish when executed on a large scale . . . With an auger, a hole of half an inch, of an inch, or even an inch and a half, may be easily bored; but by the same instrument it would be impossible to bore one of a palm in diameter; and no one would think of attempting in this way to bore one of half a foot, or larger. Thus that which may be effected on a small or moderately large scale, cannot be executed beyond certain limits of size.[20]

Renaissance engineers did not always accept the severe restrictions sketched by Vitruvius. Leonardo da Vinci, in particular, made extensive use of models for all manner of problems.[21] He was evidently acquainted with Vitruvius's statement about models, and he dismissed it. He attempted to show how one might scale up the auger for larger tasks by using a simple geometrical rule. The force required for an auger, he thought, would be proportional to the area swept out, that is, to the square of the radius.[22]

Leonardo's optimism, while premature, was founded on one of the great advances in the history of technology, the Renaissance rediscovery of linear perspective. Graphical methods were of extraordinary importance for engineering, particularly as later systematized by Gaspard Monge's descriptive geometry, first published in 1795.[23] However, the problem of relating small models to larger artifacts was far more complex than Leonardo realized. Intuition might lead technologists to correct similarity relations based on geometry. But Leonardo was wrong in assuming that this could easily be extended to include kinematic and dynamic similarity as well. Unfortunately, in the majority of cases simple geometrical arguments such as Leonardo's were not sufficient. Experimenters found, to their sorrow, that in most cases the kinematic and dynamic behavior of scale models was not an accurate guide to the analogous behavior of full-sized artifacts. Galileo reflected the ultimate failure of scale-model experiments in his rejection of "the current belief that, in the case of artificial machines the very large and the small are equally feasible."[24]

In sharp contrast to the situation in engineering, Newton had supplied rules of similitude for physics in his *Principia*.[25] But while effective for systems of particles, Newton's work on similitude failed to help technologists. Engineering artifacts are characterized by complex shapes and they cannot generally be reduced to systems of particles. Attempts by engineers to understand such devices experimentally encountered problems of scale. What was true for a model was only seldom true for full-sized devices of identical geometry.

Engineers still had to deal somehow with the inherent complexity of the design problems with which they were presented. They could not ignore messy phenomena such as friction and viscosity, however difficult they were to deal with. The ability of Galileo and his followers to simplify and idealize their problems was crucial to the rapid advance of the new experimental philosophy. Galileo, the scientist, could ignore the atmospheric drag on projectiles or the turbulent flow of fluids, but Leonardo, the engineer, could not do so.[26] The result was a parting of the ways as between engineering and physical science. Engineering plodded along a highly empirical, experimental path, as in the engineering science of hydraulics, while a physics free to linearize and unhampered by scale effects, could soar, as in the mathematical theory of hydrodynamics.[27]

Though engineering could not follow physics, graphics provided engineering with a path of development. Graphics became a way of thinking and knowing as well as a means of communication. In the course of the eighteenth and nineteenth centuries, engineers, sometimes aided by scientists, developed a group of engineering sciences that provided a rational basis for engineering design and engineering practice. There was a strong tendency for the engineering sciences to be expressed graphically and mapped into descriptive geometry, as, for example, in the cases of graphical statics,[28] velocity triangles for turbomachines,[29] and indicator diagrams and thermodynamic cycle diagrams for heat engines.[30]

The reasons that geometric methods were so successful in providing engineering with a route to the development of classical engineering sciences was that geometry carries with it -- built in, so to speak -- certain similarity relations. The angles of triangles, for example, may be treated as dimensionless quantities.[31] Thus geometric similarity is an integral part of the graphic approach. Though engineers did not fully understand why their graphical methods sometimes worked so well, they eventually discovered methods that allowed motions, forces, and stresses to be analyzed in a nondimensional fashion through graphics. Engineers still relied on intuition in developing graphical methods appropriate to their discipline. But the graphical statics of Carl Culmann and Otto Mohr, for example, were remarkable examples of how far graphical approaches could carry engineers, even in the absence of general methods for avoiding dimensional traps.[32] The graphical mind set had a particularly heavy impact upon

American engineering in the nineteenth century. Until the dimensional revolution brought with it a more universal analytical way of conceptualization, engineers were free to incorporate the values of visually oriented engineering subcultures within their sciences.[33]

The dimensional revolution produced a convergence between engineering and physics. Engineering theory in the twentieth century was recast in emulation of the analytical style of theories in physics. Experiment in engineering benefited by the high standards developed in physics, and engineers sought to emulate those standards also. The resulting convergence between physical science and engineering is precisely what one should expect by the restructuring of engineering as a "mirror image twin" of the basic sciences. This represents the latest step in what I have elsewhere termed the "scientific revolution in technology."[34] The dimensional revolution was an important part of this broad movement, which also led to rather drastic changes in the form of theories in many fields of engineering. This change sometimes took place in two stages. The first focused upon the traditional concern of engineering theory, that of gaining a scientific understanding of particular types of artifacts. The second involved constructing engineering theories that had the sort of generality long characteristic of theories in physics. This often involved recasting engineering theories in the form of differential equations describing phenomena that influence entire families of engineering artifacts. These two stages were not always clearly separated in time.

Chemical engineering provides an interesting example of the evolution of modern engineering theory. Chemical engineering in America was revolutionized in the 1920s and after by the concept of unit operations.[35] A second stage occurred after the Second World War. Olaf Hougan and others sought to recast chemical engineering theory on the model of physics. This involved a shift in emphasis from the theory of particular artifacts (e.g., from unit operations) to a new stress upon basic molecular and transport phenomena underlying virtually all chemical reactors. Hougan was somewhat put out when he discovered that physicists were not interested in helping, he and other chemical engineers had to develop their own theories.[36]

The path followed by German chemical engineering was somewhat different. Unit processing was not adopted in Germany (at least not in the 1920s). The reason was, in part, because this approach appeared to involve a retreat from a programmatic goal of German chemical engineering, that of basing engineering theory on the more "fundamental" theories of the basic sciences. Another reoason was the difference in factor endowments, which led American chemical engineers to emphasize oil refining and the petrochemical industries, while German chemical engineers emphasized their traditional concern with coal-based chemical technologies.[37]

Despite differences, in both America and Germany chemical engineering theory dealt with mathematical models of particular types of reactors as well as with the development of more general theories. The key to success lay in the development of appropriate dimensionless parameters. Gerhard Damköhler did much to found chemical engineering theory upon dimensionless parameters. He discovered four of them. Two were of great use in understanding those particular types of reactors depending upon diffusion and absorption. With the other two dimensionless parameters Damköhler laid the foundations for chemical reaction engineering, a theory underlying the design of virtually all chemical reactors. The two additional dimensionless parameters concerned, respectively, the ratio of chemical flux to physical flux and the ratio of chemical flux to diffusional flux. These discoveries were an important part of the attempt to understand chemical engineering processes in terms of basic transport and molecular properties.[38]

Electrical engineering went through a somewhat similar evolution. In electrical engineering dimensional methods were used to derive fundamental properties of particular types of electrical machinery. The French electrical engineer Félix Lucas was one of the first to have glimpsed the possibility of using dimensional analysis to derive the fundamental equations that described electrical machines.[39] This work was extended by another French electrical engineer, E. Carvallo.[40]

The early application of dimensional considerations to electrical engineering theory contributed to a major breakthrough in dimensional analysis. Another French electrical engineer, Aimé Vaschy, independently developed an early version of the theorem often credited to Edgar Buckingham.[41] Buckingham was perhaps clearer in expressing the key idea that given n parameters in k dimensions, any engineering theory can be expressed by $n-k$ dimensionless parameters. Buckingham was a physicist with the United States Bureau of Standards and was, along with his employer, perhaps somewhat more energetic in demonstrating the utility of his theorem for a variety of engineering fields. The theorem, however, should properly be called the Vaschy-Buckingham theorem.[42]

Electrical engineering science was given a unified treatment by Charles P. Steinmetz. Ronald Kline, in an outstanding study of Charles P. Steinmetz and the development of electrical engineering science, has suggested that the growing similarity of theories in physics and engineering made them more like "fraternal twins." He also correctly noted that engineering theorists such as Steinmetz sought the greatest possible scientific generality, rather than the specificity traditionally attributed to designers.[43]

Electrical engineers came to adopt much of the methodology and style of physics. But they insisted upon maintaining their basic values and the autonomy of their community. In the post World War II era electrical engineering came of age as a research discipline. Frederick Terman, for example, sought to upgrade electronic research and base its theory on fundamental physical considerations, while rejecting what he considered the limitations of traditional engineering sciences, at least as usually taught in the United States. In this he was deliberately emulating physics.[44] But it is significant that he did not want electrical engineering dependent upon physics in an area of its vital interests. "Never again," he wrote in reviewing the postwar upgrading of electrical engineering graduate curricula, "will electrical engineering have to turn to men trained in other scientific and technical disciplines when there is important work to be done."[45]

The term "general," like "fundamental," is relative. In the case of electrical engineering there were a series of steps by which theory gained greater generality. But each was modulated by advances in the relevant technology as well as by the desire of engineers to place their theories upon more scientific foundations. Thus, classical electrical engineering science was absorbed by vacuum-tube electronics in the 1950s. Electronics has been, in its turn, subsumed within microelectronics in the last twenty years.[46] That is, engineering disciplines may borrow from physics, but they strive to preserve their autonomy, and their borrowings are guided by the technical needs of the technologies that they advance.

Modern mechanical engineering theories share many of the same characteristics as engineering theories in other fields. They too are less concerned with specific types of artifacts, and tend to focus increasingly upon basic processes common to many types of artifacts. Perhaps nowhere is this trend better illustrated than in modern heat and mass transfer. For example, Ernst R. G. Eckert, one of the principal founders of modern heat transfer, helped develop the theory of transpiration cooling of turbine blades in the 1940s in relation to jet engines; but the same theory could be equally applied to the ablation cooling of the nose cones of reentry vehicles which emerged as a significant problem in the 1950s and '60s.[47]

Designers continue to be interested in specifics; namely, the ones associated with the particular artifacts they are interested in designing. But the expansion of the scientific role of engineers, both in universities and in industrial research laboratories, has brought forth a body of engineering scientists whose interests are more general and whose values are sometimes closer to those of the scientist than to those of the designer. Research engineers often adopt scientific values such as a taste for generality and elegance in mathematical theory. That is, changes in education, research, and institutional structures have made it easier for ideas, values, and personnel to move from physics to engineering and from engineering to physics, and these circumstances have helped to produce a convergence of physics and engineering. This has been achieved at a cost; there is now a tension between science and design in modern engineering that would have been unimaginable a century ago.[48]

Recasting engineering theories in emulation of those of physics had several important advantages. Engineers were able more easily to call upon a powerful body of theory and experiment developed in physics and other basic sciences. In the case of turbomachinery, for example, the theory of Euler (which had long been neglected by turbine designers) was revived in the early twentieth century, providing engineers in this area with some powerful theoretical tools.[49] This tendency to "adopt" scientific ancestors was technically beneficial, but it has undoubtedly contributed to historical confusion, particularly the misleading notion that technology is no more than applied science. The fact that engineering disciplines came to be cast into a form consistent with an applied science model served to obscure not only their historical origins but their internal dynamics as well.

The dimensional revolution in engineering was linked to the Michelson-Morley experiment and the events in physics that followed that experiment. Michelson's perfection of the interferometer as a tool of extraordinary experimental precision provided a link between the new physics and the new engineering.

The interferometer was adapted for technology. The version used in engineering was the Mach-Zehnder interferometer (sometimes called the Mach refractometer). It differed from the Michelson interferometer in several particulars, including the way that the two light beams could be separated by a large distance; transparent cells were introduced in such a way that one of the light beams travelling between mirrors could be directed through them. This allowed the interferometer to be used to measure spatial variations of the index of refraction in gases and liquids. Changes in the index of refraction are correlated with changes in density for high-speed flow and with changes in temperature for low-speed, heated flows. At first the instrumentation was somewhat limited in size. The Zeiss company, which built these devices, produced mirrors that were initially restricted to about three centimeters in diameter. By the late 1930s the Zeiss firm was able to build mirrors approximately ten times as large. This increase greatly facilitated engineering research, but even so the size was not sufficient for testing full-sized engineering artifacts. This marvelous tool was therefore limited to scale-model experiments. Thus the use of the interferometer by engineers was an integral part of the dimensional revolution in engineering.[50]

One of the engineering uses of the Mach-Zehnder interferometer was for model studies in high-speed aerodynamics.[51] But in a broad sense this use of the interferometer was linked to the coming to scientific maturity a generation earlier of mechanical and civil engineering. In particular a very sophisticated engineering science, fluid mechanics, along with its close relative, aerodynamics, served to set the path for scientific development in a number of engineering fields. The rise of fluid mechanics was associated with a distinguished group of engineers including Ludwig Prandtl, Paul Heinrich Blasius, Theodore von Kármán, Jakob Ackert, and others in the period 1904 to 1930. Fluid mechanics was characterized by a very complex and sophisticated theory. It was less idealized than previous theories of fluid behavior, and included viscosity (and hence turbulence), at the boundary between thin layers of fluid. This more exact theory became, in turn, closely linked with very precise scale-model experimentation.[52]

Fluid mechanics was perhaps the science closest to the cutting edge in the revolution in mechanical and civil engineering. It was certainly archetypical, and stimulated emulation in many other areas of engineering research. It has sometimes been assumed that Prandtl was among the first to avoid scale effects in the formulation of an engineering science by the use of the Reynolds number. Indeed, it has been claimed that Prandtl was the one who gave the Reynolds number its name.[53] This is an understandable error, given the leadership role of Prandtl and of fluid mechanics in the dimensional revolution in engineering.

The Reynolds number is probably the most fundamental dimensionless number for fluid mechanics research and hence a key to successful model experiments in this area.[54] But, while Prandtl introduced his concept of boundary layer analysis in 1904,[55] fluid mechanics was a bit slow to use the Reynolds number. Prandtl's first reference to the Reynolds number was in 1910, and it was used by Van Kármán first in 1911 and Blasius first employed it in 1912.[56] The precise order of adoption was not important. A more complex and accurate theory needed precise, scale-independent experiments in order to test its validity. Conversely, experimentation freed of scale effects revealed the inadequacy of idealized theories, and provided strong incentives for the construction of theories powerful enough to predict and explain the complex behavior of real artifacts.

Fluid mechanics has had an enormous influence upon engineering. Its basic formalism (boundary layer analysis) and the scientific spirit that it embodied influenced many engineering disciplines. Perhaps the most notable case was that of heat transfer, where these ideas inspired a major internal restructuring of the discipline. Ernst R. G. Eckert incorporated Prandtl's boundary layer theory in heat transfer. In 1934, while still a student at Prague, Eckert chanced upon a copy of Prandtl's classic introduction to fluid mechanics in a bookstore. Later Eckert was to write that, "it was an enormous revelation to me and really started up interest in fluid mechanics and convective heat transfer." Eckert's subsequent innovations inaugurated a new era combining theoretical and experimental precision in heat transfer research.[57] Eckert, of course, built upon the work of predecessors such as Osborne Reynolds, Wilhelm Nusselt, W. K. Lewis, Ludwig Prandtl, Ernst Schmidt, William H. McAdams, and others.[58]

Eckert gave modern heat transfer the characteristics of post-revolutionary engineering research: a close coupling of a sophisticated, complex theory with precise scale-model experiments. Eckert adapted the Mach-Zehnder interferometer to measure heat transfer. The changes in the refractive index in a gas can also be correlated with temperature changes. (For heat transfer studies the velocity of flow was kept well below that needed to induce density changes.)[59] Indicative of Eckert's role in establishing the experimental foundations for the use of the interferometer in heat transfer was the fact that a paper of his dating to 1961 was chosen as a "citation classic." It was a definitive paper on convective heat transfer experimentation using the Mach-Zehnder interferometer and it was cited in 110 publications between 1961 and 1983.[60]

To an important degree, modern engineering heat transfer was literally founded on precise interferometry, and the close match between measurements with this device and the very sophisticated theory also developed by Eckert. As a result Eckert could justly claim that:

> The separation between theoreticians treating ideal fluids mathematically and engineers collecting information on real fluids by experiments existed at the beginning of the century in fluid mechanics and was overcome by the famous 1904 paper of Ludwig Prandtl. The same separation between theory and experiment, however, still continued in heat transfer for a longer time and was, in my opinion, removed largely by my book, *Introduction to the Transfer of Heat and Mass* -- especially by introducing the boundary layer concept to heat transfer problems.[61]

Much the same sort of revolution was taking place in the rather different field of chemical engineering. But there were important parallels between heat transfer and chemical engineering. Both Eckert and Damköhler were influenced by Prandtl; Damköhler took a course in fluid mechanics from Prandtl, Eckert read his works and attended public lectures by Prandtl. Both Eckert and Damköhler were recruited to work under Ernst Schmidt at the aeronautical research establishment at Braunschweig in the 1930s. Both were pioneers in theory and in matching the results of experiment to powerful theories utilizing dimensionless parameters. In both cases the sciences with which they were associated reached full maturity in the postwar era.[62]

New and highly sophisticated tools for engineering research, such as the interferometer, did appear and they were important particularly after about 1940. But in the short run engineers derived equal or greater benefits by means of scale-model experiments using existing devices such as dynamometers, wind tunnels, and their hydraulic analogs, testing flumes, beginning in the late 1880s. These older devices now could be made to yield information of greater generality and utility than had been previously the case. The virtual elimination of scale effects allowed experimental results to be compared directly with theory; this necessitated new theories in many cases. The discovery of scaling laws went hand-in-hand with an enormous theoretical enrichment of engineering that has become one of the principal attributes of twentieth century technology, and that was to bring engineering into a closer symbiotic union with physics as part of a centuries-old cycle of interactions between science and technology.

The dimensional revolution in engineering served to emphasize the limits of pure empiricism, and the necessity for a fruitful interaction between theory and experiment. Purely conceptual changes permitted information of enormous value to be generated by the use of long-known experimental apparatus. Indeed, new information could be derived in some cases from older experimental data. The development of dimensional analysis then functioned in a manner analogous to a new method for extracting gold from its ore more efficiently. In such cases not only can new ore deposits be worked, but precious metal can be extracted from the tailings left behind at old mines. There was much precious engineering information that had been present all along in old experimental data, but it could not be fully understood or properly utilized until new theoretical insights made it available for use.

The Reynolds number provides a classic example. As Stokes noted, when presenting a gold medal to Reynolds, in fluid flow:

> . . . when the motion becomes eddying [i.e., turbulent] it seemed no longer to be amenable to mathematical treatment, but Professor Reynolds has shown that the same conditions of similarity hold good as to the average effect, even when the motion is of the eddying kind . . .[63]

Thus, in this and other ways, seemingly meaningless data could be reanalyzed, and even turbulence brought within the rule of mathematical theory.

The discovery and development of dimensional analysis had a revolutionary effect upon engineering. It helped to end a long-standing gap between physical theory and engineering experiment. One result was more accurate experiments, often involving scale models. Another result was a dramatic improvement in engineering theories. These theories could now be cast into the form of theories in physics and linked to general scientific principles, often with the aid of dimensionless parameters. Engineering theories came to serve as a bridge between the basic sciences and engineering practice.

NOTES

1. René Dugas, *Historie de la Mechanique* (Neuchatel: Editiones du Griffon, 1950), generally ignores dimensional issues. For example, he gives Navier sole credit for what English-speaking historians usually call the Navier-Stokes equations (pp. 393-407). But it was precisely Stokes's contribution to introduce dimensional considerations into the theory of viscous flow. Dugas also ignores Osborne Reynolds and other principal actors in the dimensional revolution.

2. The common fallacy in earlier years had been to consider only geometric similarity between model and prototype. In fact it is necessary that all the significant conditions be similar, including motion and forces. The full implications of this insight, as might be expected, took more than a generation to work out. The conditions of perfect similarity cannot always be realized. This condition can be stated in modern terms. If the mathematical expression describing a technological system is expressed in terms of dimensionless parameters (as had become standard, particularly since the discovery and development of Buckingham's theorem), then complete similarity is achieved when the independent dimensionless variables have the same value for the model and the prototype. This, of course, shows that scale-model experimentation and the use of dimensionless parameters in engineering theory are intimately related. It is not always possible to obtain perfect similarity. As one leading authority put it, "usually it is not possible to impose complete similarity in a model test. Consequently, some of the independent dimensionless variables . . . are allowed to deviate from their correct values. An important part of the work of the model engineer -- indeed the most important part-- is to justify his departures from complete similarity or to apply theoretical corrections to compensate for them." (Harry L. Langhaar, *Dimensional Analysis and Model Theory* [New York: John Wiley, 1951], p. 64.) On Buckingham, see below.

3. Osborne Reynolds, "An Experimental Investigation of the Circumstances Which Determine Whether the Motion of Water Shall be Direct or Sinuous and the Law of Resistance in Parallel Channels," *Papers on Mechanical and Physical Subjects*, 3 vols. (Cambridge: Cambridge Univ. Press, 1900-1903), II, pp. 51-105. John Smeaton had used models, but realized that the results might not be valid, so he delayed publication until he had compared the behavior of his models with that of full-sized artifacts. There were serious difficulties in this procedure, and many of his successors rejected model experiments and conducted tests at full scale. Benjamin F. Isherwood, a leading American engineer of the nineteenth century, held that, "a fact, to be of practical authority in engineering must be derived from experiments made on the scale and under the conditions of actual practice." This was, in fact, the practice at Lowell, for James B. Francis's experiments, in which he used only full-scale turbines for his tests. (On Smeaton's test, see John Smeaton, "An Experimental Enquiry concerning the Natural Powers of Water and Wind to Turn Mills, and Other Machines, Depending on a Circular Motion," *Philosophical Transactions*, 1759, *51*: 100-174, and see also Terry S. Reynolds, *Stronger Than a Hundred Men, A History of the Vertical Water Wheel* [Baltimore: Johns Hopkins, 1983], pp. 223-226. Isherwood is quoted in Edwin T. Layton, Jr., "American Ideologies of Science and Engineering," *Technology and Culture*, 1976, *17*: 693. On Francis and the Lowell tests, see Edwin T. Layton, Jr., "Scientific Technology: the Hydraulic Turbine and the Origins of American Industrial Research," *T&C, 20*: 72-78, 81.)

4. Osborne Reynolds, "On Certain Laws Relating to the Régime of Rivers and Estuaries and on the Possibility of Experiments on a Small Scale," ibid., pp. 326-335. See also his three committee reports on scale-modeling in hydraulics, ibid., pp. 380-518.

5. Osborne Reynolds, "On the Extent and Action of the Heating Surface of Steam Boilers," ibid., pp. 81-85. See also Ernst R. G. Eckert, "Heat Transfer," in *Osborne Reynolds and Engineering Science Today*, ed. D. M. McDowell and J. D. Jackson (Manchester: Manchester Univ. Press, 1970), pp. 160-175.

6. Osborne Reynolds, "On the Theory of Lubrication and the Experimental Determination of the Viscosity of Olive Oil," *Papers*, II, pp. 228-310. See also F. T. Barnwell, "The Founder of Modern Tribology," in *Osborne Reynolds*, ed. McDowell and Jackson, pp. 240-263.

7. Leonard S. Reich, *The Making of American Industrial Research: Science and Business at GE and Bell, 1876-1926* (Cambridge and New York: Cambridge Univ. Press, 1985) and George Wise, *Willis R. Whitney, General Electric, and the Origins of U.S. Industrial Research* (New York: Columbia Univ. Press, 1985). See also George Wise, "A New Role for Professional Scientists in Industry: Industrial Research at General Electric, 1900-1916," *T&C*, 1980, *21*: 408-29.

8. Edwin T. Layton, "Mirror Image Twins: The Community of Science and Technology in 19th-Century America," *T&C*, 1971, *12*: 562-580.

9. David F. Channell, "The Harmony of Theory and Practice: The Engineering Science of W. J. M. Rankine," *T&C*, 1982, *23*: 39-52.

10. Osborne Reynolds, "An Experimental Investigation," *Papers*, II, p. 51.

11. Ibid.

12. Jack Allen, "The Life and Work of Osborne Reynolds," in *Osborne Reynolds*, ed. McDowell and Jackson, pp. 11-14, *passim*.

13. I have argued in a number of papers that the underlying orientations toward knowing and doing, along with associated values and loyalties, are what distinguish engineering and the basic sciences. See, for example, my "Mirror Image Twins," cited above, as well as my "Technology as Knowledge," *T&C*, 1974, *15*: 31-41, and my "American Ideologies of Science and Engineering," *T&C*, 1976, *17*: 688-701.

14. Walter G. Vincenti, "The Air-Propeller Tests of W. F. Durand and E. P. Lesley," *T&C*, 1979, *20*: 712-51. See also Barton C. Hacker, "Greek Catapults and Catapult Technology: Science, Technology, and War in the Ancient World," *T&C*, 1968, *9*: 34-50.

15. Alexander Keller, "Mathematicians, Mechanics and Experimental Machines in Northern Italy in the Sixteenth Century," in *The Emergence of Science in Western Europe*, ed. M. P. Crossland (London: Butterworths, 1975): 15-34, and "Pneumatics, Automata and the Vacuum in the Work of Giambattista Aleotti," *British Journal for the History of Science*, 1967, *3*: 338-347.

16. Ladislao Reti, "Il moto dei prioetti e del pendolo secondo Leonardo e Galileo," *Le Machine*, 1968, *1*: 63-89. Much of the material in the preceding article is available in English in *The Unknown Leonardo*, ed. Ladislao Reti (New York: McGraw-Hill, 1974). Among Reti's many works there are many that deal with the experimental scientific work of Leonardo. See, for example, Ladislao Reti and Bern Dibner, *Leonardo Da Vinci, Technologist* (Norwalk, Conn.: Burndy Library, 1969).

17. Again, many works by Settle might be cited, but see, for example, his "Ostilio Ricci, a Bridge Between Alberti and Galileo," *XII Congrès D'Histoire des Sciences, Actes*, 1968: 121-6.

18. See, for example, Samuel V. Edgerton, *The Renaissance Rediscovery of Linear Perspective* (New York: Basic Books, 1975), and Vernard Foley, "Leonardo's Contributions to Theoretical Mechanics," *Scientific American*, 1986, *225*: 108-13.

19. Settle, "Ostilio Ricci." See also Arnaldo Masotti, "Ostilio Ricci," *Dictionary of Scientific Biography* (hereafter *DSB*).

20. Marcus Vitruvius Pollo, *The Architecture of Marcus Vitruvius Pollo in Ten Books*, trans. Joseph Gwilt (London: Priestley and Weale, 1826), pp. 342.

21. Kenneth Keele, "Leonardo Da Vinci," *DSB* See also Carlo Zammattio, "Mechanics of Water and Stone," in Reti, ed., *Unknown Leonardo*, p. 201.

22. H. A. Becker, *Dimensionless Parameters* (New York: John Wiley, 1976), pp. 13. As noted above, scale model experiments are valid only if motions and forces are also scaled properly.

23. Peter Jeffrey Booker, *A History of Engineering Drawing* (London: Chatto & Windus, 1963), pp. 18-36, 86-106.

24. Galileo Galilei, *Dialogues Concerning Two New Sciences*, trans. Henry Crew and Alfonso De Salvio (New York: Macmillan, 1914), pp. 4-5.

25. Isaac Newton, *Mathematical Principles of Natural Philosophy*, 2 vols., trans. Florian Cajori (Berkeley: Univ. California Press, 1962), II, book 2, prop. 32. Enzo O. Macagno, "Historico-Critical Review of Dimensional Analysis," *Journal of the Franklin Institute*, 1971, *292*: 391-402 is the most critical and comprehensive history of dimensional analysis I have used. (I would like to thank Professor Walter Vincenti for calling my attention to this important paper.) H. E. Huntley, *Dimensional Analysis* (London: McDonald, 1952) contains a brief history (pp. 33-44). For another historical study see Alton C. Chick, "The Principle of Similitude," in *Hydraulic Laboratory Practice*, ed. John R. Freeman (New York: The American Society of Mechanical Engineers, 1929), pp. 796-797.

26. Reti, "Il moto dei prioetti," pp. 64-71. See also Bern Dibner, "Machines and Weaponry," in *Unknown Leonardo*, ed. Reti, pp. 181-182. Leonardo's notebooks contain numerous painstaking drawings of turbulent flow. See, for example, Zammattio, "Mechanics of Water and Stone," ibid., pp. 192-193, *passim*. See also Benoit B. Mandelbrot, *The Fractal Geometry of Nature* (New York: W. H. Freeman, 1983), pp. 97, fig. C3.

27. Hunter Rouse and Simon Ince, *History of Hydraulics* (New York: Dover, 1957), pp. 73-112, *passim*. P. F. Neményi, "The Main Concepts and Ideas of Fluid Dynamics in Their Historical Development," *Archive for the History of the Exact Sciences*, 1962-66, *2*: 52-86.

28. Stephen Timoshenko, *History of Strength of Materials* (New York: McGraw-Hill, 1953), pp. 190-197, 283-288. On the significance of the graphical methods of Carl Culmann on structural thinking, see David P. Billington, *Robert Maillart's Bridges* (Princeton: Princeton Univ. Press, 1979), pp. 6-8. The graphical emphasis did not, of course, rule out simple algebraic expressions. Indeed the two were complementary. See, for example, Charles Hutton, *A Course of Mathematics*, 2 vols. (New York: Campbell and others, revised from 5th and 6th London editions by Robert Adrian, 1818). The second volume includes an extensive treatise on applied mechanics, in which geometric reasoning is simplified by the use of algebraic expression.

29. Julius Weisbach, *A Manual of the Mechanics of Engineering*, 3 vols., trans. A. Jay DuBois (New York: John Wiley & Sons, 1880), pp. 397-445. Something of the influence and limitations of the two-dimensional geometrical formalism employed by Weisbach and other turbine theorists is suggested in Edwin T. Layton, "Scientific Technology: The Hydraulic Turbine and the Origins of American Industrial Research," *T&C*, 1979, *20*: 74-76.

30. The older graphical tradition lingers in engineering handbooks. On indicator and thermodynamic diagrams, see *Mark's Standard Handbook for Mechanical Engineers*, ed. Theodore Baumeister, and others (New York: McGraw-Hill, 1978), section 4 pp. 19-21, sec. 9 pp. 36-38, 78-80.

31. Whether a magnitude is dimensional or nondimensional is a matter of convention. Angles can be measured in dimensional terms by degrees of radians or in nondimensional terms, as ratios of the sides of similar triangles. See, for example, L. I. Sedov, *Similarity and Dimensional Methods in Mechanics* (New York: Academic Press, 1959), pp. 2-3. Percy Bridgeman's great contribution to dimensional analysis was to show that there are no "fundamental" dimensions upon which all others depend. The determination of what is a "fundamental" dimension depends upon convention. See Percy Bridgeman, *Dimensional Analysis* (New Haven: Yale Univ. Press, 1931).

32. Timoshenko, *History of Strength of Materials*, pp. 190-197, 283-288.

33. On graphics as a language for American engineers see Larry Owens, "Vannevar Bush and the Differential Analyzer: the Text and Context of an Early Computer," *T&C*, 1986, *27*: 85-95. I have discussed the ways that engineering values influenced engineering sciences in America in my "Mirror Image Twins," pp. 562-580.

34. Layton, "Mirror Image Twins," p. 562.

35. Olaf Hougan, "Seven Decades of Chemical Engineering," *Chemical Engineering Progress*, 1972, *73*: 89-104, on pp. 94-96. The classic text in the revolution in unit operations was William H. Walker, Warren K. Lewis, and William H. McAdams, *Principles of Chemical Engineering* (New York: John Wiley and Sons, 1923). Only one dimensionless parameter (the Reynolds number) was mentioned in passing in this work.

36. Hougon, "Seven Decades," pp. 94-96. Hougan presents an interesting three-stage evolution, from industrial chemistry, to unit processing, to more fundamental engineering sciences, with emphasis upon transport process theories.

37. Klaus Buchholz, "Verfahrenstechnik (Chemical Engineering) --Its Development, Present State and Structure," *Social Studies of Science*, 1979, *9*: 43. There are no more "fundamental" theories than fundamental dimensions; in each case the "fundamental" entities must be described in terms of a set of human conventions. That is, the definition of "fundamental" is a cultural artifact. In the era in question (c. 1880-1960) the influence of physics was shown by the fact that engineers tended to define "fundamental" in the sense developed in physics. But such an approach might not, in fact, prove fundamental if the task were to optimize the design of a particular artifact. Indeed, it was often the experience in classical engineering sciences in the nineteenth century that the introduction of atomic concepts led to needless complications without compensating benefits. In an influential paper Dean W. R. Marshall denounced the excessive emphasis upon science in chemical engineering, which he called the "science worship syndrome," and the neglect of things like design that are more fundamental from the point of view of engineering and the industries that it serves. (See W. R. Marshall, Jr., "Science Ain't Everything," *Chem. Eng. Prog.*, 1964, *60*: 17-21, on p. 17.)

38. Ewald Wicke, "Gerhard Damköhler -- Founder of Chemical Reaction Engineering," *International Chemical Engineering*, 1985, *25*: 770-773.

39. Félix Lucas, "Sur les équations abstraites du fonctionnement des machines," *Bulletin du Societé Mathématique de France*, 1891, *19*: 152-154.

40. E. Caravallo, "Sur une Similitude des Fonctions des machines," *La Lumiere Électrique*, 1891, *42*: 506-507.

41. Aimé Vaschy, "Sur les lois de similitude en physique," *Annales Télégraphiques*, 1892, *19*: 25-28 and the same author's "Sur les lois de similitude de électricité," ibid., 1892, *19*: 189-211.

42. Edgar Buckingham, "On Physically Similar Systems, Illustrations of the Use of Dimensional Analysis," *Physical Review*, 1914, *4*: 345-376. The mathematical foundations of the Vaschy-Buckingham theorem was explored by Langhaar, *Dimensional Analysis*. The physical significance of dimensionless parameters has been explored in a remarkable book by Stephen J. Kline, *Similitude and Approximation Theory* (New York: McGraw-Hill, 1965). Langhaar's methods and Kline's physical insights have been combined and placed in historical context in Becker, *Dimensionless Parameters*.

43. Ronald R. Kline, "Charles P. Steinmetz and the Development of Electrical Engineering Science" (unpublished Ph.D. dissertation, Univ. Wisconsin, 1983), pp. 5-11. I do not think there is any conflict between Kline's work and my own. The basis of my "mirror image twins" metaphor was not the form of the theories but the orientations adopted by the communities of technology and of the basic sciences, in particular the relative weight given to "knowing" and "doing." It is true that in earlier papers I also emphasized the real differences in the forms of theories, particularly as they emerged in the nineteenth century. As noted above, this was in part an artifact of the graphical, visually oriented culture of American engineers. But the convergence between physics and engineering is a predictable outcome of the "scientific revolution in technology." I have noted elsewhere that the growing similarity in the forms of theories has been accompanied by a

reaffirmation of the traditional engineering concern for "doing." The differences in value are still there perhaps expressed somewhat differently. (See Layton, "Mirror Image Twins," p. 576.)

44. Frederick Terman, "Electrical Engineers are Going Back to Science!" *Proceedings of the IRE*, 1952: 738-40.

45. Quoted in A. Michal McMahon, *The Making of a Profession* (New York: IEEE Press, 1984), pp. 238-239.

46. Raymond W. Warner and R. L. Grung, *Transistors: Fundamentals for the Integrated-Circuit Engineer* (New York: Wiley-Interscience, 1983), pp. 1-91, is a history of micro-electronics, including its electrical engineering and electronics background. Warner and Grung make the interesting observation that "in microelectronics technology, science came first," (p. 1) and note that ". . . it [microelectronics] is the only technology that fits the popular view of a typical technology" (p. 3) in being based upon science. But Warner also finds much to say about the role of engineers in the history of electrical engineering, electronics, and microelectronics (pp. 4-17 ff.).

47. Ernst R. G. Eckert, Interview, 13 January 1987.

48. See, for example, Marshall, "Science Ain't Everything," pp. 17-21. Marshall held, among other things, that ". . . the cloak of the high priesthood of science has cast a dark shadow over the profession of engineering during the past decade and a half" (p. 17).

49. E. Brauer, "Eulers Turbinentheorie," *Zeitschrift für das Gesamte Turbinenewesen*, 1908, *2*: 21-24. I would like to thank Dip. Ing. Hans Häckert for calling my attention to this reference.

50. On the history of the use of the Mach-Zehnder interferometer, see Werner Kraus, "Messung des Termperatur-und Geschwindigkeitsfelds bei freier Konvektion," *Bücher der Messtechnik*, ed. L. Schiller (Karlsruhe: G. Braun, 1955), pp. 7-82. I would like to thank Dr. Virginia Dawson for calling this work to my attention. See also Ernst R. G. Eckert, Interview, 13 January 1987. E. R. G. Eckert, "Gas Turbine Research at the Aeronautical Research Center, Braunschweig, during 1940-1945," *Atomkernenergie*, 1978, *32*: 208-211. George B. Wood, "Interferometer Methods," in North Atlantic Treaty Organization, Advisory Group for Aeronautical Research and Development, *Optical Methods for Examining the Flow in High Speed Wind Tunnels* (Washington, D.C.: N.A.C.A., 1956), pp. 123-148.

51. Wood, "Interferometer Methods," pp. 123-148, *passim*. The Mach-Zehnder inter-ferometer was developed and first used in Germany; since 1945 it has become a standard tool for the study of supersonic flight. On the general problem of aircraft modeling, see Sedov, *Similarity in Mechanics*, pp. 43-46.

52. Rouse and Ince, *History of Hydraulics*, pp. 219-242.

53. N. Rott, "Jakob Ackert and the History of the Mach Number," *Annual Review of Fluid Mechanics*, 1985, *17*: 1-9, on p. 1. It was Arnold Sommerfeld who in 1908 named the "Reynolds" number.

54. The Reynolds number expresses the ratio between inertial and viscous forces in fluid flow and, in consequence, determines the critical point at which the flow changes from laminar flow to turbulent flow.

55. Ludwig Prandtl, "Über Flüssigkeitsbewegung bei sehr kleiner Reibung," in *Ludwig Prandtl Gessamelte Abhandlungen*, ed. F. W. Reigels, and others, 3 vols. (Berlin: Springer-Verlag, 1961), II, pp. 575-584.

56. On the use of the Reynolds number by Prandtl and his followers, see N. Rott to G. I. Barenblatt, 23 January 1986. I would like to thank Professor Walter Vincenti for providing me a copy of this letter and (at the same time) calling my attention to Rott's paper on Ackert (cited above); Ackert was Rott's predecessor at the Zürich Federal Institute of Technology.

57. Ernst R. G. Eckert, "Engineering Education in Heat Transfer, *Heat Transfer Engineering*, 1985, *6*: 35-38, on p. 35. The work by Ludwig Prandtl was his *Abriss der Strömungslehre* (Braunschweig: Freidrich Vieweg, 1931). Eckert's classic text on heat and mass transfer first appeared as E. R. G. Eckert and Robert M. Drake, Jr., *Heat and Mass Transfer* (New York: McGraw-Hill, 1959). On the historical development of heat transfer, see John H. Lienhard, "Notes on the Origins and Evolution of the Subject of Heat Transfer," *Mechanical Engineering*, 1983, *105*: 20-27, and Kraus, "Messung des Tempertur-und Geschwindigkeitsfeldes bei freier Konvektion," pp. 7, 102-117.

58. Ernst R. G. Eckert, "Pioneering Contributions to Our Knowledge in Convective Heat Transfer (Hundred Years of Heat Transfer Research)," *American Society of Mechanical Engineers*, 1980: paper *80-HT-137*: 1-9.

59. Ernst R. G. Eckert, Interview 13 January 1987.

60. "This Week's Citation Classic," *Current Contents*, 1983, *21*:317. The classic paper was E. R. G. Eckert and W. O. Carlson, "Natural Convection in an Air Layer Enclosed Between Two Vertical Plates With Different Temperatures," *International Journal of Heat and Mass Transfer*, 1961, *2*:106-20. This was the second most frequently cited paper reviewed to that date (1983) by *Current Contents*. The paper in question was one of the first detailed studies of natural convection using the Mach-Zehnder interferometer. In this paper Eckert substituted thin plates of optical glass (c. 2mm thick) for the thick glass plates (c. 25 mm thick) previously used with the Mach-Zehnder interferometer.

61. Interview with Ernst R. G. Eckert, 9 January 1987. Eckert's book was his *Introduction to the Transfer of Heat and Mass* (New York: McGraw-Hill, 1950). The work of Ludwig Prandtl referred to was "Über Flüssigkeitbewegung bei sehr kleiner Reibung." The boundary layer concept included viscosity (at the boundary layers) and eliminated the unrealistic idealizations that postulated ideal non-viscous fluids. The (near) elimination of idealization, however, went hand-in-hand with the creative application of dimensional analysis and nondimensional numbers and the extensive use of model laws and model experiments. As noted above, this was not until after 1910, when Prandtl began to use the Reynolds number. In heat transfer, in the paper in which Eckert emphasized the role of dimensional analysis in the rise of heat transfer he held that: "The fundamental laws of mechanics were, of course, already known since Isaac Newton and have been formulated for fluid flow in the Navier-Stokes equations. However, solutions to these equations could only be obtained for a few simple situations and even there they often did not agree with experimental results, for instance, with those describing the pressure

drop connected with fluid flow through pipes. Neither did many of the experimental results correlate among themselves." (Eckert, "Pioneering Contributions," p. 1.)

62. Edwin T. Layton, Jr. and Richard Goldstein, "Interviews with Ernst R. G. Eckert," five videotape interviews (1977). Available through Walter Library, University of Minnesota, Minneapolis, Minn. 55455. Wicke, "Damköhler," p. 771.

63. Joseph Lamor, ed., *Memoir and Scientific Correspondence of the Late Sir George Gabriel Stokes*, 2 vols. (Cambridge: Cambridge Univ. Press, 1907), I, pp. 233-234.

II

The Interpretation of Great Experiments

MICHELSON'S FIRST ETHER-DRIFT EXPERIMENT
IN BERLIN AND POTSDAM

Barbara Haubold, Hans Joachim Haubold, and Lewis Pyenson

PROLOGUE

More than 100 years ago, in April 1881, Albert A. Michelson (1852-1931) performed the first version of his famous experiment to measure the effect of the Earth's motion through the ether. In Michelson's own words: "the result of the hypothesis of a stationary ether is [thus] shown to be incorrect, and the necessary conclusion follows that the hypothesis is erroneous. This conclusion directly contradicts the explanation of the phenomenon of aberration which has been hitherto generally accepted, and which presupposes that the earth moves through the ether, the latter remaining at rest." However, as Michelson discussed in his paper, only the hypothesis of A. J. Fresnel (1788-1827) of a stationary ether was shown to be wrong. Not called into question was "the existence of a medium called the ether, whose vibrations produce the phenomenon of heat and light, and which is supposed to fill all space."[1] The centenary of Michelson's pioneering experiment has served as an opportunity to recall the context and the outcome of the experiment, which was prepared at the Physical Institute of the University of Berlin under the guidance of H. von Helmholtz (1821-1894) and performed at the Astrophysical Observatory in Potsdam by arrangement with the observatory director, H. C. Vogel (1841-1907).[2]

In 1931, the year Michelson died, the German translation of the 1881 paper appeared, at the suggestion of A. Berliner. Reprinting the results of Michelson's work in a German translation was an appreciation of his famous experiment "weil sie zeigt," as M. von Laue (1879-1960) wrote in an annotation: "wie Michelson dem von Maxwell erwogenen, aber mit einem 'Unmöglich' beiseitegeschobenen Gedanken [proof of effects to the order $(v/c)^2$] eine Wendung ins Positive gibt, und mit welchem jugendfrischen Mut er an die Verwirklichung geht."[3] It is beyond any doubt that Michelson's 1881 paper is a classic. It has frequently been reprinted in its original form or in translations.

Even the centenary of the birth of A. Einstein (1879-1955) was an occasion to refer extensively to Michelson as the master of light and to his important contributions to ether-drift problems, which were centrally relevant to Einstein's special theory of relativity.[4] The aim of the present paper is to present lesser-known documents and facts connected with the experiment that Michelson performed in Berlin and Potsdam.

THE FORMATIVE YEARS

Albert Abraham Michelson was born in Strelno (at the time, part of the Kingdom of Prussia), 19 December 1852, and was brought to the United States in 1855, when his parents immigrated. He spent most of his childhood in the world of California's gold miners.[5] He graduated from the United States Naval Academy in 1873 and returned as an instructor in physics and chemistry from 1875 to 1879. During that time, in accordance with his personal interests, Michelson started early measurements of the speed of light. More generally, he did work in several areas of physical optics, particularly on methods for measuring the speed of light established by A. H. L. Fizeau (1819-1896) and L. Foucault (1819-1868), respectively. Fizeau (1849) had measured the velocity of light in air and water using a machined gear wheel and found that it was greater in air, a crucial experiment in favor of the wave theory of light and against the corpuscular theory. In 1850, Foucault had determined the absolute value of the velocity of light by adopting a rotating mirror, an experiment that later was important in establishing the electromagnetic theory of light and the special theory of relativity. Foucault's technique attracted the interest of Michelson. S. Newcomb (1835-1909), Superintendent of the Nautical Almanac Office

Figure 1
Albert A. Michelson, about 1887
(Courtesy of Dorothy Michelson Livingston)

at the Washington Naval Academy, had taken measurements of the speed of light using Foucault's method. Michelson was aware of Newcomb's results.[6] In 1877 he modified and improved Foucault's method by using a still longer light path and larger light intensity; thus he was able to get much more precise measurements of the speed of light. The results appeared in the *American Journal of Science*.[7] Michelson received the good news that Newcomb was interested in his experiments, and in 1878 the two researchers began a long correspondence.[8] In 1879 Michelson became a member of the staff of the Nautical Almanac Office and began collaborating with Newcomb in his scientific work.

ENCOUNTER WITH GERMANY IN 1880-1881

Trials with the "interferential refractor" in Helmholtz's Physical Institute in Berlin

In the 1870s and 1880s, research work leading to a German-style Ph.D. could be conducted at only a small number of American universities. The federal government, accordingly, offered scientists a leave of absence for specialized education in Europe. Michelson benefitted from this common procedure for two years, when he studied in France and Germany under the great masters of optics.[9]

Michelson arrived in Europe in 1880 at the Collège de France and studied under the guidance of A. Cornu (1841-1902). Cornu had refined Fizeau's gear-wheel method for measuring the speed of light. Most probably through discussions with Cornu and influenced by Cornu's measurements of the speed of light, Michelson came to the idea for measuring the ether drift. Thereby he also became aware of Fizeau's paper about technical problems relating to measurement of the speed of light. Particularly important for Michelson's interferometer (still in the future) was the semitransparent mirror arrangement used by Fizeau. Fizeau refined the technical detail of a plane-parallel plate glass by silvering the front surface and utilizing only the half of the beam reflected from that front surface. The technique had already been developed by J. C. Jamin (1856) and E. E. Mascart (1872) in France.

Michelson's research under Cornu's guidance notwithstanding, the sources for the invention of the apparatus later known as the Michelson interferometer are open questions. In Michelson's classic paper of 1881, he refers to a letter[10] of J. C. Maxwell's (1831-1879), written just before he died, to D. P. Todd (1855-1939), at that time a colleague of Newcomb's. In this letter Maxwell discussed whether observations on the satellites of Jupiter could be made with such a high precision that speed of light determinations could be carried out to reveal changes in the speed of light caused by the Earth's motion through the luminiferous ether.[11] However, Maxwell stated that no apparatus existed capable of measuring effects of the order (v^2/c^2), the square of the ratio of the Earth's speed to that of the light. In an earlier discussion,[12] in the ninth edition of the Encyclopedia Britannica in 1878, Maxwell had considered the possibility of measuring variations of the velocity of light in different directions due to the motion of the Earth through the ether. However, Maxwell considered terrestrial experiments to be hopeless. This announced hopelessness may have been a challenge for the energetic and enthusiastic Michelson.[13]

Michelson arrived in Berlin in September 1880. Fortunately, his stay in Germany can be followed by his correspondence with Newcomb and A. G. Bell (1847-1922).[14] Michelson started his work in the Physical Institute of the Royal Friedrich-Wilhelms University of Berlin (today the Humboldt University of Berlin)[15] and particularly in Helmholtz's laboratory for optical research; both centers had the very best technical equipment for optical studies. In 1880, Helmholtz himself was already world famous for his principle of energy conservation, his hydrodynamics of vortex motion, his investigations of electrodynamic potentials, and for his contributions to physiological optics and acoustics.[16] In the area of physical optics Helmholtz had created a theory of microscopes and had founded the theoretical study of the anomalous dispersion of light. Just in the center of his research work at that time was the influence of motion on electromagnetic forces.

Figure 2
Physical Institute of H. v. Helmholtz, Berlin, destroyed during the Second World War (Courtesy of the Humboldt University Archives, Berlin)

Starting his training in Berlin, Michelson enrolled at the University of Berlin (see Fig. 3) to hear the lectures of Helmholtz. We have his certificate (see Fig. 4) and an assessment for him (see Fig. 5) as well as a written note[17] by the administration of the university that states that Michelson attended classes in theoretical physics, theory of light, and physical laboratory work in the fall semester of 1880 (see Fig. 6). Other evidence for Michelson's studies at the University of Berlin are some documents concerning his work in the Royal Library of the university.[18]

Date	Residence of Michelson	Archival Document
26 June 1880	Washington	Letter of Michelson to A. M. Meyer; cited in n. 8, p. 286.
16 October 1880	Berlin	Matriculation at the Königliche Friedrich Wilhelms Universität, Berlin; see Fig. 4.
19 October 1880	Berlin	Letter of Michelson to S. Newcomb; cited in n. 8, p. 287.
22 November 1880	Berlin	Letter of Michelson to S. Newcomb -- elementary work in the laboratory of H. Helmholtz; lectures by Dr. Helmholtz; quite a long conversation with Dr. Helmholtz; cited in n. 8, p. 287.
	Berlin	Letter of Michelson to A. G. Bell, not found; cited in n. 8, p. 289, note 35.
9 March 1881	Berlin	Proposal for getting the leaving certificate, see Fig. 4; handing of books to the university library, cf. Haubold and John, 1982; cited in n. 2, Figs. 3b, 6a and 6b.
11 March 1881	Berlin	Date of the proposal for getting the leaving certificate, cf. Haubold and John, 1982; cited in n. 2, Fig. 3b.
13 March 1881	Berlin	Remark on classes attended by Michelson; see Fig. 6.
23 March 1881	Berlin	Confidential report for Michelson; see Fig. 5.
5-15(?) April 1881	Potsdam	Michelson performed his ether-drift experiment; cited in Wempe, n. 2.
17 April 1881	Heidelberg	Letter of Michelson to A. G. Bell -- intention to attend lectures given by Quincke and Bunsen; review of the experiments made in Berlin and Potsdam; cited in n. 8, p. 288.
17 April 1881	Heidelberg	Letter of Michelson to S. Newcomb, not found, cited in n. 8, p. 290, note 38.
2 July 1881	Heidelberg	Letter of Michelson to S. Newcomb, cited in n. 8, p. 249.
8 July 1881	Heidelberg	Letter of Michelson to S. Newcomb, cited in n. 8, p. 295
29 August 1881	Schluchsee/ Schwarzwald	Letter of Michelson to S. Newcomb, cited in n. 8, p. 298.
23 October 1881	Paris	Letter of Michelson to S. Newcomb, cited in n. 8, p. 300.

Figure 3

Summary of A. A. Michelson's postgraduate studies in Germany, 1880-81 using documents still available.

Figure 4
A.A. Michelson's leaving certificate from the Königliche Friedrichs-Wilhelms Universität zu Berlin (Courtesy of the Humboldt University Archives, Berlin)

Figure 5
Confidential report for A.A. Michelson when he finished his courses at the Berlin University (Courtesy of the Humboldt University Archives, Berlin)

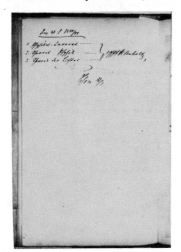

Figure 6
Administrative remark on classes attended by A.A. Michelson at the Berlin University (Courtesy of the Humboldt University Archives, Berlin)

On 22 November 1880, already fully engaged by his research work, Michelson wrote in a letter to Newcomb:

> I . . . had quite a long conversation with Dr. Helmholtz concerning my proposed method for finding the motion of the earth relative to the ether, and he said he could see no objection to it, except the difficulty of keeping a constant temperature. He said, however, that I had better wait till my return to the U.S. before attempting it, as he doubted if they had the facilities for carrying out such experi-

ments, on account of the necessity of keeping a room at a constant temperature.[19]

Indeed, it is quite interesting to note that the same objection was made many years later by Einstein in a letter to P. Ehrenfest (1880-1930) concerning the analysis of the observations of D. C. Miller (1866-1941) of a slight positive result for ether drift that he obtained in his Mount Wilson experiments in the early 1920s.[20] Miller, one of the many experimenters who repeated the Michelson experiment in a modified form, had measured a strong shift of the interference fringes in 1921 that he explained as a consequence of a relative motion of the Earth against the ether. It was R. S. Shankland[21] (1905-1982) and his collaborators who made an exhaustive statistical analysis of Miller's observations and concluded that Miller's positive signal originated in temperature changes.

During the preparations for the ether-drift experiment Michelson continued his experiments with light passing through a very narrow slit, which he also discussed with Helmholtz. Both areas were no doubt stimulated by the great interest of Helmholtz, and the discussions between Helmholtz and Michelson surely led to concrete plans for constructing the "interferential refractor" that Michelson intended to use for the determination of the relative motion of the Earth against the ether.

Obtaining funds from A. G. Bell, the founder of the telephone, Michelson had an interferometer made to his design by a well-known German instrument firm, Schmidt & Haensch,[22] which also provided instruments to other scientists, including Helmholtz.

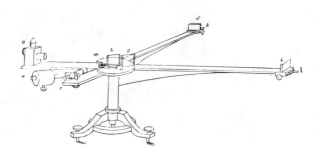

Figure 7
This drawing depicts the original "interferential refractor" that was constructed for Michelson's experiment in Berlin and Potsdam by Schmidt and Haensch, Berlin. (*American Journal of Science*, July 1881)

Michelson located his interferometer (see Fig. 7) in the basement of the Physical Institute of the University of Berlin, as his measurements had shown that the interferometer was so sensitive that the slightest motion would cause an immediate shift of the fringes.[23] Traffic on the street adjacent to the Physical Institute made it impossible to make measurements during the day and even disturbed the measurements taken during the night. Helmholtz was acquainted with the director of the Royal Astrophysical Observatory at

Potsdam, H. C. Vogel, and through mutual accord Michelson moved his apparatus to Potsdam.[24]

THE ALMOST CRUCIAL EXPERIMENT

Only the main building of the Astrophysical Observatory, begun in 1878, was available for use for experiments when Michelson arrived in Potsdam.[25] He arranged his apparatus on a stone pier in the basement under the eastern dome of the main building (see Fig. 8). In this windowless cellar the measuring equipment was screened from vibrations and kept at a relative constant temperature (see Figs. 9 and 10).[26] But, as Michelson wrote in his 1881 paper,[27] the interferometer was so extraordinarily sensitive that even the stamping of the pavement at a distance of 100 meters could cause a shifting of the interference fringes. Michelson was able to observe the fringes undisturbed, however, and he was himself astonished at the beauty of the interferometer performance.

Figure 8

The Königliche Astrophysikalische Observatorium bei Potsdam, about 1890. Michelson performed his experiment in a cellar that is in the basement of the dome seen on the far left. (Copy of a drawing; courtesy of M. Strohusch)

Michelson could measure distances in units of wavelength of light with his interferometer (see Fig. 7): The light coming from a source of light *[Laterne]; (a)* will be split at a partial transparent mirror *(b)* into a reflected *(d)* and a transmitted *(c)* part. These two light beams will be reflected at the mirrors *(d)* and *(c)*, respectively, and are reunited at *(b)* and transmitted to *(e)*, where the fringes may be observed in the eyepiece to see if the waves are in interference. The compensating plate of glass *(g)* was inserted between *(b)* and *(c)* to equalize the optical path.

Figure 9
The "Michelson cellar," which enclosed the "interferential refractor" for
Michelson's experiment in 1881, as it looked in about 1965. In the middle
of the room is the original sandstone pier. (Courtesy of H. Strohbusch)

Figure 10
The Michelson cellar as prepared for the Michelson exhibition on the
occasion of the Centennial Symposium at the Central Institute for Astrophysics of the Academy of Sciences of the GDR to celebrate the achievements of Albert Einstein, 28 February-2 March 1979. (Courtesy of H.
Strohbusch)

Early in April 1881,[28] Michelson successfully carried out his measurments for the relative
motion of the Earth against the luminiferous ether. At this time of the year the Earth's

motion in its orbit coincided roughly in longitude with the estimated direction of the motion of the solar system, namely, toward the constellation Hercules. The interferometer was arranged in such a way on the stone pier that the arms of the intrument showed to the north and east (see Figs. 7 and 9). The arm directed to the east coincided with the resulting motion. The arm directed to the north was thus oriented perpendicular to the resulting motion. If the arms were revolved through 90 degrees so that the "parallel" arm and the "perpendicular" one now interchanged direction, the interference fringes would be shifted a definite amount, following Fresnel's theory of the stationary ether. But Michelson discovered, in his experiments with the interferometer, that the interference fringes did not shift. He suggested that the hypothesis of a stationary ether was not correct and that Fresnel's ether hypothesis was erroneous.

Michelson reported his results to Bell in a letter of 17 April 1881:[29]

> ... The experiments concerning the relative motion of the earth with respect to the ether have just been brought to a successful termination. The result was however negative. . . . Thus the question is solved in the negative, showing that the ether in the vicinity of the earth is moving with the earth, a result in direct variance with the generally received theory of aberration. . . .

In 1882, shortly after Michelson's results were published,[30] M. A. Potier called Michelson's attention to a slight error in the calculations he had made in the 1881 paper: he had neglected the effect of the Earth's motion on light traveling in the arm of the interferometer at right angles to that motion.[31] In 1886, the same comment was made by H. A. Lorentz[32] (1853-1928), who pointed out that the displacement of the fringes would be reduced by half. After that correction the measured displacement of the fringes in 1881 was identical to the expected theoretical value. Michelson was naturally very disappointed.

In 1887 the Potsdam experiment was repeated by Michelson and E. W. Morley (1838-1923) but the interferometer in the Cleveland experiment had several major improvements over the one used in Potsdam. The result of the Cleveland experiment was a confirmation of the Potsdam experiment.[33]

After finishing the experiment in Potsdam, Michelson moved to Heidelberg to attend the lectures of G. H. Quincke (1834-1924) and R. H. Bunsen (1811-1899).[34] He had finished the winter semester at the University of Berlin and received his certificates (see Figs. 4 and 5). Letters of Michelson to Bell and Newcomb, respectively, show that he occupied himself with problems of further improvement of his interferometer.[35] In that connection the experience of Quincke with the production of partial silvered mirrors was quite useful for Michelson's studies in Heidelberg. At that time Michelson also planned some experiments for further precision measurements of distances in units of wavelength of the light and again for determinations of the velocity of light.

After a short trip through the Black Forest and vacation in Schluchsee, Michelson went again to Paris for consultations with French colleagues and there he met Cornu and informed him about the results of the Potsdam experiment.[36]

STILL ON THE FOREFRONT OF BASIC RESEARCH

The "negative" result of the Michelson experiment was a cornerstone for the development of the special theory of relativity.[37] The Michelson experiment is still frequently repeated, with ever-increasing precision, using electromagnetic waves from different ranges of the spectrum--every time confirming the result of the early Michelson experiments.

The expected coincidence of the experimental result with consequences of the special theory of relativity is valid as long as one does not go beyond the bounds of theory itself: The space in which the experiment is carried out has to be free of fields. Matter and external fields can strongly influence the path of light as well as the interferometer as a whole. It is possible to screen the experiment from such influences, with one important exception: the gravitational field. The effect of gravitation as a universal, long-range interaction cannot be eliminated globally. Following the principle of equivalence, there can be only local compensation for gravity. A good example of the physical situation is the movement of a terrestrial interferometer in the gravitational field of the Sun. More precisely, the interferometer is exposed to the influence of a gravitational field changing in space and in time due to the rotation of the Earth. This leads to a very small change of the optical path of the light of both arms of the interferometer.

There are many applications of the Michelson interferometer in physics and technical sciences, especially the technique of using the principles of splitting the electromagnetic wave into partial waves through reflection and refraction and uniting these partial waves after the two beams have passed equal paths. We mention only some examples concerning astrophysical research.

Michelson himself brought his interferometer to the first application in astrophysics.[38] In 1890 he showed how the diameter of a star may be measured from the visibility of the interference fringes in a two-element interferometer. Until the 1920s, he measured stellar diameters, making an important contribution to classical astrophysics. Today, measurements of angular size of several types of stars are under way using an intensity interferometer that is not limited by atmospheric scintillation effects.

One of the well-known consequences of the general theory of relativity is the deflection of light by a gravitational field. Measurements of the deflection of starlight by the solar gravitational field are considered a crucial experimental test of this theory. Recently experiments have been undertaken to measure nonlinear-order effects of the deflection of starlight by the Sun using two interferometers arranged in a geosatellite.[39] Several proposals of optical interferometers with baselines ranging from 10 meters to several kilometers have been advanced for optical interferometry in space, again with an eye to astrophysics and relativity.

The cosmic microwave background radiation is considered to be a relic of the big-bang fireball phase in the evolution of the universe. The discovery of granularity in the microwave background radiation will lead to important progress in the understanding of inhomogeneities formed in the early evolution of the universe that eventually became galaxies. The measurements of anisotropies of the background radiation are best made from space; and one piece of technical equipment for such measurements is a Michelson interferometer for the analysis of the atmospheric spectrum.

From "photon-astronomy" to "neutrino-astronomy," the Michelson interferometer is the ultimate guide to the "gravitational-wave astronomy."[40] Einstein showed more than sixty years ago that gravitational waves should exist as "ripples" in the metric. This prediction is true of a certain class of gravitational theories in which, by analogy with electromagnetic waves, one would expect moving masses to produce gravitational waves. Attempts to detect such gravitational waves are under way using a laser interferometer with an arm length of several kilometers in order to measure relative changes of length of about 10^{-21}.[41] Again the interferometer is based on the principle found by Albert A. Michelson.

NOTES

1. A. A. Michelson, "The Relative Motion of the Earth and the Luminiferous Ether," *American Journal of Science*, 1881, *22:* 120-129. The article appeared in German fifty years later: "Die Relativbewegung der Erde gegen den Lichtäther," *Die Naturwissenschaften*, 1931, *19:* 779-784. A short note of commentary was added by Max von Laue. Cf. L. S. Swenson, Jr., *The Ethereal Aether -- A History of the Michelson-Morley-Miller Ether-Drift Experiments, 1880-1930* (Austin, Texas, and London: Univ. Texas Press, 1972); V. J. Rodičev and J. I. Frankfurt (Hrsg.), *Tvorcy fizičeskoj optiki,* (Sbornik statej, Nauka Moskva 1973); H. Paul (Hrsg.), *Die Schöpfer der physikalischen Optik* (Wissenschaftliche Taschenbücher, Bd. 195, Akademie-Verlag, Berlin, 1977).

2. H. J. Haubold and R. W. John, "100 Jahre Michelsonsche Ätherdrift-Experiment," *Wissenschaft und Fortschritt,* 1981, *31:* 410-412; H. J. Haubold and R. W. John, *NTM-Schriftenreihe* (Leipzig), 1982, *19:* 31; Michelson Colloquium of the Central Institute for Astrophysics and of the Einstein Laboratory for Theoretical Physics of the Academy of Sciences of the GDR, 27-30 April 1981, in Potsdam and Caputh. Proceedings published in *Astronomische Nachrichten,* 1982, *303:* Part 1, and *NTM-Schriftenreihe fuer Geschichte der Naturwissenschaften Technik und Medizin* (Leipzig), 1982, *19:* Part 1. In a paper by U. Bleyer, S. Gottlöber, H. J. Haubold, and J. P. Mücket, "Michelson und seine Experiment," *Wissenschaft und Fortschritt,* 1979, *29:*74, the Institute of Physics of H. v. Helmholtz, where A. A. Michelson performed the first trials with his interferometer, was incorrectly identified with the building of Magnus. This error was pointed out by Professor Dr. J. Wempe (1906-1980) in *Die Sterne,* 1975, *51:* 199, who also gave the correct reference to A. Guttstadt (Hrsg.): *Die naturwissenschaftlichen und medizinischen Staatsanstalten Berlins: Festschrift für die 59. Versammlung deutscher Naturforscher und Ärzte* (Berlin: 1886).

3. M. v. Laue's annotation to the German translation of Michelson's 1881 paper, *Die Naturwissenschaften,* 1931, *19:*779-784.

4. Bleyer et al, "Michelson und seine Experiment;" U. Bleyer and H. J. Haubold, *Katalog der Michelson-Ausstellung des Zentralinstituts für Astrophysik der Akademie der Wissenschaften der DDR anlässlich des 100. Geburtstages Albert Einsteins,* 1979, pp. 9; Michelson Exhibition of the Central Institute for Astrophysics of the Academy of Sciences of the GDR, 28 February to 2 March 1979, in Potsdam.

5. D. Michelson Livingston, *The Master of Light--A Biography of Albert A. Michelson* (New York: Charles Scribner's Sons, 1973).

6. Swenson, *The Ethereal Aether.*

7. A. A. Michelson, "On a Method of Measuring the Velocity of Light," *Am. J. Sci.,* 1878, *15:*394-395; "Experimental Determination of the Velocity of Light," *Am. J. Sci.,* 1879, *17:* 324-325; "Experimental Determination of the Velocity of Light," *Am. J. Sci.,* 1879, *18:* 390-393.

8. N. Reingold, ed., *Science in Nineteenth Century America -- A Documentary History* (American Century Series) (New York: Hill and Wang, 1964).

9. Livingston, *The Master of Light;* Swenson, *The Ethereal Aether.*

10. J. C. Maxwell, "On a Possible Mode of Detecting a Motion of the Solar System through the Luminiferous Ether," *Nature*, 1879-80, *21*: 314-315. The posthumous note consists of a letter written by Maxwell to D. P. Todd, director of the *Nautical Almanac* office in Washington on 19 March 1879. The letter is commented on by Todd and the editor of *Nature* and is reprinted in Rodičev and Frankfurt, *Tvorcy fizičeskoj optiki*.

11. H. A. Lorentz, "Over den invloed dien de beweging der aarde op de lichtverschijnselen uitoefent," Amsterdam, Koninkijke Akademie van Wetenschappen, *Verslagen en mededeelingen, Afdeeling natuurkunde*, 1886, *2*: 297-372.

12. J. C. Maxwell, "Ether," *Encyclopedia Britannica*, 9th ed., Vol. III, 1878.

13. R. S. Shankland, *Michelson's Role in the Development of Relativity*, Naval Academy Lecture, Annapolis, 11 May 1973, Preprint.

14. Reingold, ed., *Science in Nineteenth Century America*.

15. K. H. Wirzberger (Hrsg.), *Die Humboldt-Universitaet zu Berlin, Bilder aus Vergangenheit und Gegenwart* (Berlin, 1973); Guttstadt (Hrsg.), *Die naturwissenschaftlichen und medicinischen Staatsanstalten Berlins . . .*

16. E. Du Bois-Reymond, "Wohlvorschlag für H. v. Helmholtz (1821-1894) zum AM [Akademiemitglied] (1870)," in *Physiker über Physiker: Zur Geschichte der Physik an der Berliner Akademie von 1870 bis 1929* (2 Vols.; Berlin Akademie-Verlag, Berlin, 1975), Vol.1, pp. 63-65.

17. This remark was written by a fellow of the administration of the "Koenigliche Friedrich Wilhelms Universitaet zu Berlin" and not--as misinterpreted in the paper by Haubold and John, *NTM-Schriftenreihe*--by Helmholtz himself.

18. See note 2.

19. Reingold, ed., *Science in Nineteenth Century America*.

20. R. S. Shankland, "Conversations with Albert Einstein," *Am. J. Phys.*, 1963, *31:* 47-57; "Conversations with Albert Einstein, II," ibid., 1973, *41:* 895-901.

21. R. S. Shankland, S. W. McCuskey, F. C. Leone, and C. Kuerti, "New Analysis of the Interferometer Observations of Dayton C. Miller," *Rev. Mod. Phys.*, 1955, *27:* 167-178. Miller's measurements are discussed in J. M. Crelinsten, *The Reception of Einstein's General Theory of Relativity among American Astronomers, 1910-1930* (Ph.D. diss., Univ. Montreal, 1982), pp. 398-410.

22. F. Schmidt and Hänsch, *75 Jahre, Festschrift*, Berlin, 1939; E. Löblich, *Deutsche Mechaniker-Zeitung*, 1914, Heft 8, 95; 1914, Heft 9, 97; Swenson, *The Ethereal Aether;* Shankland, *Michelson's Role in the Development of Relativity*.

23. A. A. Michelson, *Am. J. Sci.*, 1881; Swenson, *The Ethereal Aether;* Rodičev and Frankfurt, *Tvorcy fizičeskoj optiki;* Paul (Hrsg.), *Die Schöpfer der physikalischen Optik*.

24. E. Du Bois-Reymond, "Wohlvorschlag für H. v. Helmholtz (1821-1894) zum AM [Akademiemitglied] (1870)."

25. *Das Koenigliche Astrophysikalische Observatorium bei Potsdam*, Mayer & Mueller, Berlin 1890; J. Wempe, *Die Sterne*, 1975, *51*: 199.

26. According to Professor Dr. J. Wempe, only this cellar could have been used by Michelson for his experiment in 1881.

27. A. A. Michelson, *Am. J. Sci.*

28. As realized by Professor Dr. J. Wempe, it follows from the original series of observations given in Michelson's paper of 1881 (*Am. J. Sci.*) that he performed his measurements in the period between 5 and 15 April 1881.

29. Reingold, ed., *Science in Nineteenth Century America.*

30. Michelson, *Am. J. Sci.*

31. A. A. Michelson, *Comptes Rendus*, 1882, *94:*520.

32. Lorentz, "Over den invloed."

33. A. A. Michelson and E. W. Morley, *Am. J. Sci.*, 1887, *34:*333.

34. Livingston, *The Master of Light.*

35. *Reingold, ed., Science in Nineteenth Century America.*

36. Livingston, *The Master of Light;* Swenson, *The Ethereal Aether.*

37. G. Holton, "Einstein, Michelson, and the Crucial Experiment," *Isis,* 1969, *60:*133-197; H. J. Treder, *Astronomische Nachrichten*, 1982, *303:*91.

38. D. H. De Vorkin, "Michelson and the Problem of Stellar Diameters," *Journal for the History of Astronomy,* 1975, *6:* 1-18.

39. I. I. Shapiro,"Experimental Challenges Posed by the General Theory of Relativity," in H. Woolf, ed., *Some Strangeness in the Proportion: A Centennial Symposium to Celebrate the Achievements of Albert Einstein,* (Reading, Mass.: Addison-Wesley Publishing Company, Inc., 1980), p. 115.

40. A. De Rújula, S. L. Glashow, R. R. Wilson, and G. Charpak, "Neutrino Exploration of the Earth," *Physics Reports,* 1983, *99:*341-396.

41. P. Kafka, "Wie wichtig ist die Gravitationswellen-Astronomie?" *Die Naturwissenschaften,* 1986, *73:* 248-257.

THE MICHELSON EXPERIMENT IN THE LIGHT OF ELECTROMAGNETIC THEORY BEFORE 1900

Jed Z. Buchwald

THE OPTICS OF MOVING BODIES AS A PERIPHERAL CONCERN BEFORE THE 1890S

In 1889, two years after the Michelson ether-drift experiment was published, Henry Rowland of Johns Hopkins University concluded a New York address to the Institute of Electrical Engineers with the following words:

> ...the luminiferous ether is, to-day, a much more important factor in science than the air we breathe ... When we speak in a telephone, the vibrations of the voice are carried forward to the distant point by waves in the luminiferous ether ... When we use the electric light to illuminate our streets, it is the luminiferous ether which conveys the energy along the wires ... We step upon an electric street car and feel it driven forward with the power of many horses, and again it is the luminiferous ether, whose immense force we have brought under our control and made to serve our purpose. No longer a feeble, uncertain sort of medium, but a mighty power, extending throughout all space and binding the whole universe together, so that it becomes a living unit in which no one portion can be changed without ultimately involving every other portion.[1]

These are not the words of a doubter. For Rowland, the Michelson experiment had in no way affected his belief in the reality, indeed the very substantial reality, of the ether. And yet, since the sole explanation for Michelson's result -- FitzGerald's contraction hypothesis -- was published in *Science* only this same year,[2] Rowland was probably as yet unaware of it.

Rowland was hardly alone among British and American physicists. I do not know of a single one who, between 1887 and 1889, either dispensed with the ether as a useful tool or even downplayed its significance. Michelson's experiment, even before Fitz-Gerald's hypothesis became known, did not in any major way alter these physicists' concerns. A quick glance at the titles of papers in optics and electromagnetism published during -- and much after -- these years shows that very few indeed are at all concerned with the optical effect of motion through the ether. And this is very nearly as true for Continental investigators as for the British. Instead there are large numbers of articles on Hertzian waves, on field equations and how to use them to deal with various kinds of optical phenomena, and on related issues that do not appear to be closely tied to the kinds of questions raised by the effects of motion through the ether.

Most historians writing after 1910 or so[3] have assumed that the motion of matter through the ether was perhaps the single most important question for nineteenth-century physics. But from almost any point of view this is certainly incorrect for the last quarter of the century. Indeed, the issue was of little consequence even much earlier, since most work in theoretical and experimental optics before the 1870s concerned methods for constructing a general wave equation that could encompass known phenomena and imply new ones. Both British and Continental physicists had attempted to do this by playing with the equations of motion of the ether.[4]

By the late 1860s in Britain it was increasingly felt that the major guiding criterion to use in constructing equations for the ether was not a specific model, whether a discrete or a continuous one. Instead, attention was turning to the more abstract requirement that the ether's energy must, like the energy of material continua, be spread throughout it and that every portion of the medium, no matter how small, must obey the same kinds of general principles that govern the motion of ordinary continua.[5] This rarefied pursuit departed from the earlier, common insistence on closely following the pattern defined by the behavior of material bodies,[6] but it was no more concerned than its predecessor had been with the effect of material motion on the transmission of light through the ether.

Yet few topics in the history of physics -- excepting possibly Newton on force or Galileo on experiment -- have been so often and so extensively discussed as the optics and electrodynamics of moving bodies before relativity.[7] Nor is traditional historiography entirely mistaken. The problems raised by this issue did indeed become quite important by the mid-1890s. However, they became important in a rather different way than has usually been assumed. I shall attempt to show that what significance the issue of moving bodies did have before 1900 derives almost entirely from its stimulus to the development by H. A. Lorentz of a sophisticated, and (at the time) not widely appreciated, electromagnetic theory that was based upon the microphysics of matter. Rather than offering difficulties, I shall argue, the issue helped to stimulate this very successful development.

My goal will be to explain why, and in what ways, Lorentz was interested in the problem of material motion through the ether, and also why most of his contemporaries, particularly in Britain, were not so concerned. To do so I must begin with a brief précis of the problem's history through the late 1880s. Here I can scarcely hope to offer much novelty, but I think it is worthwhile to remind ourselves of the central issues that were involved at least through Lorentz's important work of 1892, which began to alter the tenor of the problem outside of the Maxwellian community.

Before electromagnetic concerns were brought forcefully into play by Lorentz in 1892, the optics of moving bodies raised three major issues: first, how could the various aspects of stellar aberration be explained; second, how could one explain the fact that a flowing stream of water changes the velocity of light through it by a certain fraction of its speed; and, third, how to explain the absence of any effect to the first order in the ratio of the speed of the earth through the ether to the velocity of light on terrestrial optics. Before 1886 two quite different schemes could deal reasonably well with all three of these issues.

The oldest scheme dates to 1818, in fact to the birth of the wave theory of light in France. In order to explain Arago's failure to detect any dependence of the refraction of starlight on the earth's motion, Fresnel argued that a very slight or partial dragging of the ether by a transparent body would change the light velocity in the body by a certain fraction.[8] The basic phenomenon of stellar aberration, Fresnel felt, required that the ether must remain otherwise quiescent -- that, in other words, the vast, opaque bulk of the earth must not disturb it in any substantial fashion. In 1851, Armand Fizeau was able to measure the Fresnel "drag" coefficient, and in 1873 Wilhelm Veltmann demonstrated that no optical experiment with a terrestrial source of light can, to first order, detect motion through the ether if the drag coefficient obtains. Consequently, to this degree of accuracy, Fresnel's original theory -- which required a very slight transport of the ether by transparent bodies -- was quite satisfactory.

However, in the mid-1840s, the British mathematical physicist George Stokes objected to Fresnel's requirement that the bulky, opaque earth would not disturb the ether (primarily on the basis of an analogy to the motion of a solid through a slightly viscous fluid).[9] Stokes accordingly proposed instead that, near the earth's surface, there should be no relative velocity between the ether and the earth at all -- that the vast bulk of the earth must drag the ether along completely. This at once explains why -- to any order -- optical experiments with terrestrial sources of light will always fail to detect the earth's motion.

Stellar aberration itself requires the waves of light to be skewed around in a certain fashion in the boundary region, where the ether acquires the earth's speed. Stokes easily showed that the correct skewing will occur only if the ether's velocity satisfies a certain kinematic condition everywhere -- to wit, that it have (in modern parlance) no curl. None of this explains Fizeau's experiment (which, of course, Stokes did not have to explain in 1845 since it was only performed in 1851), but in later years those who held to Stokes's hypothesis simply assumed that transparent bodies are less efficient draggers of ether than is the bulky earth, and that they act as required by Fresnel's coefficient.

We see, then, that Fresnel had already in 1818 provided a method that was sufficient to explain aberration and the absence of first-order effects of the earth's motion on optics. Stokes merely added a different theory, which was as good as, but no better than, Fresnel's. Consequently one should hardly be surprised that the optics of moving bodies was certainly not a major concern for almost anyone between 1818 and the late 1880s. The Paris Académie did offer a prize in the subject in the early 1870s, which evidently sparked an unusual amount of interest in it, but the purpose of the prize was primarily to encourage a more elaborate theoretical and experimental investigation of applications for the Fresnel coefficient, for example to birefringence. In any case even during these peak years of interest only a few articles per annum were devoted to the subject.[10] This is hardly a large number in comparison to the considerable amount of writing that was just then beginning to appear, in Germany especially, on how to provide equations for birefringence and for the newly discovered phenomenon of anomalous dispersion.[11] Some of this work was peripherally concerned with motion through the ether, but most of it was not.

One can perhaps gauge the lack of deep concern with the issue by the fact that the Maxwellian physicist Richard Tetley Glazebrook, in an elaborate 1885 report on optics to the British Association, merely reiterated the old Stokes theory and left it at that.[12] Though the report runs over a hundred pages, only four of them concern the optics of moving bodies. This proportion accurately reflects the level of interest in the area before the late 1880s. There were simply no major problems to solve here, or so it was generally thought.

The spread in Britain of Maxwellian electrodynamics during the mid to late 1870s did alter in many important ways both British and, eventually, Continental attitudes towards the ether. However, none of these changes was conducive to a thorough consideration of the optics of moving bodies, and not merely because, at least until 1886, Stokes's (or Fresnel's) theory seemed to be entirely adequate. Rather, the new Maxwellian consensus differed from previous views in its very firm, indeed irrefragable, insistence that the link between ether and matter was, and would certainly remain for some time, deeply resistant to analysis. This attitude emerged directly from the heart of electromagnetic theory as developed in Maxwell's 1873 *Treatise on Electricity and Magnetism*. It is worthwhile spending several moments on it inasmuch as, I believe, we will find here the central difference in the late 1880s and 1890s between the Maxwellians, on the one hand, and, on the other, German and Dutch physicists, most importantly Lorentz. And that difference accounts in major part for their contrasting attitudes towards the optical and electromagnetic effects of moving bodies.

MAXWELLIAN ELECTROMAGNETICS AND MOTION THROUGH THE ETHER

Maxwellian theory was a complex, highly integrated scheme that was rarely (perhaps never) presented in a thorough, step-by-step fashion. Instead, and following the pattern

Maxwell himself had set in the *Treatise*, British physicists in the 1880s tacitly assumed a familiarity with most of the Maxwellian core ideas. Consequently their letters, notes, articles and books are filled with Maxwellianism, not infrequently of the most extreme kind. More than anything else the image of a continuous ether that is subject to qualitative alteration as a result of material action upon it underlies this uniquely British scheme.

I say uniquely British because, as I have argued at some length,[13] solely in Britain did physicists insist that the only tractable problems in optics and electromagnetism were those that could be treated by altering the ether's structure according to circumstances -- that anything that could not be treated in this fashion could probably not be dealt with at all, at least in the foreseeable future. Many of the core ideas that enabled the British to integrate their scheme were so deeply embedded in their discourse that most physicists who had not been educated under the influence of British texts found it extremely difficult, and usually impossible, fully to grasp the scheme as a whole. Consequently, and despite the great influence of British theory after Hertz's observations of electromagnetic radiation in the late 1880s, its attitudes as well as many of its details never thoroughly penetrated the Continent. Primary among those attitudes, and closely bound to the core of the scheme, was the British insistence that there was simply no point in trying to determine the mechanism by which matter acted upon the ether -- even though the essence of British theory required a most precise specification of the result of that very interaction.

Attitudes and opinions do not usually produce concrete results in physics if nothing can be calculated with them. For example the fact that early in the century most French physicists firmly believed that all phenomena involved particles and forces led to singularly few concrete results -- though it did profoundly influence their understanding of nature. Maxwellian theory, in the sense I intend, contrasts markedly with this kind of comparatively fruitless conviction because it did lead to very concrete results indeed. In particular the first successful mathematical theories of two puzzling electro-optical phenomena -- the Faraday and Kerr effects -- were Maxwellian.

Elsewhere I have discussed the explosion of activity among Maxwellians that revolved about these and other, related phenomena. It is not necessary to grasp the rather intricate details of this story. One needs only to realize that the majority of British activity in optics and electromagnetism between the late 1870s and the early 1890s centered on these kinds of questions, questions that could be answered mathematically from within the confines of Maxwellian theory, without trespassing on forbidden ground.

Before 1886, few if any Maxwellians would have been at all concerned with the optics of moving bodies. First, there seemed to be no problem here at all because Stokes had solved it forty years before, and his solution could be carried over directly to the electromagnetic ether. A great advantage of Stokes's solution was its allowing one almost completely to ignore why matter in motion affects the ether in this way: since Stokes's ether moved entirely along with the earth one did not need to assume that the effect varies with the motion -- in which case whatever mechanism was responsible for it at least remained very nearly invariant. One did not have to elaborate a mechanism, it might be said, because the mechanism never changed. But even after it became necessary to abandon Stokes's theory, Maxwellian interest in the optics of moving bodies still remained negligible -- because the failure of Stokes's account merely provided another instance of how difficult it was to analyze problems that depend very closely on the mechanism that links ether and matter.

This is not to say that Maxwellians were unconcerned with the electromagnetic effects of motion through the ether. They were at least somewhat interested in the question, but for reasons that reinforce rather than undercut their lack of concern with the intractable boundary problem of precisely how matter and ether are related to one another. Since the early 1880s some Maxwellians -- especially J.J. Thomson and George FitzGerald, and, later, Oliver Heaviside -- had addressed a very specific issue that does concern motion: namely, precisely what effects are produced when a field of electric displacement or of magnetic induction moves through the ether.

In retrospect -- in the light of electron theory and then relativity -- these seem to be very important questions, in the case of electron theory because, after all, the sources of all electromagnetic actions are charges in motion. But in Maxwellian theory this was not a very important issue because Maxwellians most certainly did not think of moving charges as sources. This is a difficult, though exceedingly important, issue, and I shall say only this about it: that Maxwellians regarded charge as an epiphenomenon of more fundamental field processes, and that effects which electron theory later attributed to moving charges were for Maxwellians due to redistributions of field energies, which in the end resulted in phenomena that could be interpreted in terms of localised changes in charge densities.[14]

Consequently, J.J. Thomson, Fitzgerald and Heaviside were not addressing something that they thought to be a foundational issue in electromagnetism. Instead, they were trying to solve, from within Maxwellian theory, a particular problem, and one that they did not think to have a very wide significance. Indeed we have here a nice instance of the investigation of a rather nasty problem by a community that has a high level of internal agreement, but a problem that could be, and was, solved.

This, then, was the state of things by 1886, the year in which Michelson and Morley reconfirmed Fizeau's result that the velocity of light differs in moving water from its value in stationary water by the amount deduced long before by Fresnel. This was also the year in which H. A. Lorentz published a lengthy discussion of the issues raised by the optics of moving bodies, a paper in which he conclusively demonstrated that Stokes's solution to the problem, which was very congenial (though hardly essential) to Maxwellians,[15] must be incorrect.

LORENTZ'S CRITIQUE

Michelson and Morley's 1886 experiment, it seems, catalyzed Lorentz's concern with the optics of moving bodies. In his detailed and comprehensive treatment of the subject that same year he insisted that "the task of the theory of light is to explain the value furnished by observations for the coefficient of entrainment"[16] -- that is, to explain why Michelson and Morley got the results they did. Of course in the light of what I said above this would not seem to be much of a problem, and in a sense it still was not. Lorentz was well aware that assuming transparent bodies to "drag" a very small amount of the ether along, leaving it otherwise unaffected, satisfies the demands of experiment. Nor did Lorentz offer anything substantially new here. The importance of his paper was not for its positive results, but, instead, for the implications which Lorentz began to see in its one major negative result.

Lorentz demonstrated rigorously that Stokes's theory of aberration was internally inconsistent. One could not, to be precise, have a velocity potential and yet simultaneously satisfy the two boundary conditions that the ether must have the same velocity as the earth along the earth's normal and also along the tangent to its surface. If the ether has a velocity potential -- as it must to explain aberration if it moves at all -- then it must also move with respect to the earth. Stokes's theory cannot be correct.[17]

Again none of this is particularly difficult for anyone, including the Maxwellians, to assimilate.[18] Lorentz knew this, but he had now begun to think about the problem in a way that linked to an earlier understanding of how ether and matter affect one another that he had developed in the late 1870s for the particular problem of optical dispersion.

These reflections, together with the stimulus of Hertz's 1888 measurement of electric waves, were, I believe, in large part responsible for Lorentz's creation four years later of a new kind of electromagnetic theory.

Lorentz in 1886 distinguished two ways of understanding the connection between ether and matter: either matter never moves the ether very much (which he took to be Fresnel's position), or else a massive, opaque object like the earth pushes the ether out of its way to some extent at least. Both alternatives require that the molecules of transparent bodies must somehow alter the speed of light through the bodies that they form in accordance with the requirements of the Fresnel coefficient. Lorentz remarked that, though many people would probably prefer the second view, in which the ether is radically disturbed by the earth's motion,

> It seems to me nevertheless the other way of thinking is at least as simple, if not simpler. It is possible that what we call an atom might occupy the same place as a part of the ether, that for example an atom is nothing other than a local modification in the state of this medium, and then one could understand how an atom might move without entraining the environing ether.[19]

This way of stating the issue would have been particularly significant only to Lorentz, or, better put, only Lorentz was in a position to see that a question that nobody had ever before tried to answer could be answered if one adopted this point of view. The ether, in the view that Lorentz now wished to maintain, almost never moves (excepting the small local disturbances responsible for electromagnetic phenomena). This is conceivable, Lorentz thought, because atoms of matter may themselves be local modifications of the ether, so that a moving atom is akin to the transmission from point to point in the ether of an ether state, and this need not involve bodily motion of the medium itself.

In the traditional understanding of the Fresnel coefficient a small amount of ether is dragged along by the molecules of transparent bodies. Yet Lorentz now wished to make of atoms pure ether states, states which are passed on without -- and this is the major point -- the "environing" ether moving along with the state. Much as a wave moves along without carrying the stuff that waves with it, Lorentz was saying, so does an atom move through the ether. But what then of the usual understanding of the Fresnel coefficient? If atoms are ether states, then how can they carry any ether along with them, in which case why does the motion of transparent matter affect light in any way at all? Why, in other words, is there such a thing as the Fresnel coefficient? This was the question that Lorentz had now to answer.

To appreciate the novelty of Lorentz's position we may contrast it with a point of view that might seem quite similar to it, one that was very common in Britain (though not on the Continent) by the late 1870s: namely, the idea that atoms are vortex rings in the ether. First suggested by William Thomson (later Lord Kelvin) in the mid-1860s, the vortex atom had captured the imagination of many British physicists. It shares with Lorentz's image the notion that atoms of matter are not distinguishable from the ether because they are made up of ether. But there the similarity ends.

The essence of Thomson's atom, what makes it possible for an atom to retain its individual identity, is this: the mathematics of fluid mechanics imply that a vortex ring, no matter how it distorts as it moves along, will always be composed of the very same elements of fluid. Ether elements swim physically along with the vortex. And so the vortex atom is not properly speaking a "local modification in the state of the ether" because it carries the ether along with it from place to place. Certainly the state of the ether is modified by the presence of a vortex atom, but the modification is not transmitted from one region of ether to another; instead the ether itself moves, and, in moving, carries the modification along.

The difference between Lorentz's view and this one is very deep indeed, since it is the difference between seeing the atom as an insubstantial condition and seeing it as a thing.[20]

This has important implications. If we think of atoms in Thomson's manner, then the Fresnel coefficient does not truly raise a new problem. In fact there is a rather simple way to explain it, perhaps not using the vortex atom proper, but using at least the idea that an atom involves ether moving at its speed. And this explanation was provided by Michelson and Morley themselves in a footnote to the second page of their 1886 paper. In it they showed quite nicely that one can very nearly obtain the required alteration in the speed of light by assuming the ether to remain fixed except for a certain small amount that moves with each molecule of matter at that molecule's speed. Since (of course) the speed of light through the ether carried by the molecule will be increased by the speed of the molecule, one can (using hypothetical molecular dimensions) compute a formula for the net effective speed change.

Lorentz could have produced the same argument; he surely read it. Yet he did not. He gave no calculation at all for the Fresnel coefficient, and instead proffered, in few words, a way of thinking about ether and matter that was quite novel among both Continental and British physicists. Clearly, Michelson and Morley's solution, though assimilable to the underlying sentiment that motivated the vortex atom, did not appeal to Lorentz.

The reason for Lorentz's lack of satisfaction with their suggestion brings me back to my claim that only he now conceived of a way to explain something -- the Fresnel coefficient -- that no one before had ever thought needed explaining. Before Lorentz, as we have seen, the common view held that if moving matter affects the speed of light then the ether must itself be moved. The point was, it seems, entirely obvious and hardly worth discussing. Consequently there was no need to explain the Fresnel coefficient, but only to calculate a formula expressing (in effect) how much ether is dragged along that would square reasonably well with the requirements of experiment. Michelson and Morley had done just that.

Lorentz's new way of thinking contrasts strikingly with this because the physical basis for understanding the Fresnel coefficient has vanished along with the substantial identity of the atom. Since the ether is not moved by matter, there can be no such thing as a "drag" coefficient. The "task" of the theory of light was therefore to provide an altogether new physics, not just a slightly different calculation for a well-understood process. But why did Lorentz come to this extremely unorthodox conclusion, so unorthodox that it did not occur either to British Maxwellians or to other Continental physicists? To attempt an answer we must go back to Lorentz's earliest work in electromagnetism and optics.

LORENTZ AND THE ETHER

At least since the end of the 1870s Lorentz had diverged from the common practice of nearly all of his contemporaries in one very important respect. Before the early 1890s he alone distinguished matter from ether in constructing electromagnetic equations, even though the method that he used to do so had been introduced many years before by his intellectual mentor, the German physicist Hermann von Helmholtz. Since the peculiar character of Lorentz's views on the optics of moving bodies seems to me to reflect this unique approach to electromagnetism, I shall spend a few moments explaining what Lorentz did.

Lorentz's first work in electromagnetism in 1875 did not diverge from the foundations of prevailing sentiment on the Continent, though he certainly operated at its furthest limits by treating the ether in Helmholtz's fashion as a magnetically and electrically polarizable body. This approach was not common at that time even in Germany, where variants of the old, Weberean electrodynamics were still preferred, and where the ether was still thought of as a mechanical structure. However, Helmholtz's views were in fact considerably closer to those of his Weberean contemporaries than they were to Maxwell's, because electric charge remained for Helmholtz an accumulation of something with the same kind of identity as matter itself, and the electric current remained the flow of electric charge. For the Maxwellians, as I mentioned above, electricity was entirely insubstantial, being an epiphenomenon of more fundamental processes. And precisely because electricity remained substantial in Helmholtz's theory, it was possible for Lorentz, who enthusiastically embraced it, to connect ether and matter in a way that was later to make the optics of moving bodies more of an issue for him than it ever was for the Maxwellians.

In 1878 Lorentz took a momentous step beyond Helmholtz by uniting two of Helmholtz's own theories: he combined Helmholtz's 1870 electromagnetic theory of the ether, which he had already used in 1875, with another theory that Helmholtz had developed that same year to explain the newly discovered phenomenon of anomalous dispersion. The latter was, however, entirely non-electromagnetic. It began with two distinct mechanical equations: one for the ether, the other for matter, with terms held in common between them.[21]

Lorentz appropriated the idea from this work that two linked equations were necessary to explain phenomena that involved interactions between ether and matter. But he went further by combining it with Helmholtz's polarizable ether to produce in 1878 an electromagnetic theory of dispersion. That theory required particles of matter simultaneously to possess charge and mass, whereas the particles of Helmholtz's ether (should the ether be discrete) needed to possess only charge. On this basis Lorentz was able to provide a technique for handling the connection between ether and matter that the Maxwellians completely lacked, because it depends upon the very thing that the Maxwellians steadfastly denied: namely, the substantiality of charge. As yet (1878) Lorentz had applied it only to generate a dispersion formula, but even here he had done something that no Maxwellian ever did attempt or ever could have.

It is important to understand how Lorentz's technique worked in order to see why the optics of moving bodies were to present him with a challenge that it never presented Maxwellians. Let us begin, by way of contrast, with Maxwellian techniques for solving problems in new areas of electromagnetic research. Their procedure -- in vast simplification -- amounted to altering the very structure of the ether itself in just the right way to obtain the desired effect. The cause of the change was to be found in the action of matter upon the ether, but Maxwellians were concerned almost exclusively with the effect and not the cause. Lorentz's approach in his 1878 dispersion theory was entirely different, amounting almost to an inversion of the Maxwellian method. Instead of ignoring matter and, as it were, monkeying with the ether, Lorentz ignored the ether and monkeyed with matter, though before 1886 he did not assert that one should never play with the ether.

Despite the fact that Lorentz had mentioned that one must consider the effect of motion through the ether in 1878 he did not then do so, and his attention was, as we shall see in a moment, directed elsewhere for the next seven or so years. It was the stimulus of the 1886 Michelson-Morley confirmation of the Fresnel coefficient that brought the issue home to him in ways that returned him to his earlier ideas. Lorentz already knew that he could generate results in optics by assuming Helmholtz's electromagnetic ether to be unaltered in structure by material systems, which are linked to it through a single driving term in the material equation of motion. And he became convinced that the Fresnel coefficient would yield to this same technique.

Seen in this light, Lorentz's otherwise unusual and even odd remarks about atoms and ether reveal their proper meaning: they provide the physical underpinnings for a theory in which the ether's structure, as represented by the field equations, remains, absolutely unaffected by the presence of matter. Since matter is itself an ether state it must be associated with certain effects in the ether, but these effects are invariant because the ether-matter link -- the nature of matter as an ether state -- is as invariant as the ether itself. Only the organization of the particles that together constitute the material system changes.

It would be difficult to overestimate the novelty of Lorentz's scheme, and not only in contrast to the Maxwellian. Well into the 1890s in electromagnetism most physicists, on the Continent as well as in Britain, continued to alter the ether's structure rather than to play around with material particles. For many years Lorentz was very nearly alone, and yet, in an ironic turn of events, when his method did become common by the late 1890s, few people at the time associated it specifically with him.

But I run ahead of my story. If I am correct, then the Michelson-Morley experiment that had the most contemporary significance, the most immediate and far-reaching effect, was not the famed 1887 measurement of ether drift. That experiment, I shall shortly argue, had a rather limited influence. Instead, their 1886 confirmation of the Fresnel coefficient had the greatest impact because it turned Lorentz's attention back to a way of thinking that he had not pursued for nearly eight years. This turn led quite directly to his construction of a new, and ultimately immensely fruitful, electromagnetic theory based upon the microstructure of matter.

APPLYING ELECTRODYNAMICS TO MOVING BODIES

Now Lorentz's dispersion theory was not his only work on the connection between electromagnetism and optics before 1892. He had also discussed the implications of the Hall effect for the action of matter on light in the presence of magnetic fields (the Faraday and Kerr effects). And here he had taken a markedly new tack. Unlike his earlier dispersion theory, which did not play with the ether, Lorentz's analysis of the optical implications of the Hall effect did. In the context of Helmholtz's theory this means that dispersion did not require Lorentz to alter the constants that govern the ether's properties, whereas the Hall, Faraday and Kerr effects did require one to do so[22] -- even though both phenomena obviously concern the effect of matter on ether. Lorentz, I believe, began thoroughly to understand the implications of this inconsistency only in 1886, when it again arose in the context of understanding why such a thing as the Fresnel drag coefficient exists.

In 1886 he had two choices, at least if, as was almost certainly the case, he felt that the drag coefficient had -- like dispersion and the Faraday-Hall-Kerr effects -- to have an electromagnetic interpretation. Either it arises, like these latter, because of a change imposed by matter on the ether, or else it arises like dispersion, as a result of the way in which matter generates electromagnetic effects in the ether. To Lorentz, I believe, these alternatives came to be embodied in the difference between Stokes's and Fresnel's theories of the optical effects of motion through the ether. Stokes's analytically faulty theory was similar to Lorentz's own analysis of the Faraday-Hall-Kerr effects because both theories modified the ether (by dragging it along with a large opaque body in the case of the former or by altering the basic electromagnetic field equations in the case of the latter). Fresnel's theory, of course, had traditionally involved a partially dragged ether, but only a very slight amount of ether must move in comparison to

Stokes's massive requirement. Accordingly, Fresnel's theory shared much of the spirit of Lorentz's account of dispersion.

Lorentz realized that these kinds of differences evoked an extremely general issue: namely, whether the ether is, or is not, ever to be modified. In a radical and unprecedented step Lorentz decided sometime between 1886 and 1892 that one need not alter the ether in any way at all for any optical or electromagnetic phenomenon. Since one could understand this puzzling lack of effect by matter on the ether by thinking of atoms as ether states, what might otherwise seem to be an extreme and unwarranted speculation on Lorentz's part takes its proper place as the physical foundation for a new electrodynamics and optics -- one in which only matter, and never the ether, changes its structure.

And so Lorentz had a new task: to create a theory in which the Fresnel coefficient is due exclusively to material organization. This was an extraordinarily difficult problem because Lorentz was not willing to settle for half measures. He insisted that dispersion, and the Fresnel coefficient, must emerge from a thorough computation of the effects that a set of radiating material particles have on one another. Had Lorentz not insisted on such a precise, detailed theory, he might have achieved the same final results in a considerably simpler fashion than he did. Indeed, he could have created the very system that, unlike his own, did achieve a rather widespread influence in Germany during the 1890s, particularly as a result of Paul Drude's publications.[23] In this scheme we have a set of field equations -- the "Maxwell limit," it was often said, of Helmholtz's relations -- together with a linked equation for matter, and we manipulate the latter to produce effects. This was approximately what Lorentz had himself done for dispersion in 1878. But Lorentz did not now take this approach, no doubt because he felt that it was not sufficiently rigorous.

The theory that Lorentz did publish in 1892 generated all electromagnetic effects, most especially the "drag" coefficient, by means of what have since been called retarded forces. In this scheme charged particles accelerate and so radiate, sending out waves that propagate through the eternally invariant ether. At some moment these waves strike other particles, thereby causing them in turn to move and so to radiate. The pattern of interactions that results determines the optical properties of the system, and Lorentz was able to show in a straightforward but intricate manner that, to first order, a system of particles moving through the ether will behave in the fashion required by the drag coefficient.

In Lorentz's theory, then, the "drag" coefficient and dispersion emerge from precisely the same kinds of effects, the only difference being that one ignores a common translation of the particles in calculating dispersion. But if this motion is not ignored, then the "drag" coefficient emerges directly as an extra effect of retardation. To put it crudely, the coefficient exists because, since the particles move, the forces must travel different distances through the ether than when the particles are at rest. Motion per se has no effect except to add an additional, common term to the velocities of the atoms.

This beautiful unification was just what Lorentz had been looking for since 1886 or so: all optical and electromagnetic effects, including those attendant on motion, now, it seemed, could be dealt with on the basis of a single physical scheme with no extraneous elements at all. Lorentz's theory was, in both British and Continental contexts (though in different ways) strikingly new, and I have not found any evidence of its specific influence. Quite the contrary. In Britain, where Maxwellian views remained firmly entrenched through the mid-1890s, Lorentz's work, since it employed an electric substance, seemed antithetical to a field tradition. On the Continent, where field equations rapidly spread after Hertz's discovery of electric waves, Lorentz's central concern with retarded actions seems to have had almost no impact at all. Yet, at the time, only in this way could the "drag" coefficient actually be deduced from electromagnetic principles -- a fact that underscores the continuing unimportance of the optics of moving bodies even as late as the mid-1890s.[24]

THE MICHELSON EXPERIMENT AT THE MARGIN

In 1895, just three years after re-creating electromagnetism on the basis of retarded actions, Lorentz produced a unified and elaborate account of his theory entitled *Versuch einer Theorie der elektrischen und optischen Erscheinungen in bewegten Korpern.* Despite certain novel features to which I shall momentarily turn, Lorentz's *Versuch* is not fundamentally different from his 1892 analysis. Retarded actions underlie the system, and all electromagnetic effects emerge out of the pattern of interactions between charged particles.

The major difference between the *Versuch* and Lorentz's 1892 work is generality.[25] In the *Versuch* Lorentz detailed a mathematical technique that makes it quite simple to solve any optical problem involving motion through the ether to first order, whereas a rather large number of manipulations were necessary in 1892 to do so, in particular to deduce the "drag" coefficient. The problem was this. In a system moving through the ether the electromagnetic field equations do not have the same form as they do when the system is stationary. This is to be expected and indeed explains why such a thing as the "drag" coefficient exists. However, this does make it rather difficult to solve problems, and so Lorentz developed a technique for reducing the equations to the same form that they have when matter is at rest in the ether.

To do so Lorentz introduced a new variable in place of the usual time coordinate, a so-called local time, while leaving the spatial coordinates of the moving system untouched. Using the local time in the usual equations for the field in the moving system, and discarding terms beyond the first order, Lorentz found that the field equations remain the same in form as in a system that is not moving. Furthermore the relationships between the fields in this new fictitious system, and the true fields in the moving system, is such that the same regions which have a given field pattern in the one must have a related pattern in the other. In particular if there is no field in some place in the fictitious system, then there must also be no field in the same place in the actual system. And since the fictitious system has a familiar form, problems can be solved in it and the solutions transferred directly to the moving system.[26]

In all of this the Michelson-Morley ether drift experiment of 1887 was decidedly marginal, despite the fact that Lorentz had been puzzled by it well before 1892. It is well known how he dealt with it in that year and, again, in 1895: he assumed, as had FitzGerald before him, that the "line joining two material particles shifting together through the ether"[27] contracts in just the right amount to nullify the second-order path difference that the experiment was capable of detecting.

I do not wish to contribute to the long history of philosophical debate concerning the "ad-hocness" of the Lorentz-FitzGerald contraction, but I do wish to remind you that the effect could be assimilated to the well-known behavior of the electric force in the moving system. It had been known for quite some time that the electric field is affected by the motion of the system. This effect would in fact produce the proper second-order contraction if the material particles were held in their positions of equilibrium by electric forces, or by forces that transformed like them. Consequently there was a plausible physical basis for the contraction.

And plausibility was all that Lorentz needed at this time. The Michelson-Morley experiment was but a single fact, one which had to be accommodated but certainly not one to cast any serious doubt on the overall power of Lorentz's microphysical scheme. Both the acceptance of the contraction's plausibility as well as a conviction in the marginal character of the problem, are well illustrated by Paul Drude's 1900 comments:

However unlikely the hypothesis that the dimensions of a substance depend upon its absolute motion may at first sight seem to be, it is not so improbable if the assumption be made that the so-called molecular forces, which act between the molecules of a substance, are transmitted by the ether like the electric and magnetic forces, and that therefore a motion of translation in the ether must have an effect upon them, just as the attraction or repulsion between electrically charged bodies is modified by a motion of translation of the particles in the ether. Since v^2/c^2 has the value 10^{-8}, the diameter of the earth which lies in the direction of its motion would be shortened only 6.5 cm.

As far as Drude was concerned Lorentz had neatly solved all problems of motion, which meant that Drude could ignore the issue and get on with the major task of building a coherent theory on the basis of the field equations and a linked equation for matter. Far from posing any grave difficulties, to Drude the issue of motion through the ether was, if not a closed book, at least one hardly worth opening.

Nor was Drude alone in considering the Michelson-Morley experiment to pose at most a marginal question. One can cite many examples, but a particularly striking one will suffice. In 1896 Emil Wiechert wrote a paper entitled "On the Foundations of Electrodynamics," which gave in outline form the twin-equation scheme that I described above: unalterable field equations for the ether, linked by a single term to an equation of motion for material particles. He considered that stellar aberration as well as "the observations of Fizeau, Michelson and Morley" simply confirm that the electromagnetic field is "independent" of material motion, which, on Lorentz's 1892 and 1895 theories, is correct. Wiechert does not even mention the 1887 ether-drift experiment, much less the issues raised by it. For him the most important thing was to come to grips with the novel elementary structure of electromagnetic theory, not to examine marginal effects. This viewpoint was if anything even more pronounced in Britain than on the Continent, because until after the mid-1890s Maxwellians were for the most part unconcerned with effects that had to be due to the incalculable interaction between ether and matter. FitzGerald's short, and shortly ignored, suggestion for explaining the ether-drift experiment notwithstanding, most Maxwellians showed very little concern with this kind of problem.[28]

Both Maxwellian indifference to the problem and Continental indifference to the special character of Lorentz's work had begun markedly to alter by about the turn of the century. Maxwellian theory was by then in nearly complete decline as a result primarily of Joseph Larmor's electron theory and the support it apparently received from J. J. Thomson's experiments. In the form it had acquired by circa 1900, Larmor's theory was quite similar to Lorentz's. And Lorentz's work was acquiring intense scrutiny, particularly in Germany, in major part because physicists saw in it the possibility of founding all of physics on electrodynamics. Attention was by this time turning directly to problems that centered on the behavior of the electron proper, and here the issues raised by the Michelson-Morley 1887 experiment were more to the point.

NOTES

1. Henry A. Rowland, "Modern Views with Respect to Electric Currents," *Transactions of the American Institute of Electrical Engineers*, 1889, 6:342-357; *Physical Papers* (Baltimore: Johns Hopkins Press, 1902), pp. 653-667.

2. George F. FitzGerald, "The Ether and the Earth's Atmosphere," *Science*, 1889, *13*:390.

3. With the notable exception of E. T. Whittaker, who, however, did not regard Einstein as a significant contributor to relativity.

4. These equations were usually closely patterned on the kinds of relationships exemplified by Cauchy's stress-strain analysis of elastic solids or else on his equations for a point lattice governed by conservative forces.

5. In particular that the medium must be governed by Hamilton's principle both as to the partial differential equation which determines its evolution over time and as to the boundary conditions across any surfaces drawn within it.

6. This new British emphasis on abstract dynamical analysis is closely linked to the appearance in 1867 of Thomson and Tait's *Treatise on Natural Philosophy*, a text based directly on energy principles and associated extremum conditions. Its success marks the full ascendence to major influence of the kind of highly abstract, analogical thinking that was exemplified by Thomson and Maxwell, in contrast to the older emphasis on traditional dynamics, which Stokes continued to represent.

7. Recent accounts include Lloyd S. Swenson, *The Ethereal Aether. A History of the Michelson-Morley-Miller Aether-Drift Experiments, 1880-1930* (Austin: Univ. Texas Press, 1972); Tetu Hirosige, "The Ether Problem, the Mechanistic World View, and the Origins of the Theory of Relativity," *Historical Studies in the Physical Sciences*, 1976, 7:3-81; Arthur I. Miller, *Albert Einstein's Special Theory of Relativity* (Reading, Mass.: Addison-Wesley, 1981); Abraham Pais, *"Subtle is the Lord..." The Science and the Life of Albert Einstein* (Oxford: Oxford Univ. Press, 1982). Aspects of what I shall say are treated in greater detail by at least one of these three. Still the most comprehensive and detailed account of analyses before Lorentz is M. E. Mascart, *Traité d'Optique* (3 vols, Paris: Gauthier-Villars et Fils., 1893); see vol. 3, chap. 15.

8. Specifically, the speed of light relative to stationary ether changes from c/n if the refracting body is at rest to $c/n + u - u/n^2$ where u is the speed of the body, c is the speed of light in the ether, and n is the body's index of refraction when it is at rest.

9. On which see David B. Wilson, "George Gabriel Stokes on Stellar Aberration and the Luminiferous Ether," *British Journal for the History of Science*, 1972, *21*:57-79.

10. On this and the Parisian prize, see Hirosige, "The Ether Problem."

11. On which, see Jed Z. Buchwald, *From Maxwell to Microphysics. Aspects of Electromagnetic Theory in the Last Quarter of the Nineteenth Century* (Chicago: Chicago Univ. Press, 1985) and C. Jungnickel and R. McCormmach, *Intellectual Mastery of Nature. Theoretical Physics from Ohm to Einstein* (2 vols., Chicago: Chicago Univ. Press, 1986), vol. 2, pp. 116-118.

12. Except for a very brief mention of Doppler's principle, which Glazebrook did not consider to pose any kind of important problem.

13. Buchwald, *From Maxwell to Microphysics*, and Jed Z. Buchwald, "Modifying the Continuum: Methods of Maxwellian Electrodynamics," in *Wranglers and Physicists. Studies on Cambridge Mathematical Physics in the Nineteenth Century*, ed. P. M. Harmon (Manchester: Manchester Univ. Press, 1985.)

14. Buchwald, *From Maxwell to Microphysics*, Part I.

15. As it was to Maxwell himself, who recommended it in his famous article on ether for the *Encyclopaedia Britannica* (9th Edition), though he took care to emphasize how little we can properly say about the ether's motion:

> The theory of the motion of the aether is hardly sufficiently developed to enable us to form a strict mathematical theory of the aberration of light. Professor Stokes, however, has shewn that, on a very probable hypothesis with respect to the motion of the aether, the amount of aberration would not be sensibly affected by that motion.

[James Clerk Maxwell, *Scientific Papers*, ed. W. D. Niven (New York: Dover, 1965), p. 769.]

16. Lorentz, "De l'influence du mouvement de la terre sur les phénomènes lumineux," *Versl. K. Akad. Wet. Amsterdam*, 1887, 2:297; in H. A. Lorentz, *Collected Papers*, ed. P. Zeeman and A. D. Fokker (9 vols., The Hague: Nijhoff), vol. 4, pp. 152-214; see p. 202.

17. At least if, as Maxwellians and most others required, the ether is incompressible.

18. To whom it would merely mean, as I remarked above, that things had to be much more complicated than they had thought. There is little if any Maxwellian response to Lorentz's demonstration at the time, which again shows that Maxwellians were never deeply concerned with these kinds of issues.

19. Lorentz, "De l'influence," p. 203. Emphasis added.

20. Given the few words Lorentz devotes to this issue, one might think that he had not deeply thought about it, and that, perhaps, the distinction I am drawing was not one which he himself drew. However, earlier in his paper Lorentz had insisted that "the intermolecular ether moves in [transparent bodies] in the same way as did the free ether which occupied the same place [as this intermolecular ether before the transparent bodies were inserted]." If the molecules are vortex atoms this cannot possibly be so because vorticity necessarily alters the velocity field everywhere. But if the atoms are mere states, conditions which are passed from one part of the ether to the next, then this might be so.

21. Buchwald, *From Maxwell to Microphysics*, chap. 27. I discuss Helmholtz's analysis under the rubric of his "twin equations."

22. In Lorentz's 1884 theory of the Hall effect [see Buchwald, *From Maxwell to Microphysics*, pp. 98–99 for details], he implicitly altered the basic electromagnetic field equations, and this amounts to changing the structure of the ether itself. From this work (which was very influential) we can see that before 1886 at any rate Lorentz not only allowed one to play with the ether, he himself did so. Nevertheless Lorentz did not play with abandon. His single excursion in this regard was limited and terse almost to the point of obscurity, which contrasts markedly with his elaborate analysis six years before of dispersion. He only reluctantly, and with almost no comment, restructured the ether. I think it highly probable that Lorentz was terse and even obscure here because he was not satisfied with what he had to do in order to generate equations, but that he did not at this time see how to do it otherwise.

23. Buchwald, *From Maxwell to Microphysics*, chap. 29.

24. Consider, for example, that Paul Drude's 1894 *Physik des Aethers* does not even mention motion through the ether, though it does discuss dispersion, metallic reflection and the effects of a magnetic field on optics in great detail.

25. As remarked by Miller, *Special Theory*, sec. 1.6.

26. This made it, for example, extremely simple to deduce the "drag" coefficient because Lorentz could now analyze the problem in the fictitious system. He did not have explicitly to consider the retarded interactions which, in 1892, had given him the coefficient. This way of obtaining the coefficient made it independent of all approximations except the first-order limitation. In 1892, by contrast, it had emerged only after a series of approximations concerning the interactions between particles. In effect the 1895 deduction produced a general constraint on the approximations that could be used for explicitly calculating the effects of motion on a system of charged particles.

Nevertheless in his 1900 *Theory of Optics* Drude did not think it worthwhile to deduce the drag coefficient using the local time. He also avoided Lorentz's considerations of retarded actions. Instead he obtained the coefficient by operating directly with the field equations and the linked material equation, all referred to moving axes. He did then introduce the local time, which he called position time, and used it explicitly to show that, to first order, the earth's motion cannot be detected using terrestrial sources of light. This was actually unnecessary because Veltmann's old proof, given the drag coefficient, guaranteed this result.

27. Lorentz, "The relative motion of the earth and the ether," *Versl. K. Akad. Wet. Amsterdam* 1892, *1*:74; in *Papers*, Vol. 4, pp. 219–23.

28. This also notwithstanding Oliver Lodge's attempts in the early 1890s to measure ether "viscosity" [Bruce Hunt, "Experimenting on the Ether: Oliver J. Lodge and the Great Whirling Machine," *HSPS*, 1986, *16*:111–34.] I see no compelling evidence in correspondence or in the published materials to conclude that Larmor, FitzGerald or even Lodge himself ever considered the results of these experiments to pose significant difficulties for Maxwellian theory. In Larmor's case he was certainly disappointed that Lodge found no indication of an effect because, had Lodge detected one, then this would have provided support for Larmor's contention that magnetic force corresponds to ether velocity. But Lodge's negative results did not in the least deter Larmor from continuing to develop the theory in quite intricate ways until it foundered on an entirely different issue [Buchwald, *From Maxwell to Microphysics*]. The normal, daily practice of Maxwellians seems hardly to have been affected by problems involving ether motion. One needs

only to look at J. J. Thomson's *Notes on Recent Researches* of 1893 to see this strikingly illustrated. Thomson treats the electrodynamics of moving bodies without a mention of optical issues, concerning himself, e.g., with such problems as the induction of currents in a rotating sphere.

"AD HOC" IS NOT A FOUR-LETTER WORD:
H. A. LORENTZ AND THE MICHELSON-MORLEY EXPERIMENT*

Nancy J. Nersessian

INTRODUCTION

The null result of the Michelson-Morley experiment did not lead to the rejection of the hypothesis it seemed to disconfirm--that of a stationary ether. Although designed as an experiment to determine the relative motion of the earth and the ether, the Michelson-Morley experiment was reinterpreted by H. A. Lorentz as having demonstrated a relationship between the dimensions of moving matter and the ether. His hypothesis that matter in motion through the ether changes its dimensions has been criticized since its introduction for being ad hoc, first by H. Poincaré and later by A. Einstein. In recent times the issue of whether or not the hypothesis is ad hoc was joined in a debate over E. Zahar's characterization of Lorentz's "research program."[1] His and other attempts to show the hypothesis not to be ad hoc seem to be motivated by the assumption that the use of an ad hoc hypothesis is necessarily bad scientific method. Lorentz himself acknowledged that the hypothesis was introduced ad hoc, but argued that it was "not merely artificial."[2] I will argue here that although the hypothesis is ad hoc, its introduction was, nevertheless, warranted.

Lorentz first found a hypothesis that would explain that one experiment and then searched for a causal explanation of the phenomenon. Thus, he had not "deduced the LFC [Lorentz-Fitzgerald contraction] from a deeper theory, namely from ... the MFH [molecular force hypothesis]," as claimed by Zahar.[3] That deeper causal explanation was constructed to account for the hypothesis. However, the "change of dimensions hypothesis," as Lorentz's version of the hypothesis should be called, was not arrived at by what A. I. Miller has claimed to be "a physics of desperation" either.[4] It was found by investigating what Lorentz called the "theory of the experiment" and by recognizing that the problem could be placed in the class of transformation of coordinate problems he had been working on just prior to its introduction. As a method, Lorentz's approach was a good way of handling a specific sort of problem: a piece of anomalous data that one has adequate evidence to believe, but which, if taken as a falsifying instance, would require the abandonment of a fundamental principle of the theory.

What Lorentz did is what scientists often do when faced with a piece of data that apparently falsifies a hypothesis of their theory. In practice the response to an unexpected and unwanted experimental result depends upon several factors, among which are how central a tenet of the theory the hypothesis in question is and how coherent it is with the evidence thus far accumulated. If the hypothesis is central and/or coheres well with the other available data, the choice is usually to attempt a change elsewhere in the theory in order to produce a coherent explanatory account that includes both the hypothesis in question and the anomalous datum. This practice is supported as being good scientific method by anti-foundationalist epistemologies. According to "holist"[5] accounts of justification and explanation, an experimental result is not sufficient to require the abandonment of *any particular* theoretical hypothesis. A hypothesis is not tested in isolation, but in relation to an intricately woven network of hypotheses and

---

*I want to thank Allan Franklin for letting me expropriate the main title, which he used in a lecture on ad hoc hypotheses.

data. Faced with recalcitrant data, the scientist must change something but there is some choice as to what and how.

The obvious problem then becomes: "How to determine what changes to make?" There is no universal answer to this question. Such changes cannot be made according to some recipe. What to change will be constrained by the particulars of the case. There is a guiding principle, though, that proposed changes should conform to: they should be made in a conservative manner, so as not to "sink the ship." The more fundamental a principle, i.e., the more central it is to the explanatory account presented by the theory, the more damage its revocation will impose on the theory. Thus, it is good scientific practice not to reject a fundamental tenet of a theory on the basis of just one experimental anomaly, but rather to seek another way to accommodate the result.

LORENTZ'S THEORY IN 1892

In his major paper of 1892[6] Lorentz developed the first full version of his "theory of electrons." Although the Michelson-Morley experiment had been of concern to him since 1887, he did not attempt to deal with it in this paper, keeping only to first-order phenomena. The change of dimensions hypothesis would be put forth later in the year. Thus, this paper represents the most complete statement of his theory and the problems with which he was concerned while he was puzzling out a response to the experiment. Buchwald presents an overview of Lorentz's achievements elsewhere in this volume, so it is not necessary to do so here. Rather, I will simply highlight some details of this paper and of Lorentz's earlier work pertinent to my argument.

The central problem of Lorentz's research, as articulated in his 1878 Inaugural Address for his assumption of the Chair for Theoretical Physics at the Univ. of Leiden,[7] was to combine an ether-based theory of electric and magnetic actions with a molecular theory of matter. In his dissertation of 1875,[8] Lorentz had established the superiority of Maxwell's electromagnetic theory of light over the elastic solid theory for the solution of the problems of the reflection and refraction of light. He adopted Maxwell's continuous-action ether in 1891,[9] arguing that, given Hertz's results, accounting for electrostatic and electromagnetic energy requires including the medium in the surrounding space in the analysis. By 1878[10] he had already made a clear distinction between ether and ordinary matter, with the ether as the only true dielectric. And, in 1886,[11] he had established that Stokes's hypothesis that the earth drags the ether at its surface along with it completely is untenable, given Stokes's own premises, and that Fresnel's hypothesis of no overall motion of the ether in the direction of the earth's motion best explains the phenomenon of aberration.

The problem of "The Electromagnetic Theory of Maxwell and its Application to Moving Bodies" was "to know the laws which govern the movement of electricity in bodies which traverse the ether without carrying it along."[12] Here Lorentz makes a departure from the Fresnel ether hypothesis: Fresnel allowed for some local motion or "dragging" of the ether within matter. Lorentz's ether is immobile. The major result of this paper is the derivation of the Fresnel "dragging coefficient" without the actual dragging of the ether. The "physical hypotheses" of the paper are: (1) Material bodies are systems of microscopic charged "ions" that are in part mechanical bodies to which Newton's laws apply and (2) The ether is everywhere locally at rest. These are the fundamental principles of what was to become the theory of electrons.[13] Thus, contrary to Buchwald's claim that the result of the Michelson-Morley experiment "did not breach the basic structure of Lorentz' new physics in any significant way,"[14] it calls into question one of Lorentz's fundamental assumptions: the stationary ether. If interpreted as a test for the state of motion of the ether, the results of the Michelson-Morley experiment would seem to contradict one of the *central* tenets of the theory. Thus,

Lorentz searched for a different interpretation. While Buchwald is correct in claiming that the 1886 repeat of the Fizeau experiments by Michelson was influential on the development of Lorentz's microphysical theory, it should be pointed out that these experiments also provided corroboration for the Fresnel-Fizeau stationary ether hypothesis, and thus confirmed Lorentz in making it a fundamental hypothesis of his theory. We see just how important Lorentz considered the retention of the immobile ether hypothesis in his response to the problem that in his system the ponderomotive force with which the ether acts on the ions violates the law of action and reaction. He stated that "as far as I can see, there is nothing to compel raising that law to a fundamental law of unlimited validity."[15] A mechanical representation for electric and magnetic forces is in principle impossible for an immobile ether, so there is no reason to assume a priori the universal validity of Newton's laws. Thus, the law of action and reaction should be abandoned before the immobile ether hypothesis.

One further result of the paper under discussion needs to be recalled. After the derivation of Maxwell's equations and the new force law for an ether-based system, the problem of how to transform between moving systems and systems at rest had to be tackled. Lorentz was to work on this problem until 1904 before he would produce a full set of transformation rules, but the process begins with the 1892 monograph. In order to maintain the correct form of the wave equation for radiation in going from a system at rest to one in motion, a further transformation, after application of the standard Galilean transformations, was required. Lorentz introduced what he considered to be a purely mathematical coordinate transformation to an "imaginary" system to produce the correct form. Once he obtained the transformation rules, Lorentz then restricted the value of his results to first-order phenomena for which all observational evidence indicated that the velocity of light is independent of the motion of its source. In these circumstances, the spatial transformations retain their Galilean form, but the temporal one has "mixed" spatial and temporal coordinates· With this latter transformation, he was able to formulate a "general theorem" for first-order phenomena: The laws of electromagnetism have the same form whether formulated for a system at rest in the ether or for one in constant motion relative to it.

THE CHANGE OF DIMENSIONS HYPOTHESIS

Selective details of Lorentz's 1892 paper have been given in order to highlight part of the context in which he was thinking about the Michelson-Morley experiment. For, although he does not mention it in this paper, he proposed his solution to the problem of how to interpret the experiment in a brief paper later that year. He comments in the paper that "I have long reflected in vain upon this result and have finally been able to find only one means by which to reconcile its result with the theory of Fresnel."[16] When we couple these remarks with comments made to Rayleigh in a letter written between the two papers, viz., "I am totally at a loss to clear away this contradiction [between the experiment and Fresnel's hypothesis], and yet I believe if we were to abandon Fresnel's theory we should have no adequate theory of aberration at all...,"[17] we come away with the impression that the experimental result posed more than a "marginal"[18] problem for Lorentz. Indeed, if interpreted in the way Michelson and Morley would have it, the ether must be in motion, contrary to Lorentz's fundamental principle.

It is important to underscore the centrality of the stationary ether hypothesis, because this has direct bearing on the warrant for his adoption of the change of dimensions hypothesis. To accept the obvious interpretation of the experiment would mean abandoning this hypothesis--in short, his whole electromagnetic theory. But, this theory, incomplete as it was in 1892, had just accounted for all other electromagnetic phenomena and, most importantly, had allowed the derivation of the "dragging coefficient," which no other electromagnetic theory could do. Thus, rather than accept that the experiment provided a test for the state of motion of the ether with respect to the earth, he posed the question: "Can there be some point *in the theory of Mr. Michelson's experiment* which has as yet been overlooked?" [my italics][19] That is, he raised the possibility that the experiment was not measuring what it was supposedly designed to detect, but, rather, something else.   He would conclude that the experiment could not provide evidence for the motion of the earth with respect to the ether because all matter in motion through the ether changes its dimensions.

Lorentz does not tell us how, a few months later, he came upon his solution to the problem that had nagged him for five years. Using the classical addition of velocities theorem, it is easy to calculate the discrepancy between the expected and actual results, and it seems a short step from that to hypothesizing that matter in motion contracts by that amount. This is precisely what Fitzgerald had done some years earlier, unbeknownst to Lorentz. [20] But, until Lorentz had formulated his theory, as presented in the 1892 paper, there were no *good reasons* for making such a supposition.   I reconstruct Lorentz's path to the change of dimensions hypothesis as follows: In order to formulate his theory, Lorentz had to suspend his worries about the experiment. Although incomplete, and restricted to first-order phenomena, the success of the theory convinced him that he was on the right track; in particular, that the stationary ether hypothesis was correct. He then turned his attention once more to the experiment and realized that the problem was similar to the transformation of coordinates problems he had just been working on. Although he was working specifically on the problem of the transformation of electric and magnetic forces, formulated generally, the problem is: What happens to a system at rest when it is placed in motion through the ether? When he calculated the requisite "translation factor" for the experiment to have given a null result, he realized that it was *exactly* the same factor he had derived for the transformation of electric and magnetic forces, in particular, the Lorentz force. As an explanation, he hypothesized that motion through the ether might have an effect on molecular forces as well; i.e., that there is a causal relationship between matter in motion and the ether.

His actual presentation of the hypothesis[21] is as follows. He first argues that only Fresnel's hypothesis can account for aberration. He then states that he is convinced of the correctness of the Michelson-Morley experiment, and that the only way to accommodate it is to suppose that the perpendicular and parallel lengths of the arms of the interferometer and the dimensions of the slab do not remain constant throughout the experiment. The rest of the paper argues for why it is plausible to assume that as with electric and magnetic forces, molecular forces act through the ether. He cautions that since the hypothesis cannot be tested without further knowledge of the nature of molecular forces, one could not attach too much weight to it. Nevertheless, "it cannot be denied that a change in the molecular forces, and, consequently, in the dimensions of a body are possible."[22] From this Lorentz concludes: "The experiment of Michelson loses its demonstrative force for the question for which it was undertaken. It's significance--if one accepts Fresnel's theory--lies rather in that it can teach us something about the changes of dimensions."[23] Thus, the Michelson-Morley experiment cannot provide evidence for the motion of the earth with respect to the ether.

AD HOC BUT "NOT MERELY ARTIFICIAL"

In a letter to Einstein in 1915,[24] Lorentz discussed Einstein's contention that "to account for experiments with a negative result through hypothesis devised ad hoc [in particular, the change of dimensions hypothesis] is very unsatisfactory." Lorentz countered that one can be satisfied with such an hypothesis if its explanation is "not merely artificial." He argued that his explanation of the change of dimensions hypothesis by means of molecular forces transforming like electromagnetic forces was not artificial and that "[i]f [he] had stressed it more, the hypothesis would have made less of an impression of having been devised *ad hoc*." In a footnote to this latter statement, he did admit "I must, indeed, confess that I first made these remarks [about molecular forces] *after* I had found the hypothesis" [Lorentz's italics].

There are two points at issue here: whether or not the change of dimensions hypothesis was introduced ad hoc and whether the explanation of it is artificial. As we have seen, the change of dimensions hypothesis was devised to account for one experiment and in this sense it is ad hoc but this, in itself, does not make it "unsatisfactory." The major criticism offered by both Poincaré and Einstein is that by following such a procedure generally, we could end up inventing a new hypothesis to account for each new experimental result, which would indeed be "artificial" and "unsatisfactory." Here Lorentz was in agreement with them. However, he felt that the explanation in terms of molecular forces -- even though devised afterward -- redeemed the procedure *in this case*. His public introduction of the hypothesis was *together* with the explanation and his proposed reinterpretation of the experiment.

Combined with the molecular forces explanation, the change of dimensions hypothesis is warranted. First, the value of the change, calculated on the basis of the empirical evidence and the customary addition of velocities theorem, is exactly the same as the transformation of electric and magnetic forces requires. Second, so little was known about molecular forces that it was quite plausible that the ether could mediate them. Recall that Lorentz's goal was to combine an ether-based electromagnetic theory with a molecular theory of matter. It was far from clear how such a theory of matter would develop. Although Lorentz was not an active proponent of the electromagnetic theory of matter, he was sympathetic with the view. He, himself, entertained the possibility of matter being "local modifications" of the ether. In addition, as he said, "[w]ith respect to theory, there is nothing in opposition to the hypothesis."[25] Finally, his other option was to give up a central hypothesis of his theory -- one that had facilitated his explanation of all other electromagnetic phenomena -- because of *one* experiment. In the course of his work, both the change of dimensions hypothesis and the molecular force hypothesis were made more general: They encompassed more phenomena, including new experimental results; they allowed for the extension of the "theorem of corresponding states" to many second-order phenomena. And so he retained them.

It is a fact that the change of dimensions hypothesis is ad hoc but to use this appellation in a pejorative manner is to say that the hypothesis can be evaluated in isolation, that is, independently of the network into which it was introduced. But, the evaluation of a hypothesis depends importantly on the other hypothesis with which it furnishes explanations. Fitzgerald's "contraction hypothesis" is unwarranted because it was introduced in isolation.[26] Lorentz's change of dimensions hypothesis was not. And, it should be noted that Fitzgerald himself withdrew his objections to the hypothesis when he came to understand it in Lorentz's terms.[27] I interpret Lorentz's claim that the change of dimensions hypothesis is not "merely artificial" to be saying that if he had not found a satisfactory explanation for why matter in motion through the ether should change its dimensions -- one that could incorporate it into his theoretical system and thus provide the best total explanatory account of electromagnetic phenomena -- he would not have retained the hypothesis. What more can be required of good scientific practice?

NOTES

1. E. Zahar, "Why did Einstein's Programme Supersede Lorentz's?" *British Journal for the Philosophy of Science*, 1973, *24*: 95-123, 223-262; A. I. Miller, "On Lorentz's Methodology," *BJPS*, 1974, *25*: 29-45; K. F. Schaffner, "Einstein vs Lorentz: Research Programmes and the Logic of Comparative Theory Evaluation," *BJPS*, 1974, *25*: 45-78; J. Leplin, "The Concept of an *Ad Hoc* Hypothesis," *Studies in the History and Philosophy of Science*, 1975, *5*: 309-345.

2. Lorentz to Einstein, January 1915.

3. Zahar, "Lorentz," p. 106.

4. A. I. Miller, "On Some Other Approaches to Electrodynamics in 1905," in *Some Strangeness in the Proportion*, ed. H. Woolf (Reading, Mass.: Addison-Wesley, 1980), p. 30.

5. See, e.g., P. Duhem, *The Aim and Structure of Physical Theory* (Princeton: Princeton Univ. Press, 1962); W. V. O. Quine, *Ontological Relativity and Other Essays* (Cambridge, Mass.: M.I.T. Press, 1969); G. Harman, *Change in View* (Cambridge, Mass.: M.I.T. Press, 1986).

6. H. A. Lorentz, "La théorie électromagnétique de Maxwell et son application aux corps mouvants," in *Collected Papers* (The Hague: Martinus Nijhoff, 1935-39), vol. 2, pp. 164-343.

7. H. A. Lorentz, "De moleculaire theorieen in de natuurkunde," *Collected Papers*, vol. 9, pp. 1-25.

8. H. A. Lorentz, "Over de theorie der terugkaasting en breking van het licht," *Collected Papers*, vol. 1, pp. 1-192.

9. H. A. Lorentz, "Electriciteit en ether," *Collected Papers*, vol. 9, pp. 89-101.

10. H. A. Lorentz, "Concerning the Relation between the Velocity of Propagation of Light and the Density and Composition of Media," *Collected Papers*, vol. 2, pp.1-119.

11. H. A. Lorentz, "Over den invloed dien de beweging der aarde op de lichtverschijnselen uitoefent," *Versl. Kon. Akad. Wetensch. Amsterdam*, 1886, *2*: 297-372.

12. Lorentz, "La theorie," p. 216.

13. For a more extensive discussion of the hypotheses of Lorentz's theory, see N. J. Nersessian, "Why Wasn't Lorentz Einstein?" *Centaurus*, 1986, *29*: 205-242.

14. J. Z. Buchwald, "The Michelson Experiment in the Light of Electromagnetic Theory before 1900," in this volume.

15. H. A. Lorentz, "Versuch einer Theorie der electrischen und optischen Erscheinungen in bewegten Körperen," *Collected Papers*, vol. 5, p. 28.

16. H. A. Lorentz, "De relative beweging van de aarde en den ether," *Versl. Kon. Akad. Wetensch. Amsterdam*, 1892, *1*: 74-79, p.74.

17. Lorentz to Rayleigh, 18 August 1892, Rijksarchive, The Hague.

18. In correspondence with me Buchwald claims that what he meant by "marginal" is simply that Lorentz did not feel he had to explain the Michelson-Morley experiment before he presented his new theory. I have no problems with this interpretation. However, this does not mean that Lorentz did not see the explanation of the result of this experiment as a high priority for his theory.

19. Lorentz to Rayleigh, 18 August 1892.

20. See S. Brush, "Notes on the History of the Fitzgerald-Lorentz Contraction," *Isis*, 1967, *58*: 230-232.

21. Lorentz, "Relative."

22. Ibid., p. 78.

23. Ibid., p. 78.

24. Lorentz to Einstein, August 1915. I am indebted to Anne Cox for providing me with a copy of this letter.

25. Lorentz, "Relative," p. 78.

26. Fitzgerald himself speculated that the contraction might be due to the effect of the ether on molecular forces, but he did not pursue this because he had no theory into which he could incorporate the speculation.

27. G. Holton, "Einstein, Michelson, and the Crucial Experiment," in *Thematic Origins of Scientific Thought* (Cambridge, Mass.: Harvard Univ. Press, 1973), p. 181.

WHAT INSTRUMENTS MEASURE AND WHAT PEOPLE BELIEVE:
REFLECTIONS ON INTERPRETATIONS
OF THE MICHELSON-MORLEY EXPERIMENT

Stanley Goldberg

INTRODUCTION

Albert Michelson was a great physicist. His genius was his ability to harness state-of-the-art engineering and machining techniques in the service of measuring instruments of sophisticated sensitivity. Precision engineering techniques were there for the taking. Only a handful of those who might have taken advantage of state-of-the-art techniques by virtue of their talent and training in natural philosophy were actually equipped in other ways to do so. Michelson was one of those. He held himself and those who worked with him to the highest standards of precision attainable. And his charisma, which was probably defined by his vision of what was possible and by his admiration for precision work, allowed him to marshall and recruit patrons both in and out of the engineering and precision machining communities.[1] And he was a hard taskmaster--not just to others, but to himself as well. As Loyd Swenson has pointed out, Michelson was "never fully satisfied with the precision of former measurements. . . ."[2] Taken all together, it was these qualities that led to his world-wide recognition and the first Nobel Prize awarded to a United States scientist. The opening sentence of the presentation speech of Michelson's Nobel Prize states:

> The Royal Academy of Science has decided to award this year's Nobel Prize for Physics to Professor Albert A. Michelson of Chicago, for his optical precision instruments and the research which he carried out with their help in the fields of precision metrology and spectroscopy.[3]

But we must recognize that history is peppered with individuals with precisely this kind of drive and charisma. In an earlier time in the history of science, it was Tycho and his magnificent, state-sponsored observatories; Stephen Hales, fiendishly insisting on his servants' collecting daily measurements on the sap pressure in the trees and other plants provided by the royal Garden at Hampton Court; or Lavoisier seeing clearly that the oxides of tin were heavier than the tin itself when, for almost all others, the measurements were in the grey area of experimental error.[4] In the period since Michelson, it has been Ernest Lawrence commanding his troops and marshalling scarce resources at the height of the Great Depression; Barbara McClintock seeing a different kind of order in the chromosomes of the corn plant than most others saw; or Burton Richter and Samuel Ting and the armies they commanded in the work leading to the identification of the J-psi meson.[5]

The kinds of individual personality traits required to conceive of, organize and see such projects through to a conclusion probably transcend time, place, culture and social organization. But the practitioners' convictions concerning the epistemological and ontological significance of this kind of work certainly has varied with time and, I have argued elsewhere,[6] with culture. From near the end of the sixteenth to the beginning of the nineteenth century, most individuals who undertook serious investigations of the workings of the physical universe saw their work as free of metaphysics. Conclusions were arrived at through the kind of inductive inference outlined and advocated by Francis Bacon. The surest way to damn the work of their rivals and foes was to label the work as "un-philosophical," i.e. not stemming from observation and careful experimentation. "Theoretical," as opposed to "experimental" was not a meaningful distinction. Examining early volumes of *Transactions of the Royal Society* or the *Comptes Rendus*, one is impressed by an ambiance which is best described as combining reverence, amazement and awe. The wonder of it all! There was a world out there and human beings, by assiduous and artful application of the experimental philosophy could find out how it worked--God's plan was there to be understood. This is not to say that there were not easily identifiable national and local differences and idiosyncracies.[7] But these had more to do with subject, treatment, and ontology than with epistemology.

In the nineteenth century the emergence of academic disciplines and subdisciplines as we know them had a profound effect on the relationships not only among the arts, humanities, and the sciences, but among the sciences themselves. In Britain and the United States the word "scientist" was first introduced in about 1840, quickly displacing "natural philosopher."[8] The transition was a significant one. Scientist: "one who knows." Of the definitions of philosopher, I would argue that it is the activities associated with "one who investigates the nature of things based on logical reasoning rather than empirical methods"[9] that was being replaced.

This transition in language reflects the emerging view of the special epistemological quality attributed to some sciences in contrast to the humanities and the arts. The modern term, "hard science," is an apt description of that knowledge which is certain, sure, and unlikely to be subject to change. This is in contrast to the so called "soft sciences." What discriminates hard disciplines from soft disciplines is the degree to which the knowledge is seen to be anchored to careful, physical measurements done with precise instruments of unequivocal purpose. Philosophy and physics had gone their separate ways.

One can imagine that at any stage in history, people have been impressed with state-of-the-art craftsmanship, whatever that state-of-the-art might have been. The ability to make better and better tools and finer and finer measuring instruments is an iterative process that feeds on itself. (Imagine, for example, the problem of building the first lathe without a lathe with which to make it.) It is progressive in clearly defined, unambiguous ways. What has been termed "precision engineering practices" of the nineteenth century were nothing more than a milestone along a road defined by past practice.

It is not surprising that in the United States, where there was a marked tendency to value the practical over the theoretical and where technological innovation was often construed as science,[10] the great scientists made their reputation in the realm of measurements done with carefully crafted precision instruments of their own design.

By the end of the nineteenth century, in Europe, theoretical physics had become an established subdiscipline.[11] Max Planck could say openly that in making judgments about the relative value of different theoretical approaches, one needs to balance agreement with measurement against internal harmony.[12] At the time such a position would have been heresy in the American physics community.

Michelson was a great experimental physicist and an indefatigable gatherer of data. His forté was measurement. His life was devoted to discovering the most inner secrets of the behavior of light. In his view -- and the view of most of his colleagues -- measurement had revealed light to be a phenomenon characterized by the propagation of energy in a medium. Light was wave phenomenon and as such, required a medium, the ether. This was not romantic, idle speculation. This was not even theoretical. For physicists such as Michelson, it was a physical fact, forced on us by the results of repeated, precise, measurements -- measurements of the sort Michelson excelled in. He was able to measure the speed of light to six significant figures. He made precise and ingenious instruments for sorting out the component parts of light energy, and he led the way in making experiments of hitherto unknown sensitivity on the motion of the earth through the ether. But contrary to what seventeenth and eighteenth century natural philosophers might have propounded, an experiment such as the Michelson-Morley ether-drift experiment does not speak for itself. It must be interpreted. Exploring the variety of ways the Michelson-Morley experiment has been interpreted leads, *seriatim* to an examination of the measuring process and what it is that instruments actually measure. From this perspective, we will return once again to the question why it is that there can be so many interpretations of a single experiment.

INTERPRETING THE MICHELSON-MORLEY EXPERIMENT

The Michelson-Morley experiment has been subject to a number of interpretations. For Michelson and Morley, the fact that there was no discernible shift in fringes, taken by itself, meant that the ether must be totally dragged along by matter--in this case, the instrument itself. But one could not take this one experiment by itself. There are three possible types of interaction of ether and matter. The ether can be totally unaffected by the motion of matter, the ether can be partially dragged along by matter or it can be totally dragged by matter. By the time Michelson made his first interferometer in 1881, a host of experiments and observations, such as Bradley's observation of stellar aberration and the Fizeau measurements of the speed of light in still and moving water, had been performed. Some, for example, the observation of stellar parallax, were most easily comprehended by assuming no interaction between matter and ether. Almost all others were interpreted as evidence that the ether partially dragged matter.[13]

In 1878, H. A. Lorentz published a theory of optical dispersion in which he made a radical departure from previous assumptions about the ether. Lorentz assumed that the ether was absolutely fixed. According to Lorentz, light waves moving through the ether caused the ions in matter to oscillate, thereby creating innumerable small waves that through interference with themselves and with any incident radiation were changed in such a way that the velocity of the wave appeared to depend on the frequency of the radiation, thereby causing the phenomenon of dispersion in transparent matter. As a result of this analysis, Lorentz predicted a relationship between the density and the index of refraction for dielectric media.[14] The theory requires a strange model for the behavior of ether. As matter passes through the ether the density of the ether defined by the boundaries of matter increases. The ether itself is absolutely fixed. It is as if air was scooped in at the front end of a jet engine, but on being expelled from the back end, stopped instantaneously, without any perturbation, or movement of any kind. In this case, it was any material body that played the role of the jet engine and ether that played the role of air.

In 1892, Lorentz published what was to be the first of a series of papers leading to a comprehensive electromagnetic theory of matter and ether.[15] In this paper Lorentz restricted his analysis to electromagnetic phenomena, combining particulate views of electricity with Maxwell's field equations and the concept of an absolutely fixed ether. In Lorentz's view, then, partial drag was an appearance, a manifestation of the special character of the ether. As time went on, for Lorentz, the ether became more and more abstract--a benchmark of Newton's absolute space. The fact that the ether had special and seemingly contradictory physical properties was not a bother, any more than it is a bother today that the model used for a molecule by structural chemists is totally at odds with the model used by high energy physicists.

Three years later, in 1895, Lorentz again attacked the problem of the electrodynamics of moving bodies.[16] In this work, Lorentz sought to simplify the theory while at the same time bringing still more phenomena into account. The simplification of the theory resulted from the introduction of a new transformation equation for the relationship between temporal coordinates in different inertial frames of reference. Up to this time,

Lorentz had simply employed the familiar Galilean transformations when describing phenomena in different inertial frames of reference. The implication of using the Galilean transformation equations was that the only frame of reference in which the fundamental laws of electromagnetism applied exactly was the absolute rest frame--the frame which was at rest with respect to the ether. All attempts to measure the effects of the motion of the earth with respect to the ether to the first order in the ratio of the velocity of the earth to the velocity of light (v/c) had failed. This failure was interpreted to mean that the description of phenomena in frames of reference other than the ether frame must be the same, at least to the first order. The new temporal transformation equation reflected the fact that it took a finite amount of time for electromagnetic energy to propagate from one place to another.[17]

Though this was a very bold and radical step to take, Lorentz had surprisingly little to say about the sharp departure such a transformation represented. He designated the transformed temporal component as "local time [Ortzeit]" as opposed to the general, or true time [allgemeine Zeit]. Otherwise he made little comment at the time on the meaning of the transformation save that he clearly intended it to be little more than an aid to calculation. Nor did Lorentz indicate how he had arrived at the idea of this transformation equation. The upshot was that using the new transformation equation, the fundamental laws of electricity and magnetism applied equally in the ether frame and in inertial frames of reference exactly to the first order.

While the Lorentz theory predicted that to the first order in v/c one should expect no evidence of the motion of matter through ether, it did predict that one should see such evidence in experiments sensitive enough to detect second-order effects in v/c. The Michelson-Morley experiment was one such experiment. In fact, the last chapter of Lorentz's 1895 treatise was entitled, "Investigations Whose Results Cannot be Explained without Further Ado."[18] The experiment that received the most attention was the Michelson-Morley experiment. As Lorentz pointed out, in terms of a model of the ether, the null result of the Michelson-Morley experiment suggests that the ether is totally dragged by ordinary matter. No matter how abstract the concept of the ether became, it is not possible to conclude that the moving matter totally drags and partially drags the ether. One cannot have it both ways.[19]

It is that contradiction which is so stunning about the results of the Michelson-Morley experiment. It is no accident that before Michelson and Morley performed the 1887 interferometer experiment, they undertook a replication of the Fizeau measurements of the speed of light in still and moving water.[20] That is not an easy experiment to do and requires the kind of virtuosity in measurement that Michelson and Morley, and few others, possessed. They confirmed the Fizeau result -- that the ether appears to be partially dragged by the water.

At first, in 1892, Lorentz proposed that the null result of the Michelson-Morley experiment and of other second-order experiments that followed it could be explained by assuming that motion through the ether somehow caused the apparatus to shrink just enough in the direction of motion to foil the attempt to determine the speed of the apparatus relative to absolute space. At about the same time, G. F. FitzGerald also proposed that the apparatus shortened in the direction of motion.[21]

As will be developed below, a number of other interpretations of the 1887 Michelson-Morley experiment are possible and have been adopted by various physicists over time. Those interpretations include using the experiment as evidence to support the view that the ether does not exist, the view that the speed of light is the same in all directions for all observers, and the view that the speed of light is independent of the motion of the source.[22] Some historians have even tried to use the Michelson-Morley experiment as primary evidence for how it was that Einstein came to propose the special theory of relativity.[23]

An important and largely unexamined issue that presents itself is the relationship between how the instrument operates and the conclusions one feels justified in making based on the data provided by the instrument. Before considering the significance of the variety of interpretations given the Michelson-Morley experiment, I turn to the general question of the nature of the measuring process itself and the question of what it is that instruments are capable of measuring.

THE MEASURING PROCESS

At a minimum, an empiricist must hold that theory comes from measurement rather than vice versa. And it has generally been held that instruments serve as extensions of the senses.[24] Let's begin then with the senses -- the sense of sight, for example. It is a simple matter to extend the argument to the other senses.

In his extremely heuristic *Patterns of Discovery*, N. R. Hanson provocatively addressed the issue of the mediating role of the senses. Hanson stated that on seeing the

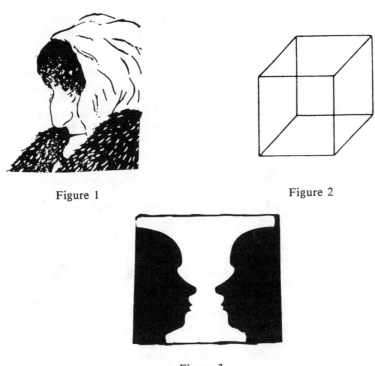

Figure 1   Figure 2

Figure 3

dawn, Tycho and Simplicius do not see the same thing as Galileo and Kepler. Tycho and Simplicius see a static sun, while Galileo and Kepler see a mobile sun:[25]

> Tycho sees the sun beginning its journey from horizon to horizon. He sees that from some celestial vantage point the sun (carrying with it the moon and planets) could be watched circling our fixed earth. Watching the sun at

> dawn through Tychonic spectacles would be to see it in
> something like this way.
> Kepler's visual field, however, has a different conceptual
> organization. Yet a drawing of what he sees at dawn
> could be a drawing of exactly what Tycho saw, and could
> be recognized as such by Tycho. But Kepler will see the
> horizon dipping, or turning away, from our fixed local
> star. The shift from sunrise to horizon-turn is analogous
> to . . . shift-of-aspect phenomena . . . . It is occasioned
> by differences between what Tycho and Kepler think they
> know.[26]

Like most of us, I have spent many hours looking at sunrises and sunsets. I have
looked at them in Cambridge, Massachusetts, in Yellow Springs, Ohio, on the high
savanna of Central Africa; and at the base of the Holyoke range in Amherst, Massachusetts. I have been a thinking, convinced Copernican since 1960. But try as I may, I
have never been able to see the horizon "dipping or turning away." I couldn't do it in
1962 when I first read Hanson, I couldn't do it this morning and no one I have ever
asked to join in has been able to do it either.

The shift-of-aspect phenomena Hanson was referring to are shown in figures 1 to 3.
They are the usual array of Gestalt figures with which we have all become so familiar.
In figure 1 the switch is between seeing the portrait of an old woman wearing a
babushka and a young woman fashionably dressed. Figure 2, the Necker cube, can be seen
in two different perspectives. Figure 3, the Koehler vase, can be seen as a vase or as
the silhouettes of two people facing each other. According to Hanson, the switch
happens instantaneously.

When first confronted with such images there is no doubt that the images seem to
shift back and forth, instantaneously. But having become acclimated to the image, by
looking at the right place, one can hold either perspective as long as one wants -- or
even better, keep both perspectives in view at the same time. In fact the best that one
can say with regard to the so-called Gestalt switch is that it happens in less time than
one has been able to measure.

Even without introducing physical instruments, a strong case could be made for the
proposition that sense impressions themselves are not simple or theory free. There are
some dramatic situations in which it is clear for at least some individuals, that one must
learn to organize what to most of us is the immediate sense experience of vision itself.
When adults, blind from birth, are given sight as a result of corneal transplants, the
world is not instantly organized. According to J. Z. Young:

> The patient on opening his eyes for the first time . . . reports only
> a spinning mass of lights and colors. He proves to be quite unable
> to pick out objects by sight. . . . His brain has not been trained in
> the rules of seeing, does not know which features are significant
> and useful for naming objects and conducting life. We are not
> conscious that there are any such rules; we think that we see, as
> we say "naturally." But we have in fact learned a whole set of rules
> during childhood.[27]

Young goes on to say that at least a month passes before the subjects can recognize
even a few objects, and ". . . if sufficiently encouraged they may after some years
develop a full visual life and be able even to read." As Leonard Nash has said of this
passage, seeing is no simple passive reception of unequivocal signals.[28]

While this kind of example calls into question the concept of sense data as raw, uninterpreted information, it is not the kind of activity I would identify as being analogous to "doing science." Rather, it is more like the behavior of an accomplished typist or an experienced driver operating an automobile. The activity becomes routinized to the point of seemingly requiring no thought.

In class after class, for twenty-five years, I have asked students what is it that they see early on clear mornings when they look to the eastern sky. Though it is clear that they know what I am talking about, the first response is stone silence. They are almost afraid to say "sunrise" because they want to appear to know better. When I reassure them that what I see is the sun coming up over the horizon, they seem almost relieved. In order to maintain my intellectual commitment to the Copernican point of view, the first thing I have to do is deny the evidence of my senses.

Consider the following remarks of Pierre Duhem:

> Enter a laboratory; approach the table crowded with an assortment of apparatus, an electric cell, silk-covered copper wire, small cups of mercury, spools, a mirror mounted on an iron core; the experimenter is inserting into small openings the metal ends of ebony-headed pins; the iron oscillates and the mirror attached to it throws a luminous band upon a celluloid scale; the forward-backward motion of this spot enables the physicist to observe the minute oscillations of the iron bar. But ask him what he is doing. Will he answer "I am studying the oscillations of an iron bar which carries a mirror?" No, he will say that he is measuring the electric resistance of the spools. If you are astonished, if you ask him what his words mean, what relation they have with the phenomena he has been observing and which you have noted at the same time as he, he will answer that your question requires a long explanation and that you take a course in electricity.

Of this quote Hanson says "The visitor must learn physics before he sees what the physicist sees."[29]

What Hanson has in mind here can be illustrated in figure 4. Those of you who recognize 4a and 4b as copies of some of Roger Price's Droodles give your age away.[30] What can we make of 4a? According to Price, it is a mermaid sliding into home plate. And 4b?--what might be called the classic Droodle? Of course we see that it is a ship coming too late to save a drowning witch.

My description does not organize the picture for you in the way you organize the Gestalt figures (figures 1-3). Yet you are able, I am sure, to understand how I have construed the images in figure 4. I have provided a theory. You may accept it, reject it, yawn at it and even provide a countertheory. There are, then, two kinds of active interpretation behind what we call sense impressions: Automatic, yet learned, organization and the deliberate intellectual interpretation of the meaning behind what the senses tell us. But even if one admits that the process of sorting out the complex array of information that stimulates the sense requires, at the beginning, active interpretation of one kind or another, it might be the case that instruments not only extend the range of the senses, but also simplify the data-gathering process. The notion of "pointer reading" that one finds referred to so often as primary data was introduced into the literature by Arthur Eddington in 1928.[31] Over the next ten years, he often referred to the concept, suggesting that the methodology of physical science is concerned with establishing correlations of those experiences that can be weighed, measured or otherwise manipulated by the tools of science. According to Eddington, the necessary data for scientific knowledge is composed solely of pointer readings.

> The eye need not have the power of measuring or graduating light and shade; I think it is sufficient if it can just discriminate two shades so as to detect whether an opaque object is in a certain position or not.[32]

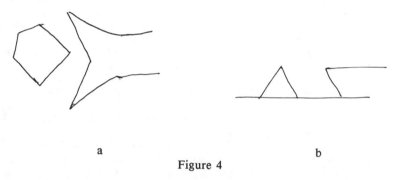

a   b

Figure 4

As John Yolton has noted,[33] one need have only one eye -- the cyclops observer -- because for Eddington, the visual sensations of pointer readings have been the data of the exact sciences. Again, whether or not one accepts Eddington's epistemological formulation, the notion of pointer readings as raw data, direct measurement, has certainly permeated the literature of science, of the history of science and of the philosophy of science. Since the middle of the nineteenth century, there has been little examination of the connection between the process whereby an instrument of high precision delivers a pointer reading, a number, or an analogue signal and the claim that is made for just what it is the reading represents.[34]

WHAT INSTRUMENTS MEASURE

It turns out that there is a very close connection between dimensional analysis and interpretations of the behavior of measuring instruments. The first attempt at dimensional analysis seems to have been by Fourier in the first decade of the nineteenth century.[35] A system of units, for example, the MKS or the SI system must satisfy the rules of formal logic. The system begins with a set of primitives or undefined quantities. In the cgs system the primitives are length (centimeter), mass (gram), and time (second). In the MKS system of more recent vintage, along with length (meter), mass (kilogram) and time (second), there is a fourth primitive, charge. And in the SI or International system of units there are five primitives: length, mass, time, electric current, and temperature. Once the primitives are chosen, all other quantities in the system are defined in terms of the primitives. Primitives serve the role of undefined postulates that are needed to get the system started.[36]

One might choose any primitives one wants in a logical system, but as I have argued elsewhere,[37] doing science is not a matter of formal logic. While being logical -- i.e. thinking straight -- is of value, being logical does not mean using formal logic. When it comes to measurement, there is a close connection between systems of units and our perceptual machinery. The primitives of the measuring system correspond to what might be called "perceptual primitives."

How does one answer the question "What is time?" There is nothing I can say. I might try to change the question to "What is a time interval?" In that case you might use analogies such as growth and decay, life and death, before and after. But unless I know what you mean when you say "time passes," no amount of arm waving is going to allow you to communicate to me the concept of time. In almost any system of units, time is a primitive. A similar argument can be made for physical extension and for mass. Either you know what I mean when I say "the distance from here to there" or you don't. Later when electrical phenomena became important and commonplace, a fourth primitive related to electric charge was added to the collection. These primitives of measuring systems are not just the result of historical accident, they reflect the workings of our perceptual machinery.

Now the curious thing is that while I cannot tell you what electric charge, or time interval, or extension is, I can specify an unequivocal operation for measuring a primitive. For example, to make a standard meter, prepare a particular alloy of platinum and iridium, cast it in a specified shape, and place two scratches at locations dictated by a specific procedure. In other words, for all of the primitives, while I cannot define them, in order to communicate with others about them, I must begin with so-called operational definitions of the sort Bridgman is famous for.

All other quantities in the logical system of units are defined. For example, weight is defined as the product of mass (a primitive) and acceleration, which itself is defined as the ratio of distance to time to time, or, as it is usually rattled off in Physics I, "acceleration is the ratio of distance traversed to time squared."

How do we measure weight? One way would be to use a spring balance. A spring is suspended from a beam. The spring has a hook on the end to which a pan is attached. A pointer is placed on the spring and a scale constructed next to the spring. Take a known weight, throw it on the pan and the pointer descends along the scale. Mark the scale where the pointer comes to rest with a number corresponding to the weight that was thrown on the pan. But how do we get a known weight? There is only one way. We calculate it by comparing the mass of the weight to our standard mass and multiplying the figure thus arrived at by the acceleration due to gravity. In other words, we do a calculation based on some theory. What the spring balance actually measures is length, a primitive.

An alternative technique for standardizing the measurement of weight would be to determine the force constant of the spring and then, using Hooke's Law, equate the change in the length of the spring to the weight of the sample. But as before, such an identification relies on a calculation based on the theoretical relationship known as Hooke's Law.

Take another example, speed. Speed is a defined quantity. In the system of units it is defined as the ratio of length traversed to time elapsed to traverse that length. One can think of laying out a track of known length and measuring the time elapsed to traverse the measured length and then calculating the average speed. What about using a speedometer? At one time, speedometers were mechanical. The length measured was related to the circumference of the wheel; the time measured was proportional to the number of revolutions the wheel makes in a given time: Each time the wheel made a complete revolution, a tab on the wheel struck and displaced a tab attached to a flexible cable which itself was connected at the other end to a spring loaded needle. The rate at which the tabs on wheel and cable interacted is related to the linear distance the vehicle traversed in a unit time and is proportional to the average force the flexible cable exerted on the spring. Given the force constant of the spring, the circumference of the wheel, one can calculate where the needle should point when the car is going five, ten, fifteen, twenty miles per hour etc. What was actually measured was length and time. And they were measured during the process of calibrating the system. Recent automobile speedometers make use of a permanent magnet rotating inside a diamagnetic cup. The cup is attached to the speedometer needle and the induced magnetic drag on the cup works against a spring. The response of the instrument is a function of electric charge and time.

In every case, the instrument responds to some combination of primitives and then a calculation is made, based on some theory, to provide a scale so that the pointer reading can be properly interpreted. Sometimes, though not always, the theory used to calibrate the instrument is the very theory the instrument is supposed to be testing or verifying.[38] Consider for example, the degree to which chemists use the results of infrared analysis as confirming evidence for correctness of the atomic theory. For example, a dip of a certain kind on the pen record is, we are told, a carbon–hydrogen bond. How do we know that? We calibrate the machine by putting through it something we know contains nothing but carbon–hydrogen bonds, methane for example, and get a pen record which can only represent the energy absorption of carbon–hydrogen bonds. But how do we know that methane contains nothing but carbon–hydrogen bonds? The answer is, of course, nineteenth century combining weight and volume analysis, and the assumptions chemists such as Dumas and Cannizzaro made in working out the gas density technique for determining atomic formulae.

Examine the workings of any instrument. No matter what the type, no matter if the output is a signal, a number, or the motion of a pointer, the meaning of the output is determined independently by a calculation based on some theory. What is actually measured is some combination of primitives. The only exceptions seem to be the fundamental comparison measurements of primitive standards wherein one must specify the operations to be followed. The notion of a pointer reading as representing primitive interpretation-free information, a notion popularized by the empiricist philosophers, simply does not stand up to careful examination.

WHAT PEOPLE BELIEVE

Let us now return to examine the Michelson-Morley experiment. What is it that the instrument actually measures? When one looks into the telescope, what one sees is a series of light and dark bands--fringes. As the instrument is slowly rotated, one measures the number of fringe shifts. In order to interpret this measurement one must perform a calculation based on one's convictions about the nature of light energy and the details required by such a theory. At the end of the nineteenth century, while not everyone accepted the electromagnetic theory of light,[39] there was no rival to the wave theory of light. But waves mean medium, so there must be a medium. And as I said earlier, the shock of the Michelson-Morley experiment was that it suggested behavior of the medium contrary to conclusions drawn from all previous ether drift experiments. The results of the experiment are so stunning, that ever since, whenever a new radiation-based technology appears that allows even more sensitive measurements, the Michelson-Morley experiment is done yet again.[40]

At the time the experiment was performed, the Lorentz hypothesis that the apparatus actually shrinks in the direction of motion not only seemed reasonable, it seemed the only way out. For example, when it came to the basic building blocks of matter, Max Abraham rejected the notion of a length contraction: Lorentz's deformable electron was unacceptable to Abraham on the grounds that such an assumption would require non-electromagnetic forces, thus violating Abraham's (and Lorentz's) conviction that electromagnetic interactions were the sole basis for the behavior of matter, radiation and their interaction. For macro-phenomena, however, Abraham invoked the very same Lorentz contraction without making any reference to his earlier critique of Lorentz's use of the application of the length contraction to the case of the electron.[41]

It is no accident that at about the same time as Lorentz, George F. FitzGerald independently also suggested that matter contracted in the direction of motion. According to Oliver Lodge, it happened in Lodge's study:

> The FitzGerald-Lorentz hypothesis I have an affection for;
> I was present at its birth. Indeed I assisted at its birth;
> for it was in my study . . . with FitzGerald in an
> armchair and while I was enlarging on the difficulty of
> reconciling the then new Michelson experiment with the
> theory of astronomical aberration and with other known
> facts that he made his brilliant surmise: "Perhaps the
> stone slab is affected by the motion. I rejoined that it
> was a 45 degree shear that was needed. To which he
> replied, "Well, that's all right--a simple distortion." And
> very soon he said, "And I believe it occurs and that the
> Michelson experiment demonstrates it.[42]

In other words, according to FitzGerald, what the Michelson-Morley interferometer measures is the degree to which the arm in the direction of motion shrinks.

Lodge and FitzGerald did not question the circularity of such an argument. Lorentz might have been highly disturbed. Having seen that one could account for the results of the experiment by proposing that the instrument shrinks in the direction of motion, Lorentz first tried to understand the proposed shortening in terms of the effects of motion on the electrical forces between the electrons in the atoms of material in the arm of the interferometer. As far as I know, he never got beyond a very qualitative account of how this might work.

Being the great physicist he was, Lorentz next turned to generalizing the result. Apparently by sheer trial and error, he figured out what the coordinate transformation equations had to be in order to give the proper contraction and preserve the applicability of the fundamental laws of electromagnetism in all inertial frames of reference.[43] An unwanted and un-understood by-product of that effort was the prediction that identical clocks in different frames of reference would not keep time at the same rate. For Lorentz, such a result could have no physical meaning. It was, he said, an "aid to calculation."[44] Lorentz's approach and his conclusions -- taken in context -- show him to be the ultimate theoretical physicist.[45]

THE EINSTEIN RESULT

In 1905, seemingly out of nowhere, Albert Einstein, a brash, 26-year-old patent examiner in Bern, Switzerland, proposed a new theory in which he assumed as premises the principle of relativity and the invariance of the speed of light. The Lorentz transformations follow as a consequence of those premises. The approach was so out of step with earlier developments in the area that virtually no one had an inkling of what Einstein was talking about and virtually no one came to his support.[46] Part of the problem was that Einstein's presentation was so sparse. Only later and to a lay audience did Einstein spell out the steps he took in producing the theory.[47] The question he apparently asked himself in 1905 was: What does it mean to say that two events are simultaneous? The significance of the question can be illustrated by the following thought experiments.[48] As we see in figure 5a, suppose we lay out a course along a railroad track marking point M to be precisely midway between A and B. I have direct contact with God and arrange that he or she send straight down, two lightning bolts at A and B, simultaneously. Now, wanting to check up on the Lord, you station yourself at M with two mirrors placed in such a way that you can look down the track in both directions at the same time. God sends down the bolts (the jagged representations are stylistic license), leaving visible

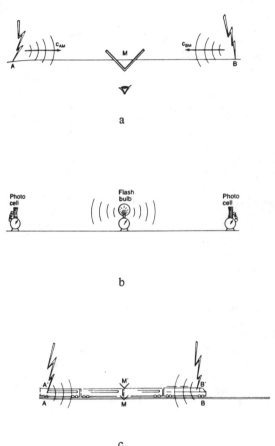

a

b

c

Figure 5

evidence on the tracks that the lightning really did come down at A and B. The light from the bolts spreads out in all directions and looking in the mirror you say, "simultaneous."

But then I ask you how you know that the speed of light from A to M is the same as the speed of light from B to M. Finally, in frustration, you arrange a second experiment (fig. 5b). You replace the mirrors with a flash bulb. At A and B you place two identical clocks set to zero connected to switches driven by photocells. You fire the flashbulb, the light spreads out in all directions and the clocks start. But then I ask you how you know that the speed of light from M to A is the same as the speed of light from M to B. There is a long silence. Finally you turn to me in triumph and say, "I know that the speed of light in the direction M to A is the same as the speed of light in the direction M to B because I *stipulate* that it is the same. Since nothing can keep up with the speed of light, I am free to stipulate anything I want about it and I stipulate isotropy of space with regard to the speed of light in order to come to a meaningful definition of simultaneity." An immediate consequence of that stipulation is that events such as the lightning strokes, which are simultaneous for one observer on the track, cannot be simultaneous for an observer on the train. I leave it to the reader to complete the argument -- illustrated in figure 5c, it being assumed that the lightning strokes occur at the instant that M and M' are adjacent to each other.[49]

Stipulate?! Postulate?! Not measure?! Indeed. Not only must there be stipulations such as the isotropy of space for the speed of light behind every theory, there have to be such stipulations lurking behind every measurement.[50] Such a notion flies in the face of late-nineteenth century pride in precision tooling, precision engineering and precision measurement. This is not to deny the vital role of measurement in the physical sciences. It simply is a rejection of the proposition that measurement is the *source* of successful physical theories. Of course, measurement is a vital tool in the confirmation of theory; and measurement is often an important heuristic for developing new theories. But without the introduction of metaphysically guided stipulations and postulates, there could be no meaningful theory; there could be no meaningful measurements.

In 1907, after giving the first of a series of lectures at Göttingen, Albert Michelson, in the company of Philipp Franck and the rest of the seminar, retired to a local coffeehouse. When they entered, those who had attended Michelson's lecture were sitting at two tables. According to Franck,[51] Michelson turned to him and with some force asked, "At which table shall I sit? Which table holds the idolaters of relativity and which table holds the physicists?" One should not be surprised. Michelson had firm convictions.

Twenty years later, a year before Lorentz died, he remarked that although he admired greatly Einstein's contributions to physics, for him the notion of simultaneity was very definite and he had no trouble distinguishing between simultaneous and non-simultaneous pairs of events.[52] Undoubtedly, Einstein, who used physics to explore basic philosophical issues, would have responded that he too had no problem distinguishing simultaneous from nonsimultaneous events. But, Einstein would have continued, that is not to say that events which are simultaneous for one inertial observer, would also be simultaneous for an observer in another inertial frame of reference. One can imagine a long, long silence. And finally, with typical grace and good will, Lorentz would have shrugged his shoulders, risen from his seat, warmly embraced his colleague, and wished him a good day. For Lorentz, as for most physicists, philosophy, if not the enemy, was completely irrelevant.

NOTES

1. Loyd S. Swenson: *The Ethereal Aether: A History of the Michelson-Morley Aether-Drift Experiments* (Austin: Univ. Texas Press, 1962), Cf. T. P. Hughes, "Science and the Instrument-maker: Michelson, Sperry, and the Speed of Light," *Smithsonian Studies in History and Technology*, No. 37 (Washington, D.C.: Smithsonian Institution Press, 1976).

2. Loyd Swenson, "Michelson, Albert Abraham," *Dictionary of Scientific Biography* (16 vols.; New York: Charles Scribner's Sons, 1970-1980), vol. 9, pp. 371-374, on p. 371; hereafter *DSB*.

3. K. B. Hasselberg, "Physics 1907," *Nobel Lectures in Physics: 1901-1921* (Amsterdam/London/New York/Tokyo: Elsevier Publishing Co., 1967), pp. 159-165, on p. 159. See also the remarks by the President of the Royal Swedish Academy to Michelson: "The Swedish Academy of Sciences has awarded you this year the Nobel Prize in Physics in recognition of the methods which you have discovered for insuring exactness in measurements . . . ." Ibid., on p. 164.

4. On Tycho, see Max Casper, *Tycho Brahe* (Stuttgart: W. Kohlhammer, 1948); on Hales, see Henry Guerlac, "Hales, Stephen," *DSB*, vol. 6, pp. 35-48; on Lavoisier, see James B. Conant, "The Overthrow of the Phlogiston Theory: The Chemical Revolution of 1775-1789" in James B. Conant, ed., *Harvard Case Studies in Experimental Science* (2 vols.; Cambridge, Mass.: Harvard Univ. Press, 1951) vol. 1.

5. On Lawrence, see Charles Suesskind, "Lawrence, Ernest Orlando," *DSB*, Vol. 8, pp. 93-96; on McClintock, see Evelyn Fox Keller, *A Feeling for the Organism: The Life and Work of Barbara McClintock* (New York/San Francisco: W. H. Freeman and Company, 1983); on Ting and Richter, see Brian McCusker, *The Quest for Quarks* (Cambridge/London/New York/Melbourne/Sidney: Cambridge Univ. Press, 1983), pp. 104-107.

6. Cf. S. Goldberg, *Understanding Relativity: Origins and Impact of a Scientific Revolution* (Boston: Berkhaeuser, 1984), esp. pp. 179-264, and S. Goldberg, "Putting Old Wine in New Bottles: The American Response to Einstein's Special Theory of Relativity," in Thomas Glick, ed., *The Comparative Reception to Relativity* (Dordrecht/Boston/Lancaster/Tokyo: D. Reidel Publishing Co., 1987).

7. See J. T. Merz, *A History of European Thought in the Nineteenth Century* (4 vols.; Chicago: William Blackwood & Sons, 1904-1912), esp. vol. 1.

8. On the introduction of the word "scientist," see Merz, *European Thought*, vol. 1, p. 89, fn.; p. 90, fn. Cf. R. E. Butts, "Whewell, William," *DSB*, vol. 14, pp. 292-295. Butts points out that not only did Whewell introduce the word "scientist," he was also responsible for the word "physicist."

9. This definition is taken from *The American Heritage Dictionary of the English Language* (Boston/New York/Atlanta/Geneva/Dallas/Palo Alto: Houghton Mifflin Co., 1971).

10. See, for example, Charles Rosenberg, "Science and American Social Thought," in David Van Tassel and Michael G. Hall, eds., *Science and Society in the United States* (Homewood, Illinois: The Dorsey Press, 1966), pp. 135-162, esp. pp. 158 ff.

11. Christa Jungnickel and Russell McCormmach, *Intellectual Mastery of Nature* (2 vols.; Chicago/London: Univ. Chicago Press, 1986), esp. vol. 2.

12. S. Goldberg, "Max Planck's Philosophy of Nature and His Elaboration of the Special Theory of Relativity," *Historical Studies in the Physical Sciences*, 1976, 7: 125-160; pp. 130-133. Cf. S. Goldberg, *Understanding Relativity*, pp. 189-190.

13. For a review of the variety of ether drift experiments, see S. Goldberg, *Understanding Relativity*, Appendix 5.

14. Max Born, "H. A. Lorentz," *Gott. Nach. Ges. Mitt.*, 1928/29, pp. 69-73 on p. 69. Cf. Max Born, *Einstein's Theory of Relativity* (revised, enlarged edition, New York: Dover Publications, 1965), pp. 204-207. Cf. S. Goldberg, "The Lorentz Theory of the Electron and Einstein's Theory of Relativity," *American Journal of Physics*, 1969, 37: 982-994.

15. H. A. Lorentz, *La Théorie électromagnétique de Maxwell et son application aux corps mouvants* (Leiden, 1892), esp. chapter 4.

16. H. A. Lorentz, *Versuch Einer Theorie Electrischen und Optischen Erscheinungen in Bewegten Körper* [*Attempt at a Theory of Electrical and Optical Appearances in Moving Matter*] (Leiden, 1895; unaltered 2nd edition, Leipzig, 1906).

17. The new transformation equation proposed by Lorentz was

$$t' = t - vx/c^2$$

where v is the velocity of the frame of reference relative to the ether, x is the distance coordinate in the moving frame at which the measurement of the time coordinate (t') is made, and t is the temporal coordinate as measured in the ether frame of reference.

18. Ibid., pp. 115-138. "Versuche deren Ergebnisse Sich Nicht Ohne Weiteres Erklären lassen." The translation is by Holton: See Gerald Holton, *Thematic Origins of Scientific Thought: Kepler to Einstein* (Cambridge: Harvard Univ. Press, 1973), on p. 301.

19. E. Cunningham proposed an ether that was at one and the same time at absolute rest but that had no velocity relative to any inertial frame of reference. For Cunningham, the ether represented nothing more than the equations that described it. His point of view was not adopted by anyone else and was sternly rejected by his mentor, Joseph Larmour. See S. Goldberg, "In Defense of Ether: The British Response to Einstein's Special Theory of Relativity, 1905-1911," *HSPS*, 1970, 2: 89-126; Cf. S. Goldberg, *The Early Response to Einstein's Special Theory of Relativity, 1905-1911: A Case Study in National Differences* (Ph.D. dissertation, Harvard Univ. 1969), pp. 318-319, 339ff.

20. A. A. Michelson and E. W. Morley, "Influence of the Motion of the Medium on the Velocity of Light," *American Journal of Science*, 1886, 31: 377-386.

21. This is not to suggest that Lorentz and FitzGerald agreed on the qualities of the ether that might result in contraction of matter in the direction of motion. As far as I know, Fitzgerald never specified a precise mechanism for the contraction. Lorentz hypothesized that the contraction was a manifestation of the ultimate electromagnetic character of all matter. Contrast that view with Poincaré's that the ether exerted a physical pressure on moving matter. See S. Goldberg, *The Early Response*, chaps. 1, 3, and 4.

22. Cf. S. Goldberg, *Understanding Relativity;* S. Goldberg, "Putting New Wine in Old Bottles," and K. Schaffner, ed., *Nineteenth Century Aether Theories* (New York: Pergamon Press, 1972).

23. A. Gruenbaum, "The Special Theory of Relativity as a Case Study in the Importance of the Philosophy of Science for the History of Science," in B. Baumrin, ed., *Philosophy of Science: The Delaware Seminar* (2 vols.; New York: Interscience, 1961-1963), vol. 2, pp. 171-204. Cf. S. Goldberg, "Being Operational vs. Operationism: Bridgman on Relativity," *Revista di Storia della Scienza*, 1984, *1*: 333-354, esp. pp. 350-352.

24. Examples include N. R. Hanson, *Patterns of Discovery: An Inquiry into the Conceptual Foundations of Science* (Cambridge: Cambridge Univ. Press, 1958); A. E. Eddington, *The Philosophy of Physical Science* (New York: MacMillan, 1939), esp. pp. 99-100; and P. W. Bridgman, *The Way Things Are* (Cambridge, Mass.: Harvard Univ. Press, 1959), esp. pp. 132-149.

25. N. R. Hanson, *Patterns of Discovery*, pp. 17-23.

26. Ibid., on p. 23.

27. J. Z. Young, *Doubt and Certainty in Science* (Oxford: Oxford Univ. Press, 1951), p. 62, quoted in Leonard K. Nash, *The Nature of the Natural Sciences* (Boston/Toronto: Little Brown, 1963), on p. 7.

28. Ibid.

29. Hanson, *Patterns of Discovery*, on pp. 16-17.

30. See, for example, Roger Price, *Oodles of Droodles* (New York: Simon and Schuster, 1965).

31. The earliest reference to "pointer reading" by Eddington seems to have been in *The Nature of the Physical World* (Ann Arbor: Univ. Michigan Press, 1958), on. p. 251. The book was originally published in 1928. The analysis of Eddington's work in this paper follows closely the work of John W. Yolton, *The Philosophy of Science of A. S. Eddington* (The Hague: Martinus Nijhoff, 1960).

32. Eddington, *The Philosophy of Physical Science*, p. 12.

33. Yolton, *Eddington's Philosophy of Science*, p. 3.

34. Here I am referring to the pathbreaking work of Gauss and Weber and the determination of magnetic and electric quantities in absolute units. See Jungnickel and McCormmach, *The Intellectual Mastery of Nature*, Vol. 1, chaps. 3 and 6. Cf. Robert P.

Multhauf and Greogory Good, *A Brief History of Geomagnetism and a Catalog of the Collections of the National Museum of American History* (Washington, D.C.: Smithsonian Institution Press, 1987), esp. pp. 16ff. It should be noted that Gauss and Weber did not consider the epistemological issues being raised here. Those issues are hinted at by N. R. Hanson in *Observation and Explanation: A Guide to Philosophy of Science* (New York: Harper and Row, 1971), on p. 17.

35. J. R. Ravetz and I. Grattan-Guinness, "Fourier, Jean Baptiste Joseph." *DSB*, Vol. 5, pp. 93–99, esp. p. 97.

36. See for example, Heinrich Hertz, *The Principles of Mechanics Presented in a New Form* (New York: Dover Publications, repr. 1956), especially the discussion of the various approaches to the science of mechanics (Introduction, on pp. 13-33). Compare that treatment to the development of the system of units in Book I, pp. 45-89, esp. "Explanation," p. 45. The first German edition was published in 1894 and the first English translation appeared in 1899. The reprint edition contains a very useful introduction by R. S. Cohen who has also provided an excellent, though now dated, bibliography. A typical modern introductory discussion of systems of units can be found in any first year college physics text, for example, F. W. Sears and M. W. Zemansky, *University Physics* (Reading Mass.: Addison-Wesley, 3rd ed., 1964), pp. 1-5.

37. S. Goldberg, *Understanding Relativity*, passim. Cf. S. Goldberg, "Being Operational vs. Operationism."

38. See the discussion by Hanson, *Observation and Explanation*, on p. 17.

39. Lord Kelvin is an example of one physicist who was skeptical of the electromagnetic theory of light. See William Thomson, *The Baltimore Lectures on Molecular Dynamics and the Wave Theory of Light* (Cambridge: Cambridge Univ. Press, 1904), esp. Lecture XX. A new edition together with historical and philosophical essays has been recently published: Robert Kargon and Peter Achinstein, eds., *Kelvin's Baltimore Lecture and Modern Theoretical Physics: Historical and Philosophical Perspectives* (Cambridge, Mass.: MIT Press, 1987).

40. J.P. Cederholm et al., "New Experimental Test of Relativity," *Phys. Rev. Letters*, 1958, *4*: 342-343; Weinberger and Mossel, "Theory for a Unidirectional Interferometric Test."

41. See S. Goldberg, "The Abraham Theory of the Electron: The Symbiosis between Theory and Experiment," *Arch. Hist. Exact Sci.*, 1970, 7: 9-25.

42. Oliver Lodge, *Continuity: The Presidential Address to the British Association at Birmingham* (London: J. M. Dant, nd.), pp. 56-57.

43. Only later did Lorentz realize that Voigt had arrived at the same transformation equations some years earlier than Lorentz himself. See H. A. Lorentz, *The Theory of Electrons and its Applications to the Phenomena of Light and Radiant Heat* (New York: Dover Publications, repr. 1952), p. 198 fn.

44. See S. Goldberg, "The Lorentz Theory of the Electron," esp. pp. 990-991.

45. See the portrait of Lorentz given by Russell McCormmach in "Lorentz, Heinrik Antoon," *DSB*, Vol. 8, pp. 487-500.

46. S. Goldberg, "Max Planck's Philosophy of Nature"; S. Goldberg, *Understanding Relativity*, esp. part II.

47. S. Goldberg, "Albert Einstein and the Creative Act: The Case of Special Relativity," in R. Aris et al., *Springs of Scientific Creativity: Essays on Founders of Modern Science* (Minneapolis: Univ. Minnesota Press, 1983), pp. 232-253.

48. The thought experiments described here follow those given by Einstein in *Relativity, The Special, the General Theory*. Cf. S. Goldberg, *Understanding Relativity*, Chapter 3.

49. One way to ensure that the speed of light is the same in all directions in all inertial frames of reference is to postulate the invariance of the speed of light -- the second postulate of the special theory of relativity. Another way is to postulate the Lorentz contraction -- corresponding to the Lorentz theory.

50. One need not stipulate isotropy of space for light if one is willing to use some other signal speed for the purpose of synchronizing clocks. For example, one might choose to use the speed of sound. But then one would have to stipulate isotropy of space for the speed of sound in order to come to a meaningful definition of simultaneity. Such a choice would, of course, have ramifications for the formulation of physical phenomena, but, having synchronized clocks using the speed of sound, one could then use two clocks to make one-way measurements of the speed of light.

51. Personal communication.

52. H. A. Lorentz, *Problems of Modern Physics* (New York: Dover Publications, repr., 1967), p. 221.

MICHELSON'S ETHER-DRIFT EXPERIMENTS AND THEIR CORRECT EQUATIONS[*]

H. Melcher

MICHELSON'S SUCCESSFUL EXPERIMENTAL RESULTS

Albert A. Michelson was an experimentalist par excellence. His instruments, built by himself, were masterpieces. He won the Nobel Prize for his famous achievements in the fields of metrology, not for ether experiments explicitly. The interferometer that he so successfully manipulated bears his name and plays an eminent role in experiments in many disciplines of physics today and certainly will continue to be important in future.

The ether-drift experiments Michelson performed -- first in 1881 in Potsdam by himself and then in 1887 together with E. W. Morley -- have their place in the history of physics. These experiments led to the end of ether physics. In both of the corresponding publications, the existence of a stationary ether was correctly disproved. The authors were absolutely certain of this result. They were convinced, because they were even able to count the number of wavelengths of light waves or in other words, they were able to measure time differences less than 10^{-15} seconds.

Michelson's statement at the end of the two publications should be regarded as the result of these experiments and should be respected as a historical invariant. That is to say, in history of science, one has to avoid all extraneous ideas and, in particular, a mixture of truth and poetry. Therefore, when interpreting the experiments of Michelson and Morley, aims that were not intended by them should be omitted. Both men learned that it was impossible to measure the velocity of the earth with respect to a stationary ether, in particular the velocity of the earth in its orbit around the sun, by observing from within a closed box (laboratory).

Michelson was not satisfied with this result. In regard to the theory, he later wrote (1903):

> It was found that there was no displacement of the interference fringes, so that the result of the experiment was negative and would, therefore, show that there is still a difficulty in the theory itself; and this difficulty, I may say, has not yet been satisfactorily explained.[1]

Which theory and which difficulty did Michelson mean? What appeared unsatisfactory to him? Even after 1905, when his book was published in German translation, these questions were still open for him.[2]

The main problem seems to be his inability to give up the ether concept. He should have been able to renounce the hypothesis of a light medium as provisionally unnecessary. But Michelson preferred to keep his "beloved ether" all his life.

Obviously, Michelson did not accept Lorentz's explanation of the null-result of his ether-drift experiments. Michelson claimed that an immovable ether does not exist but Lorentz kept the concept.

Michelson understood his problem to be the measurement of a prescribed velocity, v, of an ether in an equation for a time difference Δt. He later corrected the equation

---

* I am grateful to my friends, Professor John Stachel and Dr. Jurgen Renn, Boston University, Einstein Papers Project, for their help in translation from German to English.

for this time difference Δt with the help of independent comments of H. A. Lorentz and A. Potier. In the Cleveland experiment the following equation was used:

$$\Delta t = 1/(c-v) + 1/(c + v) - 2l/(c^2 - v^2)^{\frac{1}{2}} \neq 0 \qquad (1)$$

The time difference between two light signals running the same length l in perpendicular directions should be $\Delta t \neq 0$; but Michelson and Morley measured $\Delta t = 0$. They were disappointed and cancelled their plan for continuing the experiment.

THE EXPERIMENT WAS PERFORMED IN THE RESTFRAME S'

What was possible after 1905? The answer is to give up the concept of an ether, perhaps temporarily only, in order to avoid holding on to what has now become a prejudice. Furthermore, and this is our central point, it must be said that the equation can be set up without the presumption of an ether. If there is no (moving) medium for light, then c means the velocity of light in a vacuum and v the relative velocity between an observer (instrument) and a light source. This means Michelson and Morley did not test equation (1), because in their experiment a relative velocity v did not exist between the observer and the light source; they held very rigidly to their concept of an ether.

What does an experimenter measure inside an insulated unaccelerated box? All results of any experiment are the same, whether the box is at rest ($v = 0$) or moving with the velocity ($v \neq 0$). This is the content of the principle of special relativity. And obviously, Michelson's experiment is in accordance with this principle. The proper time $t' = l'/c'$ was measured for each arm of the interferometer. Therefore, the difference $\Delta t'$ between such times for light signals in any direction is zero; this also means, for example, for the two perpendicular arms of equal length: $\Delta t = 2l'/c' - 2l'/c' = 0$. This is the time difference measured by Michelson and Morley.

The Michelson-Morley experiment is frequently claimed to measure or confirm the principle of constancy of the velocity of light (in a vacuum). This, however, is not correct because this principle involves a relative velocity between the emitter of light and the observer (absorber of light). This principle means that the velocity of light is independent of the velocity of the emitter or absorber. Michelson's experiment confirmed the isotropy of light propagation and the homogeneity of space.

THE GEDANKENEXPERIMENT WITHOUT ETHER IN SYSTEM S

But what about equation (1) where a relative velocity appears that is not equal to zero ($v \neq 0$)? In this case let us measure Δt in an inertial system or laboratory S. The light comes out of the system S' with velocity c'. A simple Gedankenexperiment shows that we have to expect $\Delta t = 0$ in contrast to the classical equation (1). In the laboratory S' two light signals are emitted simultaneously and reflected after running the same distance l' in different directions. They arrive at the same point after the same travel time; that means there is a coincidence. The arrival of two signals of light in one and the same point is an event of absolute simultaneity. And such an event is an invariant. This means in all inertial systems only one coincidence signal will occur: $\Delta t = 0$. It would be without physical meaning and logic to expect $\Delta t \neq 0$; there is no time difference in the arrival of the two signals in one and the same point. So it becomes clear immediately that there is a contradiction between the Gedankenexperiment and equation (1).

But where is the difficulty? The answer is: in the theory, i.e. in the classical equation (1). Lorentz's solution is of no interest here: He multiplied the two first terms in equation (1) by the factor $(1 - v^2/c^2)^{\frac{1}{2}}$ and got $\Delta t = 0$. This means the arms of the interferometer in each case parallel to the direction of v should be contracted to the shorter length $l \cdot (1 - v^2/c^2)^{\frac{1}{2}}$. This explanation seemed to be artificial and strange. We only mention the possibility of arriving at $\Delta t = 0$ for equation (1) by means of the time dilation.

In the following, the contradiction shown above between the experimental value $\Delta t = 0$ and the value $\Delta t \neq 0$ according to equation (1) will be solved with respect to the addition of velocities. First of all, was the addition $c \pm v$ experimentally tested and proven? In 1886 Michelson and Morley repeated the famous experiment of Fizeau (1851) and confirmed the result with a higher degree of accuracy. They found that the velocity of light in resting water $c_M' = c/n'$ added with the velocity v (of streaming water) is not $c_M = c'_M + v$, but less, namely

$$c_M = c/n' + v\ (1 - 1/n'^2).\tag{2}$$

And for the vacuum $(M = vac)$ this equation gives $c_{vac} = c$ because $n' = 1$; this result, however, was not tested.

After 1905, however, the general formula for the composition of velocities was known. This equation was given independently by H. Poincaré and A. Einstein:

$$u = (u' + v)/(1 + u'v/c^2).\tag{3}$$

Here u' and v are parallel. In the case of light in a vacuum, the velocity in system S' is c' and in S is c. Equating $u' = c'$ the addition theorem yields $c = c'$. It is occasionally maintained that the velocity of light cannot be measured in one-way experiments. This is not relevant, because the equation for composing velocities is valid for all experiments up to now. And the test of the validity of equations rests on their verification.

Using this method more generally, $u' = c/n'$, $c = c'$ to compose velocities, the thus changed equation (1) is in correspondence with the Gedankenexperiment: $\Delta t = 0$, if $n' = 1$. Instead of writing down here the correct equation for Δt with regard to the addition of velocities in perpendicular direction, I shall give the approximate equation that follows by expanding in series:

$$\Delta t = 2ln'^3/c \cdot (1/2)\ (v^2/c^2)\ (1 - 1/n'^2)\tag{4}$$

whereas it always follows from the classical equation (1):[3]

$$\Delta t = 2l/c\ (1/2)\ (v^2/c^2) \neq 0\tag{5}$$

In order to test or verify equation (4) or the exact equation (here not written down) there must exist a relative velocity v. This is not the case in experiments of the type Michelson conducted; but it was realized in experiments using light sources outside the earth (sun, stars).

Taking the correct relation (3), it can be shown that, for $(u'v)/c^2 \ll 1$, the well-known classical equation results in $u = u' + v$. But this equation is not correct, if $u' = c$ is put in, after considering $v \ll c$. (Incidentally, Einstein never considered equation (1) in any one of his published writings, but one can find this equation in many introductory special relativity books. In those books equation (1) often is set up in a rather detailed fashion; but it remains unclear to the reader why this equation was introduced, especially if there are no references to it in the following text).

Equation (1) is not valid, and experiments cannot confirm an invalid equation. Wherever equation (1) is cited, this should be said. Furthermore, it must be mentioned that in the S'-System for n' ≠ 1, we have of course $\Delta t' = 0$, but in this case there is no isotropy in the S'-System, and it yields $\Delta t \neq 0$. This was shown, for example, by the Fizeau experiment. Isotropy only holds for the propagation of light in a vacuum.

TWO CONCLUDING REMARKS

1. Perhaps it may be pedagogically easier first to introduce the composition principle for velocities and then to ask how to measure lengths and time intervals that are necessary to determine velocities.

2. Regarding methods of finding knowledge in physics, it must be pointed out that the correct equation for the time intervals in cases of electromagnetic waves in a vacuum cannot be found by means of experiments. This fact demonstrates once more Einstein's statement: "There is no straight logical path from experiment to theory."

NOTES

1. A. A. Michelson, *Light Waves and Their Uses* (Chicago: The University Press, 1903).

2. A. A. Michelson, *Lichtwellen und Ihre Anwendungen* (Leipzig: Ambrosius Barth, 1911).

3. H. Melcher, "Koinzidenz-Experimente statt Michelson-Versuch," *Praxis der Naturwissenschaften (Physik)*, (Part 7) 1987, *36*:34-39; H. Melcher, "Versuche vom Michelson-Typ--mit und ohne Äther," *Wiss. Zeitschr.*, Päd. Hochschule Erfurt GDR (in press).

III
Michelson and His Scientific Legacy

AMERICAN PHYSICS IN 1887[*]

Albert E. Moyer

Historians of science study late nineteenth-century American physics for a variety of reasons. Some seek to identify the intellectual and institutional roots of this national scientific community that achieved prominence in world physics during the twentieth century. Others hope to find specific episodes that illuminate the conceptual development or organizational structure of science in general. Some look for events that relate to the general history of thought, culture, and society in the Victorian era. Still others are drawn to the topic simply for enjoyment, for a glimpse at how physics was done in an era of steam locomotives, high collars, and Henry James novels.

For these reasons and more, historians of science increasingly have been examining both the intellectual and the institutional aspects of American physics in the late nineteenth century. Daniel Kevles has provided a comprehensive and integrated analysis of this formative period in his 1978 book, *The Physicists*, particularly in the chapter titled "The Flaws of American Physics."[1] He explains that most late nineteenth-century American physicists, like their European colleagues, were still committed to classical concepts -- concepts that eventually would fall with the advent of relativity and quantum theory. Unlike the Europeans, however, the Americans lacked strong institutions that could set appropriate standards and provide the physicists with a clear sense of community. Specifically, the Americans had to contend with inadequate colleges and graduate schools, ineffective professional societies, and inconsequential scientific journals. In Kevles's opinion, the three American physicists who earned the respect of the international community during the period, Henry Rowland, J. Willard Gibbs, and Albert A. Michelson, were exceptions, Americans who succeeded in spite of formidable institutional obstacles and the lack of national standards. They were "geniuses" who "needed no intellectual guidance."

Kevles's characterization of physics in the United States during the late nineteenth century is generally accurate. Most physicists were committed to classical concepts and all were burdened by weak schools, professional societies, and journals. To say that the institutional supports were weak, however, is not to say that they were useless. Of course, Kevles would never endorse such an extreme appraisal. But by emphasizing American flaws, Kevles and other historians of science -- including myself -- have tended to discount the positive contributions of American institutional structures in the late nineteenth century. We have also tended to overlook the ways in which the Americans successfully compensated for institutional deficiencies by devising alternative systems of support and patterns of action -- in a sense, by establishing informal, surrogate institutions. Rowland, Gibbs, and Michelson might have been geniuses, as Kevles states, but even they did not flourish in the United States without intellectual guidance and support. These three men along with many of their lesser-known colleagues

---

[*] For their advice in first preparing this essay, I thank Dorothy Michelson Livingston, D. Theodore McAllister, and Loyd Swenson, Jr. For their comments on the final draft, I thank Gerald Holton, John Michel, Nathan Reingold, Darwin Stapleton, and various participants in the Michelson-Morley Centennial Celebration symposium. For her assistance in obtaining documents, I thank Mary Catalfamo of the Michelson Collection at the U.S. Naval Academy.

relied on American institutional structures; and, when particular structures proved inadequate, they substituted informal alternatives. What distinguished Rowland, Gibbs, and Michelson from the majority of American physicists was their ability to extract maximum benefits from existing resources. Indeed, part of their genius was their knack for taking full advantage of both formal and informal supports.

In this paper, I describe the intellectual and institutional fabric of late nineteenth-century American physics, especially the positive contribution of domestic institutions and an informal informational network. I do this by focusing on the professional life of one physicist, Albert Michelson, during one year, 1887. This not only was the year in which Michelson, assisted by Edward W. Morley, undertook the refined interferometer test for the drift of the Earth through the ether -- the test that was to bring him fame following the physics community's assimilation of Einstein's 1905 special theory of relativity.[2] It was also the year in which Michelson, again in collaboration with Morley, used the same interferometer to investigate whether wavelengths of light could provide a standard unit of length -- an ongoing investigation that, unlike the ether-drift test, contributed directly to Michelson's 1907 Nobel Prize.[3] Although I discuss both the standard-of-length and the ether-drift projects, I do not analyze them in detail; such studies already exist.[4] Rather, I show how Michelson, in performing the 1887 investigations, fit into the larger picture of physicists, physics, and physics institutions of the day, especially in the United States. In other words, I sketch the immediate professional context within which this American produced the influential results of 1887. Through this close examination of one man in a single year, I hope to illustrate the nature of American physics a century ago.[5]

NETWORKS OF PHYSICISTS

Physicists in the United States in 1887 felt cut off from the scientific capitals of Europe -- Berlin, Paris, and London. Michelson felt especially isolated. A professor since 1882 at the fledgling Case School of Applied Science in Cleveland, he felt separated from even his American colleagues, such as Henry Rowland at the decade-old Johns Hopkins University, John Trowbridge at Harvard's new Jefferson Physical Laboratory, and Willard Gibbs at Yale. In a letter to Rowland, Michelson complained of "the inconvenience of being so far removed from scientific centers"; "I would much prefer," he added, "a position farther East."[6] Other scientists shared this perception of Cleveland's isolation. When the American Association for the Advancement of Science (AAAS) met there in the summer of 1888, the Permanent Secretary excused the disappointingly low attendance by pointing to the "natural aversion to going into the interior of the country in August."[7]

The founders of Case had sought to bring science and its practical applications to the Midwestern industrial city. But the five-year-old program suffered a setback in the fall of 1886 when a fire destroyed the main academic building, including much of the physics apparatus that Michelson had painstakingly assembled. As Michelson began the year 1887, damaged facilities compounded the problem of isolation.

However, Michelson, like many of his more enterprising colleagues throughout the United States, found ways to compensate. In response to the immediate problems caused by the fire, he relocated his laboratory and salvable apparatus to a building of nearby Western Reserve University, the home institution of chemist Morley, Michelson's research partner since 1885. As for the more basic problem of isolation, Michelson followed the lead of other active researchers by participating in an informal network of scientists who facilitated the exchange of information. The network, which the scientists sustained through letters and travel, was both national and international in scope.[8]

Michelson had become well connected within the domestic branch of the network. At the beginning of his career, while investigating the speed of light as a student and instructor at the U.S. Naval Academy, he had formed a close professional bond with astronomer Simon Newcomb. Newcomb, the influential head of the government's Nautical Almanac Office, opened doors for Michelson at home and abroad. Michelson also established close ties with two of the nation's most esteemed academic physicists, Rowland and Gibbs. For example, around 1884, when Michelson began to plan for a definitive ether-drift experiment -- a refined version of the experiment he originally performed in Germany in 1881 -- he consulted with Gibbs. Through conversation and correspondence, he questioned the Yale theorist about "the feasibility of the experiment."[9] Other members of Michelson's American network included physicists George Barker and Alfred Mayer, both of whom in earlier years had tried to get Michelson a job outside the Navy. Barker, of the University of Pennsylvania, finally arranged for Michelson's appointment as a professor through an influential friend at Case. "I am sure," Barker wrote to his friend, "the Case School will never regret this step."[10]

Michelson also cultivated relationships with prominent Europeans, the anchors of the physics network. Those ties dated back to his days as a postgraduate student in Germany and France. Like other Americans faced with the shortcomings of American graduate schools, Michelson compensated through study in Europe. He was particularly attracted to the University of Berlin, the site of Hermann von Helmhotz's famous laboratory. Beginning in the fall of 1880, Michelson attended Helmholtz's lectures on theoretical physics. In addition, he turned to Helmholtz for advice on his first ether-drift test and his newly devised interferometer. While in Germany, Michelson also studied optics and spectroscopy under the masters in these fields. In the fall of 1881, he went on to Paris, expanding his list of scientific contacts to include the leaders in French optics.

Although he visited London too, Michelson became close to top British physicists only after returning to the United States. In 1884 Michelson attended the Montreal meeting of the British Association for the Advancement of Science, chaired by Lord Rayleigh, and heard the subsequent "Baltimore Lectures," delivered at Johns Hopkins by William Thomson (later Lord Kelvin). Rayleigh and Thomson quickly became two of the most consequential members of Michelson's network, providing him with perspective and encouragement on his research. While in North America, they advised him to repeat his German interferometer experiment, but only after first checking Armand Fizeau's earlier result on the speed of light in moving water. Michelson accepted this advice, reporting to Thomson in March of 1886 that he and Morley had confirmed Fizeau's result. A letter from Rayleigh then provided the incentive for Michelson and Morley to push on to the next stage of their ether research, a repeat of the interferometer test. Michelson wrote back to Rayleigh in March of 1887 that he had been distressed by the lack of interest of his "scientific friends" and the "slight attention" the German test received. He continued, "Your letter has however once more fired my enthusiasm and it has decided me to begin the work at once."[11]

Cleveland, Ohio, may have been far from the scientific centers of Europe and the United States, but Michelson was not. Through persistent letters and much travel, he maintained tight links with the national and international elite of physics, from Rowland and Gibbs to Thomson and Rayleigh.

TRENDS IN PHYSICS

European physicists in the period around 1887 made a distinction that helps us understand the relation between Michelson's research and the physics of his day. The Europeans, particularly the Germans, distinguished between experiment and measurement.[12] Whereas "experimental physicists" typically investigated new phenomena related to unsettled theories, the "measuring physicists" typically devised precision instruments to refine the quantities specified in established research programs or to update recognized physical constants. Michelson was primarily a measuring physicist, as evidenced by his attempts to make precise determinations of the Earth's motion through the ether and of a standard of length using wavelengths of light.

Michelson fit in well with other Americans[13] who favored using their laboratories for measurement. Rowland, for example, showed this propensity in his measurements of both the ohm and the mechanical equivalent of heat and his manufacture of diffraction gratings for use in his spectral studies. Most Americans, however, emphasized the experimental rather than the measuring side of laboratory practice. Well-known experimental work of the day included Mayer's study of the configurations of floating magnets; Trowbridge's electrical research on the "superficial energy" between alloys; and Edwin Hall's systematic investigation of conductors, which led to the discovery of the "Hall Effect." Rowland also complemented his measurements with experiments, including an inquiry into the magnetic effect of moving electric charges. As in Europe, very few scientists specialized in theoretical physics or treated it as a distinct area of teaching and research. Gibbs proved to be the exception, with contributions to thermodynamics, statistical mechanics, and the electromagnetic theory of light. The dearth of theorists reflected in part the lack of advanced mathematical training in the United States. Michelson himself once admitted to Mayer that he made "no pretense" of being a mathematician.[14]

To make their exacting measurements, Michelson and the other Americans needed precision instruments. For this they did not need to turn to their European colleagues for help. Rather, they drew on an indigenous tradition of first-rate engineering and manufacture. Two of the world's foremost fabricators of optical apparatus were in the United States, John Brashear in Pittsburgh and Alvan Clark in Cambridge, Massachusetts. Michelson and Morley relied on Brashear's shop for the delicate optical components required in their ether-drift and standard-of-length investigations. And for all its disadvantages, living in Cleveland allowed them to gain an edge on fellow Americans because they had immediate access to the local firm of Warner and Swasey.[15] This precision manufacturing company, for which Morley served as a consultant, had won the contract to furnish all the mechanical and structural portions of the telescope for the Lick Observatory, the largest refracting telescope in the world. Shortly after Warner and Swasey completed the bulk of the Lick project in October 1887, Michelson and Morley began working with them to devise a mechanically advanced instrument for the next phase of their standard-of-length investigation.

Whether they were experimenters or measurers, most American physicists around 1887 worked within the same general conceptual framework. Echoing many of their European colleagues, they assumed that all physical phenomena could be represented in terms of underlying atoms, molecules, and ethers that obey the laws of classical mechanics. Although they shared the basic precepts of this traditional outlook as they pertained to gases, heat, electricity, magnetism, light, and even gravity, the physicists were quite unsettled about specific mechanical theories and models. Michelson in particular coupled an orthodox trust in the existence of the luminiferous ether to an equally orthodox mechanical view of the atomic processes that caused light waves. Throughout an address to the 1888 Cleveland meeting of the AAAS, he confidently spoke of "the forces and motions of vibrating atoms and of the ether which transmits these vibrations in the form of light."[16] Although convinced of the existence of the ether, Michelson was unsure of Augustin Fresnel's specific hypothesis that the ether was essentially at rest relative to the moving Earth. Thus, he set out with Morley to make a definitive measurement of the relative Earth-ether motion.

The measurement's well-known negative result threatened Fresnel's popular theory of the ether but not the notion of an ether itself. In fact, Michelson and Morley candidly reported their perverse result and then attempted to account for the result with alternative explanations. First they considered the ether theories of George Stokes and Hendrick Lorentz, concluding, however, that both of those options had their own weaknesses. Then they rationalized that the Earth's irregular surface perhaps had trapped portions of the ether, thus obscuring the "ether wind" they aimed to detect. Michelson and Morley never questioned the basic concept of the ether, just its particular character.

The team began working in earnest on the ether project soon after Rayleigh's encouraging letter. In April 1887, Morley reported to his father: "Michelson and I have begun a new experiment. It is to see if light travels with the same velocity in all directions. We have not got the apparatus done yet, and shall not be likely to get it done for a month or two."[17] Curiously, the measurement of ether drift was not the first application of their new interferometer -- an ingenious arrangement of mirrors mounted on a sandstone megalith floating on mercury. They first employed the new device in June to explore the possibility of using the wavelength of one of the spectral lines emitted by thallium, lithium, hydrogen, or especially sodium as a standard of length. In the prior half century, especially in France, the quest for an unequivocal and invariant measure of a meter had come to preoccupy many researchers.

Michelson and Morley initially addressed the standard-of-length question for a simple reason. While preparing the interferometer for the ether test, they first had to ensure that the perpendicular paths of the two light beams were of approximately equal lengths; they did this by adjusting the position of one of the plane mirrors at the end of one of the paths. They realized that if the mirror "moved parallel with itself a measured distance by means of the micrometer screw," then the distance could be correlated with an exact number of wavelengths; they could determine this number simply by counting alterations in the circular interference fringes as the mirror was moved. Although Michelson had thought of the method "several years ago," they now confirmed that the method "seemed likely to furnish results much more accurate" than any prior determinations of length.[18]

Not until July 1887 did Michelson and Morley finally set the adjusted interferometer into slow rotation and look for shifts in the interference fringes indicating the relative motion of the Earth and ether. A month later Michelson wrote to Rayleigh that the tests "have been completed and the result is decidedly negative."[19] In the ensuing months and years, Michelson came to value the ether-drift test not for its negative finding but for its role in the creation of a refined interferometer applicable to more pressing investigations, especially the determination of a standard of length. As a measuring physicist faithful to a classical conception of atoms and ether, Michelson developed the practical application of the interferometer while explaining away the negative result with makeshift hypotheses.

INSTITUTIONAL SUPPORTS

To announce their findings, Michelson and Morley traveled in early August to the meeting of the AAAS at Columbia College in New York City. Samuel Langley, president of that large, open-door organization, captured in his welcoming remarks the sense of community that the association provided to isolated researchers. Specifically, he reminded his colleagues that many of them had come to their first meetings "as solitary workers in some subject for which they had met at home only indifference." But they discovered at the meetings other scientists "caring for what they cared for, and found among strangers a truer fellowship of spirit than their own familiar friends had afforded."[20] Michelson found his "fellowship of spirit" in Section B, Physics, one of eight disciplinary divisions available to the 729 members in attendance. Section B, established in 1882, constituted the only national coalition of physicists. During the week-long New York meeting, the association elected Michelson to be the new vice president in charge of the section. For the next twelve months, Michelson, at thirty-four, served as the spokesman for the loose-jointed but aspiring American physics community.

Michelson and Morley jointly presented two of the thirty-eight papers read in Section B. One dealt with the relative velocity of the Earth and ether; the other covered the method for establishing a standard of length. A few days after the close of the meeting, *Science* reported that "the most important paper" of Section B was Michelson and Morley's discussion of a standard of length. The magazine devoted a full paragraph to characterizing the results. As for their ether paper, *Science* labeled it "a second paper of great interest" and summarized it in only a few lines.[21]

The opinion of AAAS members on the relative value of the two research projects was apparently shared by the leadership of the National Academy of Sciences (NAS)-- the other main, but much smaller and more exclusive, scientific society in the United States. In early November 1887, the Academy invited Michelson and Morley to the annual "Scientific Session" to present their latest thoughts on establishing a standard of length. Of the twenty-two papers read at this meeting, held again at Columbia College, theirs was the only one by nonmembers. The audience was obviously impressed: Michelson was elected a member the next year.[22]

In late December, Michelson and Morley presented their standard-of-length paper to one other appropriate audience, the Civil Engineers' Club of Cleveland. Both men were members of this regionally active group, which was certain to appreciate the practical implications of the standard-of-length research and the interferometer itself. Morley actually read the paper during the evening meeting, but Michelson participated in the discussion afterwards. "Permit me to state," Michelson commented, "that Professor Morley has given me more than my due in attributing so large a share of this work to me. Without his assistance, our present results would never have been attained."[23]

The AAAS and the NAS did more than provide national forums for airing results. The two societies also financed a limited amount of new research. Faced with the expense of precision instruments, Michelson had emerged as one of the nation's most active fund-raisers for science. Dollars from the NAS's Bache Fund helped build the Cleveland interferometer. The directors on the Bache Board matched $500 they had awarded Michelson in March 1887 with an equal amount in late July. These were sizable sums in a time when a person could have his horse shod or buy a bushel of potatoes for about a dollar. Michelson and Morley supplemented the latter grant with one of the first appropriations ever made by the AAAS. During the August meeting, the council of the association allotted the two men $175 "to aid them in the establishment of a standard of length."[24]

Michelson and Morley published their interferometer findings on both ether drift and standard of length in the *American Journal of Science*. This multidisciplinary journal, issued in New Haven, Connecticut, and edited by James and Edward Dana, provided publishing physicists with their only serious domestic outlet; the *Physical Review* was still six years in the future. Through the efforts of two associate editors, physicists Trowbridge and Barker, the *Journal* managed to carry the best products of American physics. In addition, it kept the community posted on international publications. Throughout 1887, for example, Trowbridge and Barker presented abstracts of articles from foreign journals such as *Annalen der Physik und Chemie*, *Comptes Rendus*, the *Philosophical Magazine*, and *Nature*. The *Journal*'s section on "Scientific Intelligence" added reports of domestic and foreign meetings as well as reviews of physics books.

Few Europeans read the *American Journal of Science*, so to ensure that their findings reached overseas, many American physicists published or reprinted their articles in foreign journals. Their favorite outlets were British -- the *Philosophical Magazine* and *Nature*. Michelson and Morley arranged for each of their two articles to appear simultaneously in the *American Journal of Science* and the *Philosophical Magazine*. The ether-drift paper initially appeared in the two journals in November 1887, while the article on the standard of length came out in December.

MICHELSON AND AMERICAN PHYSICS

In many ways, Michelson came close to being the "complete" physicist. That is, he was involved as deeply and creatively as possible for an American of his day with all aspects of the physics enterprise -- with the ideas, institutions, and individuals of late nineteenth-century physics. Specifically, he combined a commitment to classical concepts with a dedication to laboratory practice. Shying away from abstract theory and mathematics but embracing the design and manufacture of precision instruments, he followed the measuring rather than the experimental path of laboratory work. As for institutional structures in the United States, he relied on regional and national professional societies not only for forums to air his findings but also for funds to finance his research. And with the *American Journal of Science*, he found up-to-date surveys of international developments in physics as well as an outlet for his own publications. Finally, study in Europe and involvement in an informal network of physicists allowed him to compensate for American institutional inadequacies. Through frequent travel and letters, Michelson maintained a sense of community with leading physicists at home and abroad.

Although Michelson's American colleagues were also involved with these same ideas, institutions, and individuals, Michelson was involved more fully. Why was he better able than most of his associates to use existing resources, whether formal or informal? His advantage seems to come from a particular combination of personal qualities. Michelson blended not only native intelligence, mechanical dexterity, and the ability to identify significant research projects -- what Kevles probably meant by genius -- but also personal drive and public charm. Ironically, this seemingly propitious conjunction of qualities, especially drive, sometimes contributed to Michelson's personal problems. Morley speculated, for example, that overwork on the Fizeau experiment triggered the psychological illness that led Michelson to seek full-time medical care in the fall of 1885. In a revealing description of Michelson's personality, Morley told of "the ruthless discipline with which he drove himself to a task he felt must be done with such perfection that it could never again be called into question."[25] Michelson's recent biographers, particularly Dorothy Michelson Livingston, R. S. Shankland, and Loyd Swenson, chronicle more completely the talent, drive, and charm that set Michelson apart from the majority of his colleagues.

The point remains, however, that most American physicists were involved with the same set of ideas, institutions, and individuals as Michelson. That is, they shared in the same intellectual and institutional life. As with Michelson, most carried out laboratory investigations within the framework of classical physics. Most also benefited from the domestic institutions and the informal informational network. Michelson's professional life in 1887 simply constitutes an extremely favorable case of how physicists functioned in the United States during the late nineteenth century. Michelson's positive experiences allow the historian to see beyond the all too apparent flaws of American physics and observe a relatively vigorous and resilient community of physicists. We see the base upon which American physicists, including Michelson, would build in future decades.

NOTES

1. Daniel J. Kevles, *The Physicists* (New York: Knopf, 1978), chap. 3.

2. Gerald Holton, "Einstein, Michelson, and the 'Crucial' Experiment," *Isis*, 1969, *60*: 133-197; Abraham Pais, *"Subtle is the Lord. . ."* (Oxford: Oxford Univ. Press, 1982), chaps. 6, 8.

3. R. S. Shankland, "Michelson: America's First Nobel Prize Winner in Science," *The Physics Teacher*, 1977, *15*: 19-25. D. Theodore McAllister, ed., *The Albert A. Michelson Nobel Prize and Lecture*, Publications of the Michelson Museum No. 2 (China Lake, Ca.: Naval Ordinance Test Station, 1966).

4. The following fundamental studies also contain good bibliographies: Loyd S. Swenson, *The Ethereal Aether* (Austin: Univ. Texas Press, 1972); Dorothy M. Livingston, *The Master of Light* (New York: Scribner's Sons, 1973); R. S. Shankland, "Michelson-Morley Experiment," *American Journal of Physics*, 1964, *32*: 16-35; and Jean M. Bennett, D. Theodore McAllister, Georgia M. Cabe, "Albert A. Michelson, Dean of American Optics," *Applied Optics*, 1973, *12*: 2253-79. I am grateful to these authors for much of the background information in this paper.

5. The next three sections of this paper are reprinted, with slight changes, with permission of the American Institute of Physics, from Albert E. Moyer, "Michelson in 1887," *Physics Today*, May 1987, *40*: 50-56; the original article also includes coverage of Michelson's personal problems during 1887.

6. Michelson to Rowland, 6 November 1885; reprinted in Nathan Reingold, ed., *Science in Nineteenth Century America* (New York: Hill and Wang, 1964), pp. 311-312.

7. F. W. Putnam, "Report of the Permanent Secretary," in *Proceedings of the AAAS: Thirty-Seventh Meeting, Held at Cleveland, August, 1888* (Salem, Mass.: AAAS, 1889), p. 416.

8. For general background, see Diana Crane, *Invisible Colleges: Diffusion of Knowledge in Scientific Communities* (Chicago: Univ. Chicago Press, 1972).

9. Michelson to Gibbs, 15 December 1884; reprinted in Reingold, ed., *Science in Nineteenth Century America*, pp. 307-308.

10. Barker to J. Stockwell, 5 May 1881; reprinted in Shankland, "Michelson-Morley Experiment," p. 22.

11. Michelson to Thomson, 27 March 1886, and Michelson to Rayleigh, 6 March 1887; reprinted in Livingston, *Master of Light*, pp. 117, 123-126.

12. Christa Jungnickel and Russell McCormmach, *Intellectual Mastery of Nature*, 2 vols., Vol. II: *The Now Mighty Theoretical Physics, 1870-1925* (Chicago: Univ. Chicago Press, 1986), pp. 120-121.

13. Kevles, *The Physicists*, chap. 3. Albert E. Moyer, *American Physics in Transition* (Los Angeles: Tomash, 1983).

14. Michelson to Mayer, 26 June 1880; reprinted in Reingold, ed., *Science in Nineteenth Century America*, pp. 286–287; see also John W. Servos, "Mathematics and the Physical Sciences in America 1880–1930," *Isis*, 1986, *77*: 611–629.

15. *The Leader and Herald* [Cleveland], 17 October 1887, p. 8; *Cleveland Plain Dealer*, 26 October 1887, p. 8.

16. Michelson, quoted in Moyer, *American Physics*, p. 63.

17. E. W. Morley to S. B. Morley, 17 April 1887; reprinted in Reingold, ed., *Science in Nineteenth Century America*, pp. 312–313.

18. Michelson and Morley, "On a Method of Making the Wave-Length of Sodium Light the Actual and Practical Standard of Length," *American Journal of Science* 1887, *34*: 427–430. See also Michelson and Morley, "On the Relative Motion of the Earth and the Luminiferous Ether," *Am. J. Sci.* 1887, *34*: 333–345.

19. Michelson to Rayleigh, 17 August 1887; reprinted in Shankland, "Michelson-Morley Experiment," p. 32.

20. Langley, in "Report of the General Secretary," *Proceedings of the AAAS: Thirty-Sixth Meeting, Held at New York, August, 1887* (Salem, Mass: AAAS, 1888), p. 342; see also pp. 345–355.

21. *Science*, 1887, *10*: 86.

22. "Scientific Session," in *Report of the NAS for the Year 1887* (Washington, D.C.: GPO, 1888), pp. 10–11; see also "Elections," p. 28.

23. Michelson and Morley, "On a Method for Making the Wave-Length of Sodium Light the Actual and Practical Standard of Length," *Journal of the Association of Engineering Societies* 1888, *7*: 153–156; see also "Discussion," pp. 156–160, and "Proceedings: Civil Engineers' Club of Cleveland," pp. 110–111.

24. "Report of the Bache Fund," in *Report of the NAS for the Year 1889* (Washington, D.C.: GPO, 1891), p. 41. See also Michelson to J. Billings, 15 May 1889, Michelson Collection, Nimitz Library, U.S. Naval Academy, Annapolis, Md.; "Appendix to Report of General Secretary," in *Proceedings of the AAAS: Thirty-Sixth Meeting*, p. 355. For background, see Howard Miller, *Dollars for Research: Science and Its Patrons in Nineteenth-Century America* (Seattle: Univ. Washington Press, 1970).

25. Morley, quoted in Moyer, "Michelson in 1887," p. 54. See also the studies listed in n. 4.

MICHELSON AND THE REFORM OF PHYSICS INSTRUCTION AT THE NAVAL ACADEMY IN THE 1870S

Kathryn M. Olesko

> Even if we do not always reach the truth by experiment, this is
> often not because the first ideal concept of the experiment is not
> well-suited to produce such a result, but sometimes because it may
> be the fault of the material substance and the corruptible mecha-
> nisms that have to be made use of in putting it into practice.[1]

The decade of the 1870s was a period of transition for physics instruction at both American and German institutions of higher learning. It was also the decade in which Albert A. Michelson's professional career as a physicist took shape. At the United States Naval Academy between 1869, when Michelson entered as a student, and 1879, when he left his position as instructor in the academy's Department of Physics and Chemistry, reforms brought physics teaching there closer to what was offered to students in the teaching-research laboratories and seminars of the German universities. While at Annapolis he had learned physics from Edmund Atkinson's translation of Ganot's *Elementary Treatise on Physics*[2] and had probably been among the few students chosen to participate in elementary instruction in physical laboratory methods before he left in 1873. Three years later, he returned as an instructor and was immediately called upon to execute reforms, masterminded largely by his department chairman, William T. Sampson, that combined German and American components of physics instruction. By the time Michelson left Annapolis in 1879 he had demonstrated that he could practice an exact experimental physics.

Pedagogy and practice interacted over those years. Michelson took up the measurement of the velocity of light in 1878 after having reviewed classical French investigations on the matter in preparation for a lecture on optics in the fall of 1877. Until late 1879, when he published the last of his first set of investigations on the velocity of light, his teaching continued to shape his investigative techniques and his quantitative determination of the velocity of light.

AMERICAN VS. GERMAN PHYSICS INSTRUCTION IN THE 1870S

By the 1870s the boundaries that at an earlier time had separated physics instruction in German and American institutions of higher learning became somewhat less distinct. To be sure there remained strong differences between them, the most prominent being the role assigned to mathematics. The institutionalization of the mathematical methods of physics that had been worked out over four decades in German seminars and lecture courses strengthened in the 1870s with the establishment of chairs in mathematical physics; the opening of new seminars that combined mathematical and physical instruction; and the renewed need for secondary school teachers capable of teaching both subjects.[3] Whereas German physics students might be expected to work out elliptical integrals and transcendental equations (including their numerical solutions) as a matter of course in canonical exercises that engaged them in the manipulation of mathematical methods, American students generally did not go beyond elementary differential and integral calculus, and most often they deployed simple algebraic calculations.[4] John Servos has recently argued that even at the beginning of the twentieth century, Americans "feared demon mathematics, an idol that could pervert an unprejudiced understanding of nature." He concluded that "inadequate training in mathematics was an important reason for the experimental style of American science during the decades around the turn of the century."[5]   Quite naturally, then, topics that played a prominent role in the German physics curriculum in the 1870s -- for example, potential theory, Fourier's thermal equations, and phenomena represented by partial differential equations -- were rarely found in the American curriculum.

A second difference marked the orientation of physics instruction. The problem sets and laboratory exercises in physics undertaken by German students disciplined them in the professional standards governing the practice of physics. Problems and exercises involved more than the memorization and recapitulation of knowledge learned. They were encapsulations of the investigative techniques of physics. One did them to learn how to do physics. In Germany, both basic and advanced physics instruction was accomplished in this way in the 1870s, and instructors hoped that the better students would be able to go beyond exercises constructed largely from unanswered issues in the literature to undertake their own original investigations.[6] The fact that few actually did was not the point; the opportunity was in principle there for those who wanted to take advantage of it.   The fact that research entered the classroom did not mean, however, that it dominated the form and content of teaching or the lives of instructors. The emphasis was on teaching and research combined, feeding off one another in symbiotic ways.

With few exceptions, physics instruction in America, like instruction in other sciences, shaped citizens, not professionals. Science instruction, like instruction in any other subject, was generally part of a liberal education. By the end of the 1860s, physics instruction embraced the major divisions of physics -- mechanics, light, heat, electricity and magnetism. The worth of physics was found in its practical and applied nature, not in its abstract and theoretical formulation. Graduate programs were just beginning in America in the 1870s, and consequently higher physics instruction was rare. Generally the most informed physics instructors were either the young educated entirely or partly abroad or professionals from other fields, especially astronomy. There was a decided emphasis upon teaching, not research.[7]

These differences are well known. They are invoked so often that they are likely to become caricatures of physics instruction in both countries during the 1870s: a mathematical and professional instruction in Germany and a nonmathematical and unprofessional instruction in America. Yet neither description can be universally applied, and local studies belie conclusions drawn from the aggregate. This is true especially for the decade of the 1870s, when physics instruction in both countries was changing institutionally and intellectually. The practical experience in laboratory methods for which Germany was well known was only beginning to reach a broad audience in the 1870s, when institutes at Berlin, Leipzig, and Würzburg opened (notably all three were controlled by different educational administrations with different objectives). Earlier laboratories at Göttingen and Marburg had been used primarily for training secondary school teachers. The student laboratory at Königsberg was owned and controlled by Franz Neumann, who used it for the practical component of his exemplary seminar exercises in precision measurement. With the death of Gustav Magnus in 1870, Berlin's laboratory changed hands -- and focus -- when Hermann von Helmholtz arrived, while Gustav Kirchhoff's laboratory exercises at Heidelberg were but appendages to his course in experimental physics and his seminar on mathematical physics. Laboratories at other German universities were either nonexistent or small affairs. When new laboratories were built along with physics institutes in the last quarter of the century, teaching needs, not research, shaped their form and function.[8]

A more mature institutional setting for physics instruction -- and perhaps a more relevant example for an international comparison -- was the science seminar of the German universities. By the 1870s seminars operated routinely, but were far from being the research institutes their founders had intended them to be. The oldest, Bonn's natural science seminar, founded in 1825, included instruction in physics and its laboratory methods, but not in mathematics or the mathematical methods of physics. Although by the 1870s students generally registered for only one division of the seminar, thereby studying only one science, physics instruction did not become as comprehensive or as specialized as it could have. Halle's science seminar also covered all the natural sciences, but unlike Bonn's seminar, Halle's included mathematical instruction. Although it too by the 1870s had not helped to shape a comprehensive and systematic curriculum in physics, it was still active in promoting the review of original work in all of the sciences. Moreover, it provided students with support for publication and experimental investigations. Generally in universities south of the Main -- Heidelberg, Freiburg, Tübingen, Würzburg, Erlangen, and Munich -- physics instruction in lecture courses and seminars was not overwhelmingly mathematical, in part because the examinations for teaching secondary schools in those regions did not promote as strong a connection between mathematics and physics as the examination in Prussia did[9]

The two seminars that promoted the strongest connection between mathematics and physics were at Königsberg and Göttingen. Established in 1850, Göttingen's seminar was established for the explicit purpose of promoting a "systematic, connected" study of physics in conjunction with mathematics, although the mathematical physics eventually taught in it emphasized precision measurement rather than the deployment of the niceties of partial differential equations. Königsberg's seminar was older -- it had been founded in 1834 -- and in it a comprehensive curriculum in mathematical physics and exercises in precision measurement took shape by trial and error. These two seminars probably epitomize what are believed to be the general characteristics of German physics instruction as a whole, yet they were quite local affairs, emulated but not always successfully imitated. They, too, underwent changes in the 1870s. Wilhelm Weber retired from Göttingen in 1870, handing over the direction of the seminar to Friedrich Kohlrausch; Neumann from Königsberg in 1876, when his former student Woldemar Voigt took over. Neither seminar regained the vitality it had had at an earlier time, and in both a service clientele not particularly interested in research was now the rule rather than the exception.[10]

For German physics instruction the decade of the 1870s was a period of standardization, when physics courses became accessible to more students than they earlier had been. It is generally maintained that the contents of German physics courses were shaped by the frontiers of research. To a certain extent they were and they continued to be. But physics courses were also fitted to student needs. Choosing which issues and techniques were most appropriate for introductory courses and deciding upon the proper sequence of topics in physics courses contributed to the "curricular tightening" that characterized the entire German university system after 1870.[11] *Lehrfreiheit*, the professor's freedom to teach what he chose, was inoperable in an environment in which basic courses were created to meet the needs of a student clientele. *Lernfreiheit*, the student's freedom to choose his courses, was also compromised as students learned that certain physics courses were based on prerequisites. The pursuit of research, of knowledge or *Wissenschaft* for its own sake, was certainly realized, but only after submitting oneself to intensive training in the skills of investigation, which for physics were mathematical methods and those of an exact experimental physics. By the 1870s in Germany scholarly learning was metamorphosing into stepwise professional training, and *Wissenschaft* vied with *Beruf*. The incipient paradox of German science instruction was that it was eroding the very ideals upon which the German university was believed to rest.

The standardization of physics instruction in Germany in the 1870s offers an interesting contrast to the reforms in America during the same decade. From the 1820s through the 1860s or so, German physics professors tended not to use textbooks and instead created anew the pedagogical structure of physics. This instructional style was responsible in part for the vitality of German physics teaching in the first half of the century. As was occurring in other disciplines, however, teaching patterns in physics eventually led to a required curriculum of courses in the separate branches of physics, the lecture notes for which sometimes became the basis of textbooks. In America, by contrast, physics instruction had its origins in fixed format textbooks. By the 1870s physics instruction in America was still textbook-oriented but at some institutions a variety of textbooks began to appear, one for each special branch of physics, replacing the single textbook, the sections of which had been utilized in different courses. Whereas in Germany physics instruction began in the context of what might be called an elective system but evolved into a de facto curriculum, in America during the 1870s electives were just beginning to supplement the fixed curriculum. And whereas in Germany physics instruction began with the lecture format, adding recitations under the various names of student societies, exercises sessions, and seminars from the 1820s on, in America physics instruction began with the recitation method and did not generally include lectures until after the Civil War. Finally, whereas in Germany laboratory instruction in physics had been present since the 1830s but was not formally institution-alized (and then not everywhere) until the 1870s, in America laboratory instruction was just beginning in the 1870s, spurred only slightly by the Congressional Act "authorizing schools and colleges to import apparatus free of duty."[12] In both countries, however, the primary function of physical laboratories was pedagogical.

As is the case for the German universities, it is difficult to generalize on the character of physics instruction in American colleges and universities. In 1871 the American astronomer Edward Pickering estimated that there were about five physical laboratories for students in American colleges and universities.[13] Frank W. Clarke reported in his 1878 survey of instruction in physics and chemistry in the United States that thirty-five American colleges and universities had laboratory instruction in physics.[14] The most meaningful comparison with German university physics instruction, however, is probably made with institutions such as Harvard, the Massachusetts Institute of Technology, Cornell and Johns Hopkins. MIT's laboratory, "the first of its kind . . . in this country," opened in 1865.[15] Pickering, its director, remarked in 1871 that "for many years physicists have been in the habit of instructing their special students and assistants [in laboratory methods], yet it is only recently that the same plan has been tried with large classes in physics."[16] In the 1870s, sophomore courses in physics at MIT consisted principally of lectures based on Atkinson's translation of Ganot, modified "to keep pace with the progress of science." In their junior year students learned "physical manipulation" from Pickering's own laboratory manual, in part to learn how instruments were "of direct practical value in the arts" and in part to learn the art of physical investigation. In their senior year students learned advanced mathematical techniques that went beyond calculus; used textbooks in foreign languages "wherever it is desirable"; read "original memoirs and scientific periodicals"; and engaged in original research.[17] But MIT was a school practically oriented, and perhaps compared better with German technical institutes, not universities. Yet, as we shall see, Pickering's laboratory methods were used elsewhere.

At Harvard, physics was required in the freshman year, after which students could register for elective courses that covered the major branches of physics and for a laboratory course that included "the use of instruments of precision." Mathematical physics was offered only to graduate students, and then in the form of James Clerk Maxwell's *Treatise on Electricity and Magnetism*, a work rarely covered in German physics courses.[18] Instruction at Cornell was divided between lecture experiments and recitations, with laboratory work available in the senior year. But the students' work was "mostly of a quantitative nature, intended to give the student an idea of the methods employed and the care to be exercised in making physical measurements."[19] The most mathematically oriented physics instruction was at Johns Hopkins, where courses in thermodynamics, electricity and magnetism, spherical harmonics, theoretical dynamics, hydrodynamics, and the theory of the telescope were offered. Practical work was required, and its laboratory contained "some of the rarest and most valuable instruments of precision."[20]

If we were to compare the physics discipline in America and Germany at the level of research productivity, frequency of deployment of mathematical methods, and professional organizations including periodicals, we would find significant differences. But at the level of physics instruction, I believe, a different picture emerges. If we were to take the handful of institutions in both America and Germany considered to offer the best instruction in physics, then we would find that the differences were ones of degree, not kind. Advanced courses in America were close to the first half (and perhaps more) of physics instruction in Germany in all respects except perhaps degree of mathematization, *insofar as mathematization is taken to mean the use of partial differential equations*, and except in the relative opportunities available for students to work their way up to original research. And if we were to consider the texture of physics in Germany as it was *practiced* and *taught* in the 1870s without projecting forward to later developments, two conclusions would emerge. The first is that mathematical or theoretical physics was not a dominant mode of investigation or instruction. Second, the viability of mathematical physics was in part dependent upon its deployment in an exact experimental physics that emphasized precision measurement, another form of quantification.

The more general issue of *quantification* as it relates to an exact experimental physics offers a final way to compare American and German physics instruction. Despite the very different evolution of physics instruction in America and Germany, the first instructional manuals for precision measurement appeared in both countries almost simultaneously in the 1870s. Friedrich Kohlrausch's *Leitfaden der praktischen Physik* was first published in 1870 and very soon thereafter a second edition appeared, which was translated into English in 1873.[21] Based upon his experiences in teaching physical laboratory methods between 1866 and 1870 at Göttingen to a mixed student clientele of physics majors and students of chemistry, medicine and other subjects who were required to take a physics course, his textbook was popular and went through several subsequent editions. In America, Pickering's *Elements of Physical Manipulation* appeared in 1873, followed by a second volume in 1876. Pickering's texts were based upon the laboratory course he had offered at MIT since 1869. Both Kohlrausch and Pickering intended that their manuals be used in part to teach students the methods of original research. Impressed by the "excellence of the work done by many of the students," Pickering in fact had hoped that "valuable results might be obtained by assigning to different students the experiments in a research, taking care that each should be repeated at different times by different individuals. Their results, if concordant, would be much more conclusive than those obtained by a single experimenter, since they would be free from all personal bias."[22] To the extent that textbooks of this type were used in America, Pickering's was the standard.

The manuals by Pickering and Kohlrausch taught students experimental protocol, including the analysis of data. But their approach to measurement differed substantially. In the 1873 English translation of his manual, Kohlrausch began with a discussion of the errors of observation (including the mean and most probable error) and distinguished errors of observation from the constant errors stemming from instruments or the particular way in which experimental protocol was executed. As for the errors of observation "which [are] introduced by the true uncertainty of the observation -- that is, by such errors as give too great a value as often as too small a value,"[23] Kohlrausch demonstrated how they could be calculated by the use of the method of least squares and how those calculations could be aided by approximation methods and other rules for the numerical analysis of data.

Errors of observation that arose during the act of measuring and that could not be eliminated in the experiment provided for Kohlrausch an indication of the range of accuracy of the results overall. Important as these errors were, though, they were not Kohlrausch's primary concern. Instead, he explained in greater detail the constant errors that influenced the outcome of an experiment. Although constant errors had to be eliminated, accounted for numerically, or their effect diminished as much as possible, Kohlrausch warned that

> how far we may go in taking account of corrections depends of course upon the limit which is here also imposed upon us by the deficiences of the observations, as well as by our incomplete knowledge of the laws of nature and of the numerical values which they involve. But, on the other hand, it is frequently unnecessary to carry the accuracy of the correction to this limit. It is very much oftener sufficient to attain such a degree of accuracy that the neglected part of the corrections is materially less than the possible influence of the errors of observation on the result.[24]

The calculations for constant errors were laborious, and Kohlrausch tried to lessen the burden on the investigator by indicating that judgment calls could be made in their assessment. Some corrections -- for example, for temperature -- were obvious, but for more sensitive or unusual corrections he seemed to rely upon the intuition of the investigator to determine "what degree of accuracy" was required and what other corrections were essential in the determination of the final result.

Pickering also distinguished two kinds of errors. There were "constant errors, such as a wrong length of our scale, incorrect rate of our clock, or the natural tendency of the observer always to estimate certain quantities too great and others too small." He explained that "when we know their magnitude they do no harm, since we can allow for them, and thus obtain a value as accurate as if they did not exist." But unlike Kohlrausch, who identified several types of constant errors and how to determine them mathematically, Pickering did not discuss at all the conditions that influenced an experiment and how they affected the final outcome. Instead for him the analysis of data consisted in calculating a second class of errors, "accidental errors" (Kohlraush's errors of observation), which were "unavoidable" and which were due to such factors as the "looseness of the joints of our instruments, [the] impossibility of reading very small distances by the eye, &c., which sometimes render the result too large, sometimes too small."[25] Pickering distinguished two methods that could be used to determine accidental errors.

The first, the analytical method, involved the application of the method of least squares and the determination of both the most probable error and the relative weights of observations. He advocated winnowing data by applying to it C. S. Peirce's criterion for rejecting certain observations. Pickering also demonstrated how values intermediate between observations could be calculated. So he outlined the method of differences and its application in interpolation when the measured values were equidistant from one another.[26] When measurements were not so taken, the method of least squares could be used, or one could apply a second method, which Pickering called the graphical method.

In contrast to the analytical method, which represented data in a tabular form, the graphical method represented data geometrically, as a smooth curve connecting the coordinate pairs (x,y). Pickering did not say how to draw the smooth curve but only recommended that one should "draw a smooth curve as nearly as possible through them, and then see if it coincides with any common curve, or if its form can be defined in any simple way." He did indicate that the measured points would differ from the smooth curve by the accidental errors in the experiment, and he explained how the accuracy of the curve could be increased (although generally the graphical method was inaccurate beyond three significant figures). Interpolation was, of course, more easily accomplished with the use of this method, and he considered graphical interpolation to be more accurate than analytical interpolation because "by drawing a smooth curve the accidental errors are in a great measure corrected."[27]

In the manuals by Kohlrausch and Pickering, then, we have not just a contrast between the *types* of errors each author chose to describe in more detail -- Kohlrausch the constant errors and Pickering the accidental errors or errors of observation -- but different mathematical approaches to the errors of observation and to the purpose of error analysis overall. Kohlrausch was concerned less with the functional relationship between variables than with the limits within which a particular result was reliable. Pickering, by contrast, was less concerned with the precise determination of the limits of accuracy than with obtaining a reliable set of measurements that could demonstrate a physical relation, and hence the identification of causes. "The object of all Physical Investigation," he explained to his readers, "is to determine the effects of certain natural forces." He was motivated to include the graphical method because it had "the advantage of quickness, and of enabling us to see at a glance the accuracy of our results."[28] He explained that in his laboratory exercises the graphical method was also a convenient pedagogical device because it "enables the student to take in all his observations at one glance, while the instructor can constantly tell how carefully the work has been done."[29]

The treatments of errors by Kohlrausch and Pickering isolate, I believe, important differences between German and American physics instruction on the issue of how measuring operations were taught, how numerical results were achieved from measurements, and how a sensitivity to exactitude was inculcated in physical instruction. Remarking that the graphical method did "not seem to have attracted the attention that it deserves," Pickering knew that it was a novel feature of his textbook.[30] To what extent his textbook actually did promote graphical analysis is difficult to measure. He mentioned that in his own laboratory course, he distributed graph paper to his students. Undoubtedly instructors at other institutions did as well, but the number was certainly not overwhelming. Clarke reported only eleven or twelve universities and colleges, most of them in the Northeast, that assigned Pickering in 1878.[31] Yet although it was not until the end of the century that graphical analysis became fully integrated into mathematical and physical instruction in America, first and strongest in engineering schools, by the 1870s Pickering's manual introduced graphical methods to American physics students in a way that did not inhibit or compromise the formulation of functional relations, especially the relation between cause and effect.[32]

The positive attitude of Pickering and presumably other American physics instructors toward graphical analysis was not found in Germany in the 1870s. In the second edition of his manual, the basis of the 1873 English edition, Kohlrausch mentioned neither graphical analysis nor interpolation. The omission was not peculiar to him. The graphical analysis of data was infrequently and poorly practiced in Germany during the 1870s. Part of the reason had to do with the nature of mathematical education in German secondary schools, where the graphical representation of functions was not introduced until about 1870, and generally not until the 1880s and later.[33] Complaints that the graphical method was underappreciated and inadequately utilized in science came from many quarters even in the 1890s.[34] Significantly, Kohlrausch himself did not introduce interpolation until 1880 in the fourth edition of his work, and he did so in a very simple way for data recorded in a tabular form. None of his students employed graphical analysis in their first publications until 1878 and only two did between 1878 and 1881. After 1882, his students seem to have utilized the graphical representation of data wherever it could be used in their publications.[35] By 1892, in the seventh edition of his manual, Kohlrausch expanded his coverage of interpolation and introduced graphical representation, but again in a very simple way. Both topics were given expanded treatments in the ninth edition of 1901, when Kohlrausch finally created a separate section for graphical representation.[36]

In the mathematico-physical seminar at Königsberg -- one of the leading institutes in Germany in which students could learn the mathematical methods of physics -- graphical analysis and interpolation were not especially emphasized in the years between 1834 and 1876, when Neumann was the director of the seminar. In the first publications by his seminar students, graphs occasionally appeared but they were used poorly. For the most part students connected data points on a graph as a child would, by drawing straight lines between them, rather than by trying to construct a smooth curve. Moreover, both Neumann and his students maintained that mathematically expressed relations incorporating physical variables that were drawn in part from measured values and in part from interpolated values could not be granted the status of physical laws, and instead they called such relations (derogatorily) "interpolation formulas," viewing them as only approximations of actual functional relations. So obsessed were they with the determination of errors that they could not place much faith in the graphical representation of data for which one was in essence compelled to overlook errors and to consider in general terms the functional relation overall.[37]

There is an irony in this difference between Kohlrausch and Pickering on the issue of the graphical representation of data, and more generally between the methods of quantification taught and used in American and German physics. We think of German physics instruction as exceptionally mathematically oriented, yet graphical analysis -- the technique that best bridged the gap between the mathematical or theoretical and the experimental -- was underused, despite its potential advantages in formulating a physical law. We think of American physics instruction as largely unmathematical, yet cultivated in the most popular manual on physical manipulation was the powerful method of graphical analysis that could lead to the mathematical formulation of a physical relation. Such paradoxes belie simple representations of physical instruction in America and Germany, especially where the deployment of techniques of quantification in the construction of physical knowledge is concerned.

TEACHING PHYSICS AND MEASURING THE VELOCITY OF LIGHT

In broad outline, then, this was the character of physics instruction in Germany and America in the 1870s, when Michelson was a student and an instructor. Michelson recalled that when he was reassigned to the Naval Academy in December 1875 he "had no special knowledge of physics, and had never felt particularly interested in it," despite the high standing he had achieved in all of the physics courses he had taken in his last two years at Annapolis between 1871 and 1873.[38] Yet even before he left the academy, the reform of instruction in mathematics and physics had already begun. It was not so much that a pure scholarly, research-oriented instruction was introduced, but rather that the "professional" nature of science instruction at Annapolis -- by which was meant the application of science to military ends -- was believed to be enhanced by a more intense study of science and mathematics.

Although by 1870 chemistry was taught at the academy by using a mixed format of lectures, recitations, and to a limited extent, laboratory work, physics instruction was accomplished almost entirely through recitations. By the time Michelson began his physics courses, daily marks in physics recitations had been abolished, thus permitting in principle the development of more complex methods and ideas over a number of classes rather than requiring that at each class meeting a specific goal be reached.[39] Mechanics was the only subject in physics taught mathematically. Other methods of quantification were not used in the remaining physics courses. So although six of the ten questions on heat and four of the ten on light in the final examination in physics for the first class in 1873 -- Michelson's class -- concerned measuring instruments and specific experimental methods, such as the method of mixtures for determining specific heats, students need not have worked with the instruments in question or have performed the experiments cited to answer the questions. They only had to explain *how* they were done, and that was described in Ganot's textbook, which was not mathematically oriented. Ganot did discuss several measuring instruments and mentioned for some of them the errors that arose in observation, but he did not explain how those errors could be computed. Almost all of Ganot's textbook was covered in the second and first classes, or the final two years, at the academy.[40]

In 1871 instructors of physics and chemistry at the academy argued for the creation of a Department of Mechanics and Applied Mathematics, claiming that due to the addition to mechanics of the "study and application of the calculus, the duties [in teaching the course] are increased beyond the capability of a single individual to direct it."[41] The department was not immediately approved, but the calculus course was moved to the Department of Physics and Chemistry, where it was taught through lectures and in conjunction with mechanics.[42] The academy's Board of Visitors, a committee of outside evaluators that met annually, was sufficiently impressed by the proposed changes for calculus instruction that in 1873 they "strongly urged the importance of obtaining the best talents in the country for the headship of departments embracing such subjects as mathematics, physics and chemistry." "These ought to be men of great ability," they argued, "whose lives are devoted to the sole business of investigation and instruction in their particular subjects."[43] In the following year, Captain William T. Sampson returned as chairman of the Department of Physics and Chemistry. While Michelson was aboard the USS *Monongahela* and other ships between 1873 and 1875, physics instruction at the academy changed dramatically. Mechanics and the calculus were incorporated into the new Department of Mechanics and Applied Mathematics in 1874-75. In the same year, elective courses in higher mathematics were introduced to meet the needs of the "five or six" more talented students who appeared each year. These elective courses were not strictly separate courses on different subjects but were designed as supplemental meetings to existing courses that enabled the student to go "beyond the limits fixed by the usual course." By the fall of 1875 Sampson recommended that elective courses be offered in all departments. With the addition of elective courses, existing courses became more sophisticated. The level of instruction attained by mathematics courses is in part indicated by the high proportion of failures that now occurred in them.[44]

Cadets now found themselves taking more physics courses and covering the topics in them in greater detail. In 1875-76 an elementary physics course was created for the third class, and it was intended as a preparatory course to the more specialized ones that followed in the next two years. Recommendations were made to rearrange the sequence of the physics courses, especially to ensure that the cadets covered thermodynamics before learning the properties and operation of the steam engine. Sampson called for a review of physics textbooks, and by 1876, Ganot's textbook was made a reference book and several more specialized treatments replaced it as textbooks. Jenkin's *Electricity and Magnetism* was introduced first for cadet-engineers in 1874 and a short time later for first class cadet-midshipmen. Part IV of Deschanel's textbook was introduced for the first class course on optics in 1874. Maxwell's *Treatise on Heat* was used by the second class for a year but it was found to be too difficult. It was retained as a reference book and replaced by Balfour Stewart's simpler textbook on heat. Separate textbooks on mechanics and hydrostatics were used as well.[45]

In 1874-75 Sampson introduced experimental or demonstration lectures in chemistry, and probably a year or so later, in physics. Laboratory instruction in physics at Annapolis was enhanced considerably in 1875, when a new chemistry laboratory was built, freeing up a slightly older building, completed in 1869, for physics instruction alone. The older building could only accommodate ten students for both chemical and physical exercises. Newly designed for physics, the building was more efficiently utilized. Its first floor housed a galvanometer room. Thermal radiation and absorption experiments could be conducted in a room at the center of the building, strategically located to maximize temperature control. The second floor of this physics "institute" was taken up entirely by a lecture room 40 feet by 50 feet and a physical laboratory 30 feet by 40 feet. Exactitude was the order of the day in the laboratory, which had gained over the past few years "many valuable instruments of precision" and "facilities for more precise instruction in physical measurements and for original research." Accordingly, Sampson considered the copies of the standard yard and meter stored in the academy's observatory to be essential tools in physical laboratory exercises.[46]

Laboratory work could now begin as early as the third class, but not for everyone. The program of physics instruction at the academy was intended to produce an elite group of students who, from their second to fourth years, became increasingly competent in methods of physical measurement and who surpassed the regularly prescribed course of study. Sampson even wanted to encourage graduates still present on the grounds of the academy to use the laboratories "for further instruction in physics and chemistry." Specially chosen second class students spent four to six hours a week in the laboratory. First class students, "having had previous experience in the laboratory... [were] left very much to their own resources." The goal of their investigations being more precise measurements, Sampson considered it essential that "in almost every measurement some portion of the apparatus has been improved by each [first class] student."[47] Student investigations in 1876 included the determination of the wave length of a ray of light; specific heat by the method of mixtures; the latent heat of steam; the horizontal component of the earth's magnetism; the constant of a galvanometer. Students improvised upon apparatus, making them more precise; were sensitive to the accuracy (or the lack of) on scales; compared readings from two different instruments; calculated corrections to readings; and showed how to eliminate errors from experiments. They did not at this time, however, apply the method of least squares or discuss in any way the reliability of their readings.[48] As laboratory instruction was enhanced, the orientation of the annual examination in physics changed. Between 1873 and 1876, cadets were required to discuss the theoretical principles behind instruments; to show how the scales on instruments were determined; to show how instrumental readings could be corrected; and to supply the corrections for certain experimental procedures, such as the method of mixtures. In 1874-75 a required part of the cadet-engineer examination included practical exercises in laboratory work.[49]

The rationale for these pedagogical changes was couched in terms relevant to the professional life of the naval officer. Sampson believed that experimental observation was important for the naval officer who, on his tour of duty

> has peculiar opportunities of aiding scientific investigation by noting and recording those natural phenomena with which he comes in daily contact. It is, therefore, of the greatest consequence that he should have trained powers of observation, and be able to make a scientific record of what he sees.[50]

More time was given to electrical and thermal studies in the physics curriculum because those subjects had a "direct professional bearing," especially in "the study of the different methods of exploding submarine mines and torpedoes by electricity." Many contemporary ships being built of iron, it was essential to be able to compute the effect of a ship's iron on a compass needle, and to know what kinds of corrections had to be made to a mariner's compass in general. Yet even though the study of heat was "intended as an introduction to the study of the steam engine," "the more difficult problems of thermodynamics [were] not neglected."[51] For these reasons principally Sampson modernized physics instruction at the academy.

Hence with good cause Michelson could recall that when he began teaching at the academy in the fall of 1876 he "knew very little more than the cadets [he] was teaching." Sampson apparently initiated him into teaching gradually, because for the first year Michelson taught only recitation sections. He later explained that he could "keep his ignorance hidden by reading up in the textbook about ten pages in advance of the class."[52] We do not know to which textbook Michelson was referring, for the department continued to assign a variety of them. In 1876 Sampson even reported having added Maxwell's *Treatise on Electricity and Magnetism* as a reference book.[53]

Michelson's teaching assignment changed in 1877-78, his second year at the academy. As is well known, Sampson asked him to prepare a lecture in optics (which Michelson then considered a novel feature of physics instruction, never having been taught that way himself), for which Michelson chose Foucault's experiment on the measurement of the velocity of light. "It occurred to me," he later recalled, that "it would be interesting to show the class how it was done."[54] In preparation for his lecture he probably consulted first the 1875 edition of Ganot and the recently introduced textbook by Deschanel, both of which gave extended discussions of Foucault's experiment. Michelson would have read in Ganot that Foucault's result, 185,157 miles per second, was "less than that ordinarily assumed," but in agreement with values determined from solar parallax.[55] Deschanel reported that Foucault's experiment was "much better fitted [than Fizeau's] for rigorous demonstration," a feature that would have appealed to Michelson had he by then become acquainted with the investigations undertaken by the cadets. Deschanel also detailed the improvements that had been made between 1850 when Foucault's experiment was first performed, and 1873, when his textbook was published, and he even identified as the principal element in Foucault's calculation the speed of the mirror's rotation.[56] Probably from these sources Michelson constructed his demonstration apparatus. As he later reported, "in November 1877 a modification of Foucault's arrangement suggested itself," and he embarked upon an investigation of the measurement of the velocity of light.[57] His principal modification consisted of dispensing with Foucault's concave reflector, which resulted in being able to allow the light to travel over a longer distance.[58]

The inauguration of his investigation was not only occasioned by the reform in physics instruction at Annapolis; it also coincided with the material improvement of laboratory courses there. The Board of Visitors in June 1878 judged the department of physics "in excellent condition." They noted that "the additions to the lecture-rooms and laboratories have been made with judgment and exquisite taste" and that "the apparatus included many pieces of the latest design and the most finished construction." They even had examined the laboratory reports of students, and judged them to be "exceedingly careful and thorough."[59]

During the months between November 1877, when he first hit upon the modification of Foucault's experiment, and March 1878, Michelson performed a "number of preliminary trials" (he did not say how many) with his apparatus "in order to familiarize [himself] with the optical arrangement."[60] As he worked with materials that he could find in the Naval Academy Laboratory (he called them "crude pieces . . . which suited his purpose"),[61] adding but a revolving mirror that he had purchased himself for ten dollars, several failures occurred.[62] But in his first three investigations on the velocity of light published that year, Michelson reported only ten numerical results.[63] Data seemed to occupy him less, as between May and August 1878 we find Michelson trying to understand further his apparatus and his experimental protocol. First, in May 1878, he announced his modification and the formula he would use to compute the velocity of light from the experiment.[64] The following month he republished the description of his modification, adding "another plan" that would "probably give" a greater intensity of light.[65] Although in July 1878 he received $2,000 from his father-in-law "for carrying out these experiments on a large scale," he did not immediately do anything with the money, believing that it was necessary first "to find whether or not it was practicable to use a large distance."[66]

Whereas in his first two papers he had been concerned with his apparatus, in a third paper, the longest, presented to the St. Louis meeting of the American Association for the Advancement of Science in August 1878, Michelson turned his attention toward understanding the protocol of his experiment. For the first time he addressed the issue of accuracy in the experiment by identifying "the displacement, a quantity which enters directly into the formula" as the variable that in Foucault's experiment was "difficult to measure accurately" and that he would try to increase. For the first time, Michelson expressed his desire to obtain more precise readings of the variables he had to measure and in general to improve the accuracy of his protocol. Yet despite the fact that his initial modification had been designed to decrease errors and despite his intention to increase the precision of his apparatus and the accuracy of his protocol, at this point he described his instruments in qualitative terms only. Although he identified a few of the constant errors that afflicted the experiment, he did not make any quantitative determinations of them. Near the end of his report he gave the results of ten trials and their average for the velocity of light, 186,508 miles per second. One of his arguments for continuing to improve the experiment was that his data -- the ten trials he reported -- were in "accordance with each other." But we do not know if these were the only trials he performed (I suspect that they were not), and if they were not, what criteria he used for choosing the ones that he did. Nevertheless, the consistency of his data along with the agreement between their average and "the generally accepted result" justified in Michelson's mind "the expectation of obtaining *with proper appliances and under more favorable conditions* the correct result within a few miles."[67]

Without the quantitative determinations of the constant errors that afflicted his apparatus, Michelson's analysis of his experimental design and protocol did not even measure up to the student investigations he was presumably directing. But like his students, he expected better results solely through the improvement of the material conditions of the experiment. His expectation was not unwarranted but at some point he would have to admit that it would be counterproductive to continue to improve the material conditions of his experimental design. The criteria for making such a judgment were to be found in the standards of accuracy that Michelson deployed, but because he gave no indication of what were in his mind tolerable and intolerable limits for the reliability of his data, we just do not know how Michelson made that judgment. Michelson later recalled that he "had no idea of securing any very accurate results," and that "very much to [his] surprise, [he] discovered that with our homemade apparatus we were measuring the velocity of light with considerably greater accuracy than anybody had measured before."[68] How he had assessed that accuracy cannot be determined from his first set of papers. Either he did not think of computing the constant errors of the experiment, or he assumed that they cancelled one another out. He certainly did not think of computing the errors of observation. His results were at this point more than first approximations, but less than precise quantitative determinations.

It was not long before Michelson's conceptions of degree of accuracy and the limits of reliability of data sharpened. In the following academic year, 1878-79, Sampson added another required textbook to the upper level physics courses for cadet-engineers and cadet-midshipmen: the 1873 English translation of Friedrich Kohlrausch's *Introduction to Physical Measurements*.[69] Sampson's choice of Kohlrausch placed the Naval Academy among the minority of those institutions that assigned a textbook on physical measurements, for in 1878, although eleven or twelve colleges and universities in the United States used Pickering, only four -- Cornell, Washington University, the Polytechnic School at Washington University, and the Naval Academy -- used Kohlrausch.[70] Sampson reported that Kohlrausch's textbook had been used for reference at the Naval Academy since 1876-77. However, Michelson, who was undoubtedly busy familiarizing himself with several other newly required textbooks every year, seems not to have read Kohlrausch until late 1878 or early 1879, when he had to teach it.[71]

Concurrently with the introduction of Kohlrausch as a required textbook in 1878-79, laboratory instruction for the cadet-engineers was again considerably upgraded. For the first time a course on "Physical Measurements" was offered for them.[72] In addition, the Department of Mechanics and Applied Mathematics listed for the first time a course for the cadet-engineers dedicated exclusively to "the theory of the method of least squares and the application of the method to results derived from experiments" and for which they used Merriman's textbook on least squares.[73] The department appears to have offered the course unofficially one semester earlier, in the spring of 1878, because the second class cadet-engineers were required to answer ten questions concerning the method of least squares on their final examination in June 1878. (Members of the Academic Board had proposed the course in February 1878.)[74] Included among the questions on the examination were ones on the "errors likely to occur in ordinary observations" and the equations for certain probability curves. The cadet-engineers were also asked to "derive the formula for determining the probable error of an observation of weight unity in the case of direct observations of unequal precision upon a single quantity, and thence deduce the formula for the probable error of the general mean," and "in the case of independent observations, of unequal weight, upon several quantities," to "develop the method of finding the most probable values of the quantities and of determining the probable errors of the results."[75]

As a result of using Kohlrausch and Merriman, student laboratory investigations changed in 1878-79. For the most part, students continued to examine the kinds of problems they had in the past, but now error analysis -- the determination of constant errors and the errors of observation -- was incorporated into their reports. Also in 1878-79 for the first time Sampson was able to report that two independent investigations had been undertaken in the physical laboratory: "an original investigation of the changes of electrical resistance of some substances due to change in pressure" and "an original investigation of the electrical resistance of boiler scale."[76] Laboratory work was encouraged beyond regular classroom offerings for cadet-midshipmen "who display the greatest ability in the required courses." Measuring operations dominated this extra-curricular laboratory work, with first class cadet-midshipmen continuing electrical measurements and second class cadet-midshipmen continuing measurements in light and heat. The spring 1879 final examinations in physics for all cadets incorporated more questions concerning measuring operations and the evaluation of measurements than they had in the past. In that semester the final examination in heat and light for first class cadet-midshipmen -- the last that Michelson graded -- included a question that encapsulated the changes that had been introduced into physics instruction that year: "To what classes of errors is one liable in physical research? How may they be eliminated? Describe the method of graphical representation of the results of experiment. What are the formulas for interpolation?"[77] The changes in physics instruction did not go unnoticed. After their annual visit of the physics department in June 1879, the Board of Visitors recommended that "in view of the importance of this branch, it is recommended that the appliances for purposes of illustration and investigation be made as complete as possible, by continued addition of carefully selected apparatus by the best makers."[78]

Drawn to techniques of error analysis through his teaching, Michelson now incorporated them into his investigations. As late as December 1878 he continued to define the accuracy of his experiment in terms of the construction of its apparatus. With part of the $2,000 he had received from his father-in-law, he constructed a small house where he could execute his experiment on the velocity of light and where he could have greater control over the conditions of his experiment, especially temperature. He reported that the distance of 200 feet then travelled by the light ray "could not be measured with much greater accuracy, neither could the speed of the mirror nor the 'radius of measurement,'" but he did not say how those measurements were taken. His understanding of his instruments and his protocol gave him a sense, however, that the spread of his final measurements would be not more than .1 percent.[79] In April 1879 he published a synopsis of his presentation to the AAAS of the previous August; nothing had changed from the August version.[80]

Then in August 1879, at the Saratoga Springs meeting of the AAAS, Michelson presented a report of his findings in radically different terms. His August 1879 report was a beautifully detailed paper written in the tradition of an exact experimental physics exemplified by Bessel's seconds pendulum paper of more than fifty years earlier.[81] Now, finally, Michelson applied quantitative methods of error analysis to his instruments and his data. To a great extent, he still framed his investigation in terms of the material conditions of the experiment -- and these were now considerable -- but now he gave an extensive discussion of the errors of his experiment and the *quantitative* corrections he made for them, all as Kohlrausch had directed his readers to do. All constant errors in his experiment were accounted for mathematically and numerically. Michelson even went beyond Kohlrausch and used graphical analysis and interpolation to measure the divisions on his micrometer. Finally, he applied the method of least squares to his data and supplied the limits to the reliability of his final result. He altered one of the ten measurements he had previously reported, reducing its value by 40 miles, and recomputed his final result as 186,500 ± 300 miles per second (between .1 percent and .2 percent error). He then reported 100 additional measurements, the corrections for them, and the conditions under which they were taken, obtaining from them a final result of 299,944 ± 51 kilometers per second (.02 percent error), or rounded off, 299,940 kilometers per second (186,380 miles per second).[82]

Thereafter, the accuracy of his data, not the precision of his instruments, became an obsession. In November of 1879 he published in the *American Journal of Science* a synopsis of his AAAS paper with data drawn this time from 100 sets of 10 measurements each. He reported a result of 299,820 kilometers per second. Shortly after that article was published, but before it was distributed, Michelson recomputed those 1,000 measure-.nents on the basis of what must have been a more detailed error analysis. The recorrected data were published as an insert to his article, and they led to the result of 299,930 kilometers per second.[83] He sent a report of his investigation to *Nature* in November 1879 as well, and for it he recomputed the correction for reducing his result to vacuum conditions. In December, the second half of his article appeared in *Nature*. and the result he reported was 299,828 kilometers per second, as it had been in his *American Journal of Science* article. But when in January he submitted the corrections for that measurement, he supplied not the *American Journal of Science* corrections, but the results he had achieved the prior August: 299,944 ± 50 kilometers per second (lowering his error somewhat), or 186,380 ± 33 miles per second. It is this second set of the seven articles constituting Michelson's fledgling investigation as a physicist that displays his achievement in, as well as his learning of, the techniques of an exact experimental physics.[84]

Stanley Guralnick has warned us of the dangers of using the German model of science instruction as a measure of superiority.[85] My comparison of the Kohlrausch and Pickering manuals for practical physics was intended in part to show the wants of the German model insofar as it embraced techniques for the analysis of data and errors. The changes in physics instruction in both Germany and America during the 1870s would in some respects seem to make notions of a "model" physics instruction less and less meaningful. When Michelson visited Germany in 1880, he was not at first impressed with German physics instruction, finding the initial part of his laboratory work at Berlin to be "very elementary."[86] His reaction is not surprising. The shaping of Michelson as a physicist took place while he was an instructor at Annapolis and had been completed before he left for Germany.

I would argue, too, that it was completed before he went to work at the United States Naval Observatory in the fall of 1879, when he joined the distinguished American astronomer Simon Newcomb on government experiments to measure the velocity of light. Michelson had in fact been in contact with Newcomb since April 1878, before the publication of his first paper on the velocity of light. In letters they exchanged over the next year and a half, they discussed parts of Michelson's experiment.[87] But Newcomb's concern was largely for the material conditions of the experiment, and by the time Newcomb mentioned the constant errors of the experiment, Michelson was already teaching Kohlrausch and immersed in an environment in which the determination of errors was considered essential to a physical investigation. Michelson no doubt found support in his exchange with Newcomb, but far more important is the fact that in order to keep up with the reforms at Annapolis, Michelson would have had to immerse himself in his teaching during the academic year, learning what the cadets were learning.

Among the flaws of American physics in the final quarter of the nineteenth century, Daniel Kevles has cited a lack of institutions and communities to set standards, among them investigative standards, in physics. To be sure, Kevles considered Michelson a "distinguished American physicist" and an "exception to the rule."[88] The early years of Michelson's career at the Naval Academy both confirm and qualify Kevles's observations. For the young instructor Albert A. Michelson at the United States Naval Academy, the investigative standards of physics came by way of the reform of physics teaching. The selection of his research topic came not by way of disciplinary needs, but from teaching. The evolution of his investigative style -- its two stages as well as its ultimate constitution --was determined by the imperative to teach physics, and to teach it in a certain way with particular textbooks. The seven papers that he published between May 1878 and late 1879 map an investigative trail toward measuring more accurately the speed of light. When viewed from the perspective of his workaday world they also offer a glimpse of how, with each step taken along the way, Michelson became a professional physicist.

In his later years, Michelson revelled in the fact that when he had begun his teaching and his first investigations as a physicist, he had been ignorant and untrained. He felt that this had been his advantage, and he believed that it continued to be his advantage in the selection of problems and the execution of their solutions. He had a contempt for authority in science, asserting that "men of ability, highly trained, often have a little too much respect for tradition."[89] Yet when all was said and done, after Michelson had taught Kohlrausch and in the process learned for the first time methods of error and data analysis, he had absorbed some of the most traditional methods of the exact experimental physics of the nineteenth century.

NOTES

1. *Examples of Experiments in Natural Philosophy. Made in the Academy del Cimento under the Protection of the Most Serene Prince Leopold of Tuscany and Described by the Secretary of that Academy* [1667], trans. in W. E. Knowles Middleton, *The Experimenters. A Study of the Accademia del Cimento* (Baltimore and London: Johns Hopkins Univ. Press, 1971), pp. 83–254, on p. 217.

2. Adolphe Ganot, *Elementary Treatise on Physics Experimental and Applied for Use in Colleges and Schools.* 3rd ed., trans. Edmund Atkinson (London: Longmans, Green, & Co., 1868).

3. The earliest official institutional setting in the German universities for the combined teaching of mathematics and physics was the Königsberg mathematico-physical seminar, established in 1834. On the role of seminar instruction in promoting mathematical methods in physics and in training secondary school teachers, see Kathryn M. Olesko, *Physics as a Calling: Discipline and Practice in the Königsberg Seminar for Physics* (Ithaca, N.Y.: Cornell Univ. Press, in press).

4. Sophisticated mathematical exercises in physics were common in German mathematico-physical seminars. Although the mathematical methods of physics were introduced in lecture courses on mathematical or theoretical physics at German universities, exercises were not generally assigned in them. A sense of the level of mathematical instruction in American physics courses can be gained from Frank W. Clarke, *A Report on the Teaching of Chemistry and Physics in the United States* (Washington, D.C.: Government Printing Office, 1880).

5. John W. Servos, "Mathematics and the Physical Sciences in America, 1880–1930," *Isis*, 1986, *77*: 611–629; on pp. 614 and 622.

6. On the professional orientation of physics problems at Königsberg, see Olesko, *Physics as a Calling*, chaps. 3–8. The incorporation into physics problems of the process of scientific discovery and of mathematical and measuring methods was also cultivated at the secondary school level in Prussia; see Kathryn M. Olesko, "Physical Laboratory Instruction in Prussian Gymnasien and Realschulen: Sources of Teaching Measurement before 1860," *Osiris*, 1989, *5* (in press).

7. On physics instruction in American colleges and universities, see Clarke, *Report*; Stanley M. Guralnick, *Science and the Ante-Bellum American College*, American Philosophical Society Memoirs, vol. 109 (Philadelphia: American Philosophical Society, 1975); Daniel J. Kevles, "On the Flaws of American Physics: A Social and Institutional Analysis," in *Nineteenth Century American Science. A Reappraisal*, ed. George H. Daniels (Evanston, Ill: Northwestern Univ. Press, 1972), pp. 133–151, esp. pp. 136–138; Daniel J. Kevles, *The Physicists: The History of a Scientific Community in Modern America* (N.Y.: Random House, 1979), pp. 7, 9, 12, 25; Albert E. Moyer, *American Physics in Transition. A History of Conceptual Change in the Late Nineteenth Century* (Los Angeles and San Francisco: Tomash Publishers, 1983), esp. pp. 59–67; Melba Phillips, "Early History of Physics Laboratories for Students at the College Level," *American Journal of Physics*, 1981, *49*: 522–527; Servos, "Mathematics and the Physical Sciences."

8. Surveys of physical laboratories in German universities are found in David Cahan, "The Institutional Revolution in German Physics, 1865-1914," *Historical Studies in the Physical Sciences*, 1985, *15*(2): 1-66; and in Paul Forman, John L. Heilbron, and Spencer Weart, "Physics *circa* 1900. Personnel, Funding and Productivity of Academic Establishments," *HSPS*, 1975, *5*: 1-185.

9. German archival sources on physics teaching at these universities include: *BONN*: Acta betreffend die Einrichtung eines Seminars für die naturwissenschaftlichen Studien auf der Universität Bonn, Rep. 76Va Sekt. 3 Tit. X Nr. 4 Bd. I: 1823-1831, Zentrales Staatsarchiv, Merseburg, DDR [ZSM]; Acta betreffend das naturwissenschaftliche Seminarium an Universität zu Bonn, Rep. 76Va Sekt. 3 Tit. X Nr. 4 Bd. II: 1832-1858, ZSM; Acta betreffend das naturwissenschaftliche Seminarium der Universität Bonn, Bd. III: 1859-1888, Nordrhein-Westfälisches Haupstaatsarchiv, Bestand NW5, Nr. 483, Düsseldorf, BRD. *ERLANGEN*: Acta der königlichen Universität Erlangen. Das neugegrundete mathematisch-physikalische Seminar betr. 1874 und math. Cabinets, nun mathematisches Seminar, 1914, Universitätsarchiv Erlangen Nbg. I/20V/9. *FREIBURG*: Universität Freiburg. Das mathematisch-naturwissenschaftliche Seminarium und (1874) mathematisches Seminarium, GLA 235/7766; Das physikalisches Institut der Universität Freiburg, 1861-1944, GLA 235/7771. *HALLE*: Acta betreffend die Einrichtung eines Seminar für Mathematik und die gesammten Naturwissenschaften an der Universität zu Halle, Rep. 76Va Sekt. 8 Tit. X Nr. 36 Bd. I: 1837-1889, ZSM. *HEIDELBERG*: Das physikalische Cabinet der Universität Heidelberg, 1807-1830, Generallandesarchiv Karlsruhe [GLA] 235/3057; Universität Heidelberg. Physikalisches Institut, 1804-1920, GLA 235/3253; Universität Heidelberg. Mathematisches Seminar, GLA 235/3228. *MUNICH*: Universität München. Mathematisch-physikalisches Kabinett, Sammlung, 1853-1927. Bayerisches Hauptstaatsarchiv München, MK 11317; Acta des k. akad. Senats der Ludwig-Maximilians-Universität München. Errichtung von 1856 bis 1925/26, Universitätsarchiv München [UAM] Y/IX/3/209; Acta des k. akad. Senats der Ludwig-Maximilians-Universität München vom 1891 bis 1901 betreffend mathematisch-physikalisches Seminar. Jahresberichte. UAM, Y/IX/4²/211. *TÜBINGEN*: Acten betreffend das mathematisch-physikalische Seminar, 1869-[1930], Universitätsarchiv Tübingen [UAT] 117/892-893. *WÜRZBURG*: Mathematisches Seminar, 1872-1905, Universitätsarchiv Würzburg [UAW] 1640 III 11/9; Physikalisches Laboratorium, 1870-1922, UAW 3233 III 8/3.
On teaching examinations, see: *BADEN*: Die Prüfungen für das höhere Lehramt hier: mit den Hauptfächern aus dem mathematisch-naturwissenschaft-lichen Gebiete, 1835-64, GLA 235/19616-19620. *BAVARIA*: Karl Neuerer, *Das höhere Lehramt in Bayern im 19. Jahrhundert* (Berlin: Duncker & Humblot, 1978). *PRUSSIA*: *Die preußischen Universitäten. Eine Sammlung der Verordnungen*, ed. J. F. W. Koch, 2 vols. (Berlin: Mittler, 1839); L. M. P. von Rönne, *Das Unterrichtswesen des preußischen Staats in seiner geschichtlichen Entwicklung*, 2 vols. (Berlin: Veit & Co., 1854); L. Wiese, *Das höhere Schulwesen in Preußen. Historisch-statistische Darstellung.* 4 vols. (Berlin: Wiegandt & Grieben, 1864-1902); and *Centralblatt für die gesammte Unterrichtsverwaltung in Preußen*, 1859-1934. *TÜBINGEN*: Prüfungen im Allgemeinen, 1829-1928, UAT 117/948; Lehramtsprüfungen, 1878-1931, UAT 117/951.

10. On the Königsberg seminar, see Olesko, *Physics as a Calling*; on Göttingen's, see *Göttingische Gelehrte Anzeigen. Nachrichten von der Georg-Augusts Universität und der Kgl. Gesellschaft zu Wissenschaften zu Göttingen*. no. 6 (11 März 1850), pp. 73-79; Königliches Universitäts-Kuratorium, Göttingen. XVI IV C b. Math.-phys. Seminar 4Vk/20; Akta betr. das mathematisch-physikalische Seminar 1861, VI C/468(3).

11. Charles E. McClelland, "Structural Change and Social Reproduction in German Universities, 1870-1920," *History of Education*, 1986, *15*(3): 177-193; on p. 183. Another view of that curricular tightening is given in Kathryn M. Olesko, "On Institutes, Investigations, and Scientific Training," in *The Investigative Enterprise*, ed. William Coleman and F. L. Holmes (Berkeley & Los Angeles: University of California Press, 1988), pp. 295-332.

12. Clarke, *Report*, p. 11.

13. Edward C. Pickering, "Physical Laboratories, *Nature*, 1871, *3*: 241.

14. Clarke, *Report*, pp. 200-215.

15. Ibid., pp. 13, 42. Presumably the separate entries for each school in Clarke's *Report* were drawn, in all likelihood verbatim, from the survey forms that were returned to him.

16. Pickering, "Physical Laboratories."

17. Clarke, *Report*, pp. 41-42; Edward C. Pickering, *Elements of Physical Manipulation*. 2 vols. (N.Y.: Hurd & Houghton, 1873-1876).

18. Clarke, *Report*. p. 46; James Clerk Maxwell, *Treatise on Electricity and Magnetism*. 2 vols. (Oxford: Clarendon, 1873).

19. Clarke, *Report*, pp. 61-62.

20. Ibid., p. 88.

21. Friedrich Kohlrausch, *Leitfaden der praktischen Physik* (Leipzig: B. G. Teubner, 1870); Friedrich Kohlrausch, *Introduction to Physical Measurements*. trans. Dr. Thomas Hutchinson Waller and Henry Richardson Proctor (London: J. & A. Churchill, 1873). A schematic history of the increase in size of Kohlrausch's textbook is given in Cahan, "Institutional Revolution," pp. 48-50.

22. Pickering, "Physical Laboratories."

23. Kohlrausch, *Introduction to Physical Measurements*. p. 2.

24. Ibid., pp. 16-17.

25. Pickering, *Physical Manipulation*, *1*: 2.

26. Ibid., pp. 3-11.

27. Ibid., pp. 11-16; on p. 12.

28. Ibid., p. 3.

29. Pickering, "Physical Laboratories."

30. Pickering, *Physical Manipulation*, *1*: v.

31. Clarke, *Report*, pp. 200-215.

32. According to John Servos, graphical methods and techniques and the analysis of functional relations fostered by them became increasingly prominent in American physics and mathematics in the early part of the twentieth century (Servos, "Mathematics and the Physical Sciences," pp. 623, 627). To what extent Pickering's textbook was instrumental in promoting the use of graphical methods remains to be determined.

33. Heide Inhetveen, *Die Reform des gymnasialen Mathematikunterrichts zwischen 1890 und 1914* (Bad Heilbrunn: Klinkhardt, 1976), p. 192.

34. See, for example, E. Wiedemann, "Die Wechselbeziehungen zwischen dem physikalischen Hochschulunterricht und dem physikalischen Unterricht an höheren Lehranstalten," *Zeitschrift für math. und naturw. Unterricht*, 1895, *26*: 127-140; on 131.

35. Based on the forty-one *published* journal articles (excluding republications, dissertations, *Habilitationsschriften,* and school programs) stemming from work done by Kohlrausch's students under his direction. A list of these articles is found in *Gesammelte Abhandlungen von Friedrich Kohlrausch*, ed. W. Hallwachs, A. Heydwiller, K. Streker, and O. Wiener, 2 vols. (Leipzig: J. A. Barth, 1911), 2:lxix-lxxi.

36. Friedrich Kohlrausch, *Leitfaden der praktischen Physik*. 4th ed. (Leipzig: B. G. Teubner, 1880), p. 21; 7th ed. (1892), pp. 24-26; 9th ed. (1901), pp. 25-27.

37. Olesko, *Physics as a Calling*, chaps. 6-8.

38. Quoted in Neil M. Clark, "Michelson Holds the Stop-Watch on a Ray of Light," *American Magazine.* 1926, *101*: 24-25, 152-154; on p. 25.

39. On physics instruction at the Naval Academy in the 1860s and 1870s, see the *Annual Register of the United States Naval Academy, Annapolis, Md.* (published by academic year and providing information on instructors, courses, textbooks, and semi-annual and annual examinations of the previous year for the cadets); *U.S. Naval Institute Proceedings* (October 1835): 1425-1427; Thomas G. Ford [History of the U. S. Naval Academy] (unpublished manuscript, c. 1885) chaps. 21 and 22, USNA Special Collections, Nimitz Library, Annapolis; and James Russell Soley, *Historical Sketch of the United States Naval Academy* (Washington, D.C.: Government Printing Office, 1876), pp. 170-216. Soley wrote his history in part on the basis of questionnaires he circulated to department chairmen in the late spring of 1876. He seems to have paraphrased or quoted directly the responses he received. Therefore I have assumed that the "author" speaking in the section on physics was the department chairman in 1876, Captain William T. Sampson, Michelson's superior officer. See *Minutes of the Academic Board,* vol. 5 (1874-1876), p. 130 (12 April 1876), USNA Archives, Nimitz Library, Annapolis.

40. For the final examination questions, see the *Annual Register for 1872-1873 [s.b. 1873-1874]*, pp. 78-79. Ganot, *Elementary Treatise*. 3rd. ed. Cadets at the academy were classified by class, the fourth class corresponding to the freshman year, the third to the sophomore year, and so on.

41. *Minutes of the Academic Board.* vol. 5 (1874-1876), pp. 270-271; on p. 271 (4 May 1871).

42. Ford, *History*, chap. 21, p. 16.

43. Quoted in ibid., p. 18.

44. *Minutes of the Academic Board*, vol. 4 (1867-1874), pp. 445-446 (24 & 29 January 1874); vol. 5 (1874-1876), pp. 20 (19 January 1875), 22-23 (4 January 1875), 50 (16 February 1875), 91-92 (20 September 1875), 95 (9 October 1875), 104 (12 November 1875), 131 (17 April 1876); vol. 6 (1877-1880), p. 88 (29 June 1878); *Annual Register for 1874-1875*, pp. 62-63; Ford, *[History]*, chap. 22, p. 7.

45. *Minutes of the Academic Board*, vol. 5 (1874-1876), pp. 52-53 (1 June 1875), 55 (12 June 1875), 134 (5 June 1876), 165-166 (24 June 1876); Soley, *Historical Sketch*, pp. 196-201. By the time Ganot became a reference book, a new edition had come out: Adolphe Ganot, *Elementary Treatise on Physics Experimental and Applied*, 6th ed., trans. Edmund Atkinson (N.Y.: William Wood & Co., 1875). Some of the remaining textbooks were: Fleeming Jenkin, *Electricity and Magnetism* (London: Longmans, Green & Co., 1873); Augustin Privat-Deschanel, *Elementary Treatise on Natural Philosophy* (N.Y.: D. Appleton & Co., 1873); and James Clerk Maxwell, *Theory of Heat*, 3rd ed. (London: Longmans, Green & Co., 1872).
     The first cadre of cadet-engineers entered the academy in 1866 and graduated in 1868. Their course of study was designed to be more practically oriented than that of the cadet-midshipmen, although they took most of the basic science courses that the cadet-midshipmen did. The next class of cadet-engineers did not enter until 1871; they graduated in 1873. Two other classes entered in 1872 and 1873; both had a two-year course of study. On 24 February 1874 the cadet-engineers were given a four-year course of study. (E. H. Hart, *United States Naval Academy* [N.Y.: E. H. Hart, 1887; unpaginated]). Hence, by the time Michelson returned to the academy as an instructor, his student clientele was divided and each part had different educational goals. It seems plausible to infer that the needs of these two different clienteles contributed to the reforms taking place in the physics department.

46. *Annual Register for 1874-1875*, p. 62; Clarke, *Report*, p. 86; Soley, *Historical Sketch*, pp. 199-201.

47. Sampson's reporting in Soley, *Historical Sketch*, pp. 197-199.

48. Ibid., p. 198. With his reply to Soley's questionnaire, Sampson sent eight student investigations in physics, which Soley included as an appendix to his book. Ibid., pp. 246-274.

49. *Annual Register for 1873-1874*, pp. 91-92, 102-104, 111; *1875-1876*, pp. 77-78, 87-88, 92-95; *1876-1877*, pp. 76-77, 86, 96-99.

50. Sampson's reporting in Soley, *Historical Sketch*, p. 197.

51. Ibid., p. 199.

52. Quoted in Clark, "Michelson Holds the Stop-Watch," p. 25.

53. Sampson's reporting in Soley, *Historical Sketch*, p. 198.

54. Quoted in Clark, "Michelson Holds the Stop-Watch," p. 25.

55. Ganot, *Elementary Treatise*, 6th ed., pp. 404-406; on p. 406.

56. Deschanel, *Elementary Treatise*, pp. 875-879; on p. 875.

57. Albert A. Michelson, "Experimental Determination of the Velocity of Light," *Proc. AAAS*, 1879, *28*: 124-160; on p. 124.

58. Albert A. Michelson, "On a Method of Measuring the Velocity of Light," *American Journal of Science*, 1878, *15*: 394-395.

59. *Report of the Board of Visitors* (20 June 1878), p. 3, USNA Archives, Nimitz Library, Annapolis.

60. Albert A. Michelson, "Experimental Determination of the Velocity of Light," *Proc. AAAS*, 1879, *28*: 124-160; on p. 124.

61. Quoted in Clark, "Michelson Holds the Stop-Watch," p. 25.

62. Albert A. Michelson, "Experimental Determination of the Velocity of Light," *Proc. AAAS*, 1879, *28*: 124-160; on p. 124. Michelson wrote to the American astronomer Simon Newcomb on 26 April 1878: "Unfortunately, as I was about to make an accurate observation the mirror flew out of its bearings and broke." Michelson's correspondence with Newcomb is reprinted in Nathan Reingold, ed., *Science in Nineteenth Century America: A Documentary History* (N.Y.: Hill and Wang, 1964), pp. 278-291.

63. Albert A. Michelson, "Experimental Determination of the Velocity of Light," *Proc. AAAS*, 1878, *27*: 71-77; on p. 76.

64. Michelson, "Method of Measuring."

65. Michelson, "Velocity of Light," *Nature*, 1878, *18*: 195.

66. Michelson, "Experimental Determination of the Velocity of Light," *Proc. AAAS*, 1879, *28*: 124-160; on p. 125.

67. Michelson, "Experimental Determination of the Velocity of Light," *Proc. AAAS*, 1878, *27*: 71-77; on pp. 72, 76, 77, 77 (emphasis added). Clark, "Michelson Holds the Stop-Watch," p. 152.

68. Quoted in Clark, "Michelson Holds the Stop-Watch," p. 152.

69. *Minutes of the Academic Board*, vol. 6 (1877-1880), pp. 19 (4 February 1878), 40 (12 February 1878).

70. Clarke, *Report*, pp. 200-215.

71. *Annual Register for 1878-1879*, p. 64.

72. Ibid., p. 63.

73. Ibid., p. 64. Mansfield Merriman, *Elements of the Method of Least Squares* (London, 1877).

74. *Minutes of the Academic Board*, vol. 6 (1877-1880), p. 18 (4 February 1878).

75. *Annual Register for 1878-1879*, pp. 105-106.

76. *Annual Register for 1879-1880*, pp. 63-64.

77. *Annual Register for 1878-1879*, p. 64; *1879-1880,* p. 99.

78. *Report of the Board of Visitors* (10 June 1879), p. 3.

79. Michelson to Newcomb, 18 December 1878, in Reingold, ed., *Science in America*, pp. 280-281.

80. Albert A. Michelson, "Experimental Determination of the Velocity of Light," *Am. J. Sci.*, 1879, *17*: 324-325.

81. Friedrich Wilhelm Bessel, *Untersuchungen über die Länge des einfachen Secunden-pendels* [Besonders abgedruckt aus den Abhandlungen der Akademie zu Berlin für 1826] (Berlin: Akademie der Wissenschaften, 1828). Although it appeared in the academy's publication for 1826, Bessel's completed investigation did not arrive at the academy until 5 January 1828.

82. Albert A. Michelson, "Experimental Determination of the Velocity of Light," *Proc. AAAS*, 1879, *28*: 124-160. The evolution of Michelson's investigative style in the treatment of data and errors that I have outlined here is also evident in a laboratory notebook containing entries from January to August 1879, a copy of which is on deposit at the USNA Special Collections, Nimitz Library, Annapolis.

83. Albert A. Michelson, "Experimental Determination of the Velocity of Light," *Am. J. Sci.*, 1879, *18*: 390-393.

84. Albert A. Michelson, "Experimental Determination of the Velocity of Light," *Nature*, 1879, *21*: 94-96, 120-122; Michelson, "Velocity of Light," *Nature*, 1880, *21*: 226.

85. Stanley M. Guralnick, "Sources of Misconception on the Role of Science in the Nineteenth-Century American College," *Isis*, 1974, *65*: 352-366; esp. pp. 361-362.

86. Michelson to Newcomb, 22 November 1880, in Reingold, ed., *Science in America*, p. 287.

87. Ibid., pp. 278-291.

88. Kevles, *The Physicists*, pp. 38-44; Kevles, "Flaws in American Physics," p. 151.

89. Quoted in Clark, "Michelson Holds the Stop-Watch," p. 152.

THE MICHELSON ERA AT CLARK, 1889-1892

William A. Koelsch

Albert A. Michelson was a man of many firsts. He was, of course, the first American scientist to be awarded the Nobel Prize. He was also the first Professor of Physics in three institutions of higher education: the Case School of Applied Science, Clark University, and the University of Chicago.[1] His career at Clark was the briefest of these institutional affiliations, but it is noteworthy for two reasons. First, Clark was the site for most of the laboratory research for two of Michelson's major contributions to physical measurement, the successful search for a natural constant to replace the standard meter bar, and the first application of techniques of interferometry to astronomical measurements.[2] Second, Clark was the site of a less successful attempt to create a research and training center in pure physics along the lines successfully institutionalized at Johns Hopkins by Henry A. Rowland, which would feed an emerging national community of research-minded physicists.

Clark University was chartered in March 1887 as a comprehensive university that, like Johns Hopkins and Cornell, would stress graduate study and research.[3] By early 1888 the founder, Jonas G. Clark, and his trustees had agreed in principle to limit the scope of graduate work. After the election of the psychologist G. Stanley Hall of Johns Hopkins as president in April 1888, and following advice proffered by Daniel Coit Gilman, Woolcott Gibbs, and Andrew D. White, among others, the Clark trustees elected to begin operations as a graduate school of science, with initial specialties in mathematics, physics, chemistry, biology and psychology. This strategy, Hall argued, would bring the new institution immediate recognition as well as meet an urgent national need hitherto filled adequately only by Johns Hopkins in this country or by foreign universities. As at Hopkins, the initial plan anticipated expansion into other research areas as additional funding became available.

In order to assess the state of the art in the various disciplines as well as to get evaluations of promising academic scientists, President Hall spent much of the last half of 1888 and the first quarter of 1889 on the road. He visited a number of American institutions in the summer of 1888, attended the Cleveland meeting of the American Association for the Advancement of Science in August, and embarked in September on a seven months' tour of European universities and other institutions, partly to recruit two or three leading German professors for short-term posts, partly to tour scientific facilities, and partly to get assessments of young Americans who had worked with distinguished European scientists.

Hall's laboratory at Hopkins had been housed in Rowland's building,[4] so he had a fair working knowledge of what might be required in physics; additionally, Hall had solicited a plan of organization for an ideal physics department from "Professor [Willard?] Gibbs." Hall regarded physics both as prospectively his most expensive department and as the most difficult to organize, especially since he hoped to excel Rowland's program. After an unsuccessful attempt in Germany to recruit Heinrich Hertz, Hall decided to defer nominating a physicist to head the new department until the end of his trip abroad. This would give an opportunity to consult scientists in England, the country to which most Americans of that time looked for guidance and publication outlets in physics.[5]

Hall had met Michelson in Baltimore during the Johns Hopkins lectures of William Thomson (later Lord Kelvin) in October 1884. Both Willard Gibbs and Rowland had recommended Michelson highly to Hall. (Probably Simon Newcomb, whom Hall is known to have consulted on other plans for Clark, did so as well.) Hall had talked with

Michelson in Cleveland during the AAAS meetings and had determined that he was anxious to leave Case. It is not unlikely that Hall was among the large crowd of hearers for Michelson's vice-presidential address as head of the physics section, at the Cleveland meetings, during which he outlined an agenda for research in his new field of physics.[6]

On his European tour Hall must also have heard praise for Michelson's work both in Germany and in England, particularly from his attempts to measure the velocity of light. Hall consulted two of Michelson's greatest admirers, Thomson and Hermann von Helmholtz, on staffing issues, and may have talked with a friendly critic, H. A. Lorentz of the University of Leiden, as well. By mid-April, 1889, Hall was back in America and met with the Clark trustees on the seventeenth. At that meeting the board formally approved opening the university in October with the five graduate departments previously planned. At its next meeting, on May 23, on Hall's recommendation the board unanimously elected Michelson (Acting) Professor of Physics for a five-year term at an annual salary of $3,500, beginning 1 September 1889.[7]

When in 1881 Michelson had been chosen Professor of Physics at Case (to begin on his return from Europe in 1882), he was promised a salary of $2,000 per year and the then-substantial sum of $7,500 for laboratory equipment. But as early as 1885 Michelson had been ready to leave Case to secure a position at a research-minded Eastern university.[8] The Case students were not of the caliber he demanded, and during his seven years there, only two of them concentrated on physics. There were conflicts with the Case administration over Michelson's high-handedness concerning budgets and laboratory purchases. He was already using some of the superior laboratory facilities of his collaborator Edward Morley at Western Reserve before a fire in October 1886 wiped out his own laboratory.[9] The Case trustees were slow to rebuild and slower to re-equip, so when the formal Clark offer came in late May, 1889, Michelson accepted it promptly. Morley wrote his father on 2 June that Michelson was going to Clark, adding that the Case authorities were not sorry to see Michelson go.[10]

Michelson came to Worcester in June to spend several weeks in organizing the new laboratory and requisitioning equipment for it, as well as in preparing a series of lectures that he had been invited to deliver before the Lowell Institute in Boston the following winter. Michelson had previously warned Hall that he had "sadly understated" his department's start-up costs, backing his estimates with a letter from Rowland, who also recommended one of his own students as Michelson's assistant.[11] By the end of the month, Michelson was already weary of academic administration, though he wrote Morley that he thought that "within a year I'll be moderately comfortable."[12] A large amount of equipment was ordered in July through the president's office, a system Michelson saw as an infringement on his prerogatives, and a source of problems to come.

Michelson was also busy approving student applications and making recommendations for staff. In late September, Hall reported to the trustees that up to that point forty-one "applications" (probably mostly inquiries) had been received for advanced study in physics. That was, however, the lowest number of those specifying any of Clark's five original departments, and indeed the number was below that of those inquirers interested in modern languages (seventy-two) and classics (fifty-seven), neither of which Clark offered.[13]

Although Michelson himself, like most American and British physicists of his day, is often described as "a man of no mathematical turn of mind" (in Daniel Kevles's phrase), early on he urged Hall to secure a second man in the department with competence in mathematical physics. Michelson's first suggestion was that Hall attempt to attract Benjamin Osgood Peirce of Harvard, a Leipzig Ph.D. His second choices were either Arthur L. Kimball of Johns Hopkins or Edwin Hall of Harvard, both Rowland Ph.D.s, either being able to handle both mathematical and experimental physics.[14]

That post was to be filled only in Michelson's second year, when Arthur Gordon Webster was appointed Docent in Mathematical Physics in September 1890. Webster had studied mathematical physics under Helmholtz and had completed an experimental dissertation under August Kundt at Berlin. He also had attended lectures in physics and mathematics at Stockholm and Paris. Webster was to pick up the pieces of Michelson's department at Clark after the latter's departure in 1892 and to carry on a graduate program virtually single-handedly until his suicide in 1923.[15]

The physics department at Clark in 1889-90 was the smallest of the five founding departments, and at thirty-six Michelson was the youngest department head. But Michelson's salary of $3,500 per year was higher than that of any other member of the faculty save Arthur Michael, head of chemistry before his sudden resignation in November 1889, who was paid at the same rate. First year expenditures in the department were nearly $26,000, exclusive of fellowship support.

The remainder of Michelson's department in 1889-90 consisted of his personal assistant, twenty-one-year-old Alfred Goldsborough Mayer, at $500 per year; two fellows at $200 each, twenty-two-year-old Frank Wadsworth and twenty-eight-year-old Alexander McAdie, a meteorologist; and nineteen-year-old Arthur Warner, appointed scholar (which meant tuition remission only). McAdie had a Harvard M.A. in physics and several years of research experience with the Signal Service; the others came directly from their undergraduate schools.[16]

In addition, the physics community at Clark in 1889-90 included a senior researcher, William F. Durand, Professor of Mechanical Engineering at the Michigan State Agricultural College. Durand proposed to work on the sympathetic vibrations of membranes as well as on optical problems, and made no bones about the fact that he hoped to be called to Clark as a member of Michelson's department. Illness, first his wife's and then his own, cut short Durand's research and he returned to Michigan State in December after only two months in residence.[17]

Efforts were also made in the fall of 1889 to accommodate the research needs of another senior scholar, DeWitt Brace, Professor of Physics at the University of Nebraska. Brace, frustrated by his lack of research time and laboratory resources at Nebraska, had hoped to come to Clark in January 1890 to lecture and do research on the group velocity of light waves in liquid media. But plans began to fall apart when Brace could neither specify a budget nor come east to discuss the problems with Michelson in advance, and Brace withdrew from what was probably a docent's appointment in late December or early January.[18]

At the end of the first year Mayer, McAdie and Warner all left Clark. Mayer, son of the well-known physicist Alfred M. Mayer of Stevens Institute, had been pushed to enter physics by his father and by his own testimony was just not very good at physics because he lacked any interest in it. Michelson discharged him as his assistant at the end of the year and, after a brief period at the University of Kansas as assistant in physics, Mayer left abruptly to pursue graduate study in biological science at Harvard, where he worked with Alexander Agassiz at the Museum of Comparative Zoology. Taking an Sc.D. at Harvard in 1897, Mayer later became a leading American marine biologist.[19] McAdie left for another post in the Signal Service; though he had been promoted to a senior fellowship (at $400 per year) as of 1 February, he lacked the financial resources to stay on.[20] Warner had been reappointed on condition that he take a summer course in experimental physics elsewhere in order to prepare himself for advanced work with Michelson in the fall, but financial constraints prevented that and he withdrew from the program in July.[21]

Except for Michelson and Wadsworth, now appointed assistant in Mayer's place (a move of critical importance both for Michelson's research and Wadsworth's career), the physics department at Clark in 1890-91 was an entirely different group of people.

Webster had come as docent in September, and four new scholars were appointed: Louis W. Austin, T. Proctor Hall, Gustav Ravené, and Allison W. Slocum. Austin had applied after a year of graduate research at Strassburg; his father wished him to come home and take his Ph.D. on this side of the Atlantic.[22] Hall was a Canadian with background in mathematics and chemistry and several years' teaching experience. He already possessed a Ph.D., earned by correspondence study, from Illinois Wesleyan University.[23] Ravené and Slocum, the latter a Haverford graduate recommended by Henry Crew, appear to have left during or after their first year, for reasons not clear from the surviving records. Slocum, indeed, was a fellow in physics at Harvard during 1891-92 and may not have worked with Michelson more than peripherally, if at all.

Jonas Clark fully understood the importance of research apparatus and gave it first priority in expenditures, while urging the faculty at the same time to limit its ordering of apparatus intended for demonstration purposes only. During the first year of the university's operation, President Hall had tried to secure the well-known optical in- strument-maker James A. Brashear as the university's instrument-maker. Although Brashear refused, he did make a number of instruments for Clark, including a 6-inch grating, the optical parts for Michelson's meter-bar experiments, an interferometer for Michelson's work and a multiple-quadrant electrometer designed by McAdie and ap- proved by Michelson for McAdie's work with atmospheric electricity.[24]

By May 1890 Michelson was tired of makeshift arrangements and shared time and insisted on a full-time "mechanician" for the exclusive use of his department.[25] Although financial clouds were beginning to appear on the Clark horizon, Hall granted the request and both a mechanic, Charles A. Francis, and a mechanic's helper were hired. The total costs of the department during its second year of operation, including the mechanic ($900) and helper ($175) were about $8,500. Expenses for the third year remained about the same, though no new students were admitted, and the department as of 1891-92 consisted only of Michelson, Webster, Wadsworth, and Hall, plus the mechanic and helper. The total expenditures for physics at Clark in the Michelson era (furniture, books, salaries, and equipment) amounted to just over $43,500.[26]

A separate laboratory building, opened for use on 6 November 1889, had been built solely for the use of the chemistry department, and foundations had been laid (though never built upon) for one in biology, reflecting the anticipated appeal of those depart- ments. In a letter to an inquirer, Hall described Clark facilities in chemistry as "unsur- passed," but those in physics only as "good." As at Case, Michelson had to be content at Clark with space in the multi-departmental main university building. His department was assigned three well-lighted rooms, each 21 by 40 feet, on the main floor: one for his private office and laboratory, another research room that also housed cases of apparatus, and a third room for a physics classroom. On the ground floor Michelson was allotted about 2,500 square feet of specialized laboratory space, divided into several rooms for environmentally sensitive research in electricity and magnetism, spectroscopy, optical measurements, and other work requiring special conditions of stability and the maintenance of constant temperatures. Adjacent to these was a large machine shop, and during the first year another 21-by-40-foot room was assigned to McAdie for his meteo- rological laboratory.[27]

In physics at the level Michelson envisioned, teaching and research were hardly to be separated; indeed, Hall described the Clark system as one of "elbow-teaching" in the laboratory, assigned readings and individual conferences. All faculty were required to lecture weekly, however, and the senior fellows and docents also gave lectures in the fields of their research interest. During his first year Michelson gave two courses of lectures. He began lecturing in the fall term once a week on wave motion and on the undulatory theory of light, He also gave a so-called "minor course" (for nonphysics students) in electricity and magnetism in the spring of 1890. McAdie gave a few lectures on meteorology, probably also in the spring term.[28]

In 1890-91 Michelson lectured through the year on the theory of light, giving one series on "Interference and Diffraction," including the mathematical theory of diffraction and its impact on astronomy, and another on the "Velocity of Light," including a history of attempts to measure it, the bearing of the experimental work on the undulatory theory of light, and the effects of the medium through which light travels. In his final year at Clark, Michelson lectured on optical theories, again reviewing past and current work on the problems of refraction and reflection. Clearly his lectures closely reflected his research interests of those years.

Michelson's lectures were supplemented during 1890-92 by a more broadly based course of lectures in mathematical physics given by Webster. During his first year Webster emphasized the theory of the potential function and its application to electrostatics the area of his recent dissertation research, and during the second he lectured on the dynamics of particles. In addition, most of the physics students appear to have attended lectures in mathematics, and some may have attended Morris Loeb's pioneering lectures in physical chemistry as well. Students from mathematics and chemistry appear to have attended the minor course of Michelson in 1889-90 and those of Webster in 1890-92.[29]

It was in the laboratory, however, that staff and students were expected to spend most of their waking hours. As the first official announcement put it, "Dr. Michelson will strive, by advice and example, to encourage a spirit of diligent investigation and original research, particularly in those intending to find their life work in this department" (of knowledge).[30] Research in physics was hampered during the first year, however, by frustrating delays in the preparation of the laboratory space and in delivery of equipment. Most physics apparatus was either ordered from the stock of, or in the case of European-made equipment ordered through, James Queen and Company of Philadelphia. As luck would have it, this firm proved to be the most inept and error-prone of the university's major suppliers. Equipment that Michelson had ordered in July for a mid-September delivery was still not in hand in December. Shipments arrived broken, or incomplete, or with the wrong items. Frequently the bills did not arrive with the item, and a university rule forbade the use of an item until it was paid for. Michelson's level of tolerance for centralized ordering procedures, never high, fell even more precipitously than it had at Case.

The harried clerk, Louis Wilson, wrote to Queen on 24 November inquiring about the platinum foil and wire (intended for Michelson's astrophysical measurements) ordered in July, asking when it might be expected, and noting plaintively that "the office is questioned upon these matters every day."[31] McAdie's research in atmospheric electricity was also hindered by the late arrival of specialized electrometers and by defective insulators; he had to delay the beginning of continuous observations of the electrical potential of the atmosphere until May 1890. Webster designed a new drop-chronograph and the university ordered it in January 1891; Hall noted in his annual report at the end of September that it had not yet arrived.[32] Even Brashear's equipment made to Michelson's specifications could go awry; Michelson wrote Morley in March 1890 that the new interferometer ordered in November from that firm had arrived in perfect shape, except the adjusting screws were at the wrong end of the instrument![33]

Michelson complained vociferously about his lack of progress the first year owing to the fact that the promised equipment and facilities had been "fearfully slow in materializing," but reported that "there is lots of time for research."[34] These normal first-year hazards were partially offset by the loan of equipment and books from government agencies and nearby institutions. Newcomb's phototachometer, a cumbersome and expensive piece of apparatus, had been borrowed for Brace from the Navy's Nautical

Almanac Office, and a modified Mascart electrometer and other equipment was secured for McAdie's use from the Signal Service.[35] Using the new Brashear interferometer and a 4-inch telescope on loan from Worcester Polytechnic Institute, in the spring of 1890 Michelson began the preliminary work toward the application of interference methods to astronomical problems. By covering the lens with an adjustable slotted cap and viewing two minute holes in a piece of platinum foil, Michelson found he could increase the "resolving power" of the instrument in the laboratory by fifty to one hundred times. In July 1890, thanks to the cooperation of Edward Pickering (who later nominated Michelson for the Nobel Prize), Michelson was able to arrange for tests of his new method on the telescope at the Harvard College Observatory, though atmospheric disturbances appear to have spoiled the observations there.[36]

In January 1891, Great Britain's Astronomer Royal visited Clark to study Michelson's new apparatus. Beginning in April, additional apparatus was constructed in the physics machine shop under Wadsworth's direction and shipped to the Lick Observatory on Mt. Hamilton, California. Michelson went out to California in July to test it on the Lick 12-inch equatorial telescope in order to measure the diameters of the four satellites of the planet Jupiter. The resulting observations over four nights in August varied by only 1 or 2 percent, compared with 10 to 20 percent variations by earlier observers. In October, Michelson had Warner and Swasey in Cleveland construct, at Clark's expense, equipment to be used with the Lick 36-inch telescope to continue such measurements for the satellites of Saturn and smaller heavenly bodies.[37] The ecstatic Lick astronomers received the new equipment on Christmas Day and began to make plans for even larger apparatus to measure the diameters of fixed stars.

The International Bureau of Weights and Measures, established in Paris in May 1875, had long been interested in Michelson's work, especially in the possibility of using his interferometer to establish a universal and more accurate accepted measure of length. The Bureau hoped to replace the standard platinum-iridium meter bar with a figure derived from the light wave emitted by the heated vapor of some naturally occurring element. The meter bar itself had originally been calculated from a natural phenomenon, the circumference of the earth's surface. Attempts to substitute for it a standard based on the vibrations of a pendulum of specific length had also proven theoretically and practically unsatisfactory. A standard based on a light wave posed the problem of relating the minute measurement involved to the length of the standard meter bar without distorting errors.[38]

Michelson's preliminary work in Cleveland with the interferometer built for the more famous ether-drift experiment had shown that this might be done. At the same time the results of the latter test were published, he and Morley had also published a paper discussing their tests of various possibilities for a natural standard, incidentally revealing the phenomena now known as "fine structure" and "hyperfine structure."[39] Although Morley had expected to continue this work with Michelson in the summer of 1890, in December 1889 Michelson wrote Morley that "I am sorry that I have to work alone at the wave lengths." When the surprised Morley made a special trip to Worcester to talk over the problems, he was rather brusquely told that Michelson would pursue these investigations without his help.[40]

By 1890 the resources were fully in place at Clark to make the next moves toward establishing a natural standard. The new Brashear interferometer (or "refractometer") and other equipment had arrived, and there was a well-equipped physics workshop and full-time technical help under Michelson's supervision. More importantly, in his new research assistant, Frank Wadsworth, Michelson had a man who had an extraordinary talent for translating Michelson's ideas into plans for and production of the necessary apparatus.

The problem was to be solved in two stages, First there was the laboratory phase, begun in Michelson's last years in Cleveland and continued after things shook down at Clark. This was largely the problem of determining with precision, using interference methods, the length of the light waves emitted by various substances under specified conditions. Sodium, lithium, hydrogen, and thallium had all been tested in Cleveland but ultimately found wanting. Other substances were tested at Clark beginning in the spring of 1890 and reported by Michelson at the National Academy of Sciences meeting in Washington in April 1891.[41]

For a while either the green or the yellow line of mercury was the tentative standard of choice. Most of the radiations considered exhibited too complex a spectral line for use in the projected second phase of the work, the comparison with the standard meter bar. Finally the choice settled on cadmium light, which, emitting in the red, yellow, and blue states, exhibited narrow, simple spectral lines, The most promising, Michelson later wrote, was "the red radiation from cadmium vapor made luminous by the electric discharge." His analysis of the so-called "visibility curves" had shown that the red cadmium line was extraordinarily monochromatic and thus highly suitable for such a measurement.[42]

Since the Bureau could finance only the cost of the apparatus, Hall recommended and the Clark trustees unanimously approved a leave of absence with full salary for Michelson for the period from 1 April 1892 until the new university term opened in early October. The only condition was that "his title of Professor at Clark University be used by him in any official action or communication made by him in connection with this work." Hall put this in writing for Michelson on 12 February, but the specified credit line was subsequently to be ignored, both by Michelson and by historians.

Brashear was first asked to construct the equipment but declined, except for the optical work. The new apparatus was constructed by the American Watch Tool Company of Waltham, and assembled and tested in the Clark machine shop under Wadsworth's supervision, though some additional apparatus may have been constructed by Taylor and Francis in London for the Paris comparisons.[43] After some delays, Michelson finally sailed for Paris in July, taking his family with him. When the shipment of apparatus was opened, much of the optical equipment was found broken, and there was an additional delay until French replacements could be made and Wadsworth could join Michelson in Paris in August to supervise the reconstruction of the instruments. The final measurements were made during the spring of 1893 and the successful outcome of this long effort announced: the length of the standard meter bar equalled 1,553,163.5 wavelengths of the red line of the cadmium wave, with a probable error of one in ten million. It was a truly remarkable achievement. Unfortunately for Clark, by the time Michelson announced it he was Professor of Physics (on leave) at the University of Chicago.[44]

During his year at Clark, McAdie had carried on research in atmospheric electricity, with particular interest in the use of changes in the electrical potential of the earth's atmosphere as a predictor of the approach of storms. He also submitted a major facilities plan to President Hall in April 1890, proposing the establishment at Clark of a major modern meteorological observatory, "on a par with those of the first rank in Europe." Such a facility would be equipped with modern, self-registering instruments and other equipment to the amount of $5,000 worth at the start, and become the center of a network of cooperating weather stations from Philadelphia to the St. Lawrence valley. Neither this nor McAdie's plan for research in "the question of nervous excitement, depression, hysteria, etc., so prevalent, especially among women, during thunder-storms" would ever materialize, however.[45]

During his two years as docent, Webster continued the research in electrical measurements he had begun in John Trowbridge's laboratory at Harvard and continued through his Berlin dissertation, work that was to earn him the Elihu Thomson prize in

1895 for his experimental verification of the period of electrical oscillations in a discharging condenser. In the Michelson years Webster developed methods of determining the ratio between the electromagnetic and electrostatic units of electricity, experimental work of a very delicate order that contributed to the theory of electricity. In order to accomplish this empirically, Webster designed a new drop chronograph that was more accurate than the pendulum-interruptor of the great Helmholtz himself. He also made new refinements of the electrometer and galvanometer, experimenting with the use of quartz fibers. His early research output was not large, however, for he spent much time in developing a two-year cycle of lectures in mathematical physics, which eventually bore fruit in three influential textbooks, recently described as playing "an extremely important role in advancing physics education in America."[46]

Most of the degree candidates worked closely within Michelson's own research framework. The first full catalogue, published in May 1890, had announced that in addition to the problem of defining the standard meter (on which Wadsworth worked principally, as assistant), special attention would be paid to the "application of interference methods to the measurements of length and angles," including its use to replace mirrors in galvanometers and similar instruments, and in measurements of coefficients of elasticity and expansion and of the index of refraction of solids, liquids and gases.

The two new graduate students of 1890–91, Louis Austin and T. Proctor Hall, worked on problems of this sort. Austin applied interference methods to the measurement of the angular displacement of swinging needles, attempting to produce a galvanometer especially sensitive to very small electric currents. Later in his first year Austin applied interference methods to the measurement of the displacement of suspended bodies, with a view toward more accurate determination of the pull of gravity. Hall, who had an advanced background in chemistry, developed a new method for investigating the surface tensions of liquids by determining the values of angles of contact of various liquids with platinum and glass, claiming results ten times as accurate as those reached with earlier methods. The theory and experimental results were later written up in formal fashion to earn Hall the first Clark Ph.D. in physics.[47]

During his three years at Clark, Michelson took a surprisingly active role in university governance, becoming a key figure in the internal disputes concerning President Hall's administrative practices that led ultimately to Michelson's resignation and those of all but two men of faculty rank in the spring of 1892. He was an active member of the University Senate, composed of the president and the three other full professors (Michelson, W. E. Story in mathematics, and Charles O. Whitman in biology), which met eleven times during Michelson's tenure, largely to award fellowships and approve degree candidates. At the first meeting of the general faculty, in February 1890, it was Michelson who moved adoption of the policy that "the first consideration in making new appointments, re-appointments or promotions, should be devotion to and success in research and in stimulating the same in others."[48]

After June 1890, however, Hall never called a faculty meeting until he was forced to do so in February 1892. Despite the formal mechanisms, Hall's personal insecurities led him to try to keep all power in his own hands. He personally appointed Story, a Hopkins colleague and Hall loyalist, to be Secretary of the Faculty; he appointed a library committee that never met, so Hall personally approved all expenditures for books and did most of the ordering; he insisted on approving every piece of apparatus ordered by department heads. The docents were made directly responsible to the president, not to the heads of their respective departments. Hall also tried to use his financial powers to control the assistants and fellows.[49]

Most of Michelson's frustrations, and those of other department heads, arose out of these circumstances. Michelson objected to his lack of authority to expend his budgeted allocation without further item approval, and to having to detour all orders through the president's office. In making the same point about the budgetary authority of department heads to President Harper of Chicago later, he called such a circumstance "anomalous and embarrassing." This institutional background is probably the source of the famous anecdote, often told by Michelson while at Chicago, that he had told Hall "if he wanted to keep a first-rate physicist like himself at Clark University he would have to treat him like a first-rate physicist."[50]

A series of minor disputes over Hall's practices and statements had arisen in various departments in the first two years, and there were gradually increasing signals that the university's financial prospects were not as rosy as Hall had led his faculty to believe while recruiting them. In the spring of 1891, while making arrangements with E. S. Holden to carry out his astronomical measurements, Michelson asked Holden to submit his name to Stanford University for consideration as their first Professor of Physics. In the light of Stanford's later financial troubles, it may have been fortunate for Michelson that nothing came of it.[51]

In the fall of 1891, however, Hall's deprivation of part of the $20-per-month stipend of a fellow in biology who had returned to the campus late because of the illness of his mother triggered a powder train that blew up in Hall's face about three months later. In January several members of the faculty began to meet and to compare notes across departments; discerning a pattern of behavior where only isolated incidents had previously been seen, they concluded with some justice that Hall was not to be trusted. In order to draw the attention of the trustees to the internal problems caused by Hall's inadequacies as an academic leader, on 21 January 1892 Michelson, six other faculty members and two docents (one of them Franz Boas) submitted their resignations, citing their "lack of confidence in the President of Clark University."

This set off what Hall's biographer, Dorothy Ross, has called "a bizarre series of encounters" among Hall, faculty members, and trustees, during which the faculty tried to establish some ground rules in the face of what they perceived to be Hall's arbitrary and evasive administrative conduct. Michelson took a leading role as spokesman for the faculty in ultimately unsuccessful negotiations with the president and members of the board. Indeed, one list of faculty demands, presented to Hall on 4 February, seems in large measure a listing of Michelson's own grievances; they largely concern the authority of department heads in matters relating to ordering equipment, hiring assistants, and governing fellows and assistants in their own departments.[52]

For some months Charles Otis Whitman, head of biology, had been negotiating privately with President William Rainey Harper of the new University of Chicago, but Harper and he had come to no agreement concerning Whitman's demands for a new biology department there. As late as December 1891, Harper had been unable to attract "head professors" for any of the departments of his planned university, which was supposed to be opened in the fall of 1892, largely stressing humanistic subjects. When Whitman and others saw that no reconciliation with Hall was possible, Whitman accepted Harper's final offer and notified him of the opportunity to secure a number of disaffected Clark scientists. At a meeting held in Whitman's home in late April, Harper spoke with Michelson and others, ultimately signing up from among the Clark scientific community several mathematicians, nearly all of Chicago's first biology staff, and department heads in biology, chemistry and physics, the latter Michelson himself.[53]

Hall's explanation to the local press was that faculty were leaving for financial reasons: Jonas Clark's diminishing contributions combined with Harper's generous offers of departmental budgets and salaries ($7,000 per year for department heads) were too much, in Hall's view, for his ambitious faculty to resist. Thanks to Hall's later repetition of this theme, the financial interpretation of Clark's troubles in 1891-92 has become somewhat standard, but it was both a disingenuous and an inaccurate appraisal. Michelson and Whitman answered it directly in a public letter printed in the *Worcester Evening*

Gazette on 6 May 1892. In it they stated that, in justice to Jonas Clark as well as to those leaving, the financial aspects were not the reasons for the resignations; indeed, several people had resigned without being sure of jobs elsewhere. Nor was there anything but praise among the faculty for the high ideals and the research commitment that had characterized the institution from the beginning. The major reason was left deliberately unstated in the letter, though some newspapers began to get it by leak: President Hall's administrative mismanagement and untrustworthiness. Or, as the phrasing of the January letter of resignation had it, "lack of [faculty] confidence in the President of Clark University."[54]

Although the brouhaha at Clark must have been especially taxing to Michelson because of his need to finish up the preliminary research for the meter bar experiment (and there is some evidence that the timing for this had slipped badly), he could at least look forward to a number of months of pure research in Paris, far from the internal problems of Clark. When the formal invitation from the President and Secretary of the International Committee had arrived in late January or early February, Michelson's first resignation was still on the table. But it seems to have been assumed at that point by all parties that the internal difficulties might yet be solved, and the final confirmation of his leave assumes that Michelson would return to Clark by October. As it turned out, Michelson was notified of his election as Professor of Physics at Chicago in late April and accepted it by letter to Harper on 2 May, though he did not formally resign at Clark until June. Clark University continued to pay Michelson's salary through September 1892, and the University of Chicago continued his leave until the Paris experiments were completed, late the following spring.[55]

Michelson's resignation appears to have derailed the career plans of several Clark Ph.D. candidates in physics. It appears from sketchy surviving records that the doctoral candidacies of four of Michelson's eight Clark pre-doctoral students were still alive in 1891-92. Austin, having been reappointed Fellow for that year, returned to Germany. He wrote Wilson in July 1892 that he had heard Michelson would be "on this side" for the coming year, requested his address, and indicated Austin himself would finish his research at Strassburg "about Christmas" and hoped to return to Clark. In the end Austin took his Ph.D. at Strassburg, in 1893. McAdie had done further research for his dissertation at the Blue Hill Meteorological Observatory in Milton, Massachusetts, during the summers of 1890 and 1891. He wrote President Hall in June 1892 that his draft had been approved by Michelson and others, and in September Hall wrote him to send it on for Webster's approval. But the next month McAdie withdrew the dissertation and dropped out, either because of the high cost of publishing the plates or for some other, unknown reason.[56]

Wadsworth too appears to have regarded himself as a Ph.D. candidate in good standing, writing President Hall in July 1892 of his hopes to get the degree "next year." Hall and Michelson had agreed the year before to put up $500 in personal funds for Wadsworth to make a trip to Europe to study and report on laboratories and methods in physics, and Clark held that sum for him for some years after 1892. But the trip to Europe was never made because of Wadsworth's many other commitments, first at the Smithsonian and later at the University of Chicago. His later involvements probably also account for his failure to complete a Ph.D. Of the eight physics graduate students of the Michelson era at Clark, all of them originally candidates for the Ph.D. degree (the only one offered at Clark at the time), only T. Proctor Hall ever completed the doctorate there. His topic had been suggested and his early research directed by Michelson, but the dissertation was actually completed under Webster in 1893.[57]

Of the Michelson era graduate students at Clark whose subsequent careers can be traced, only Austin and Wadsworth secured physics posts at research universities, and both left academic life after a few years. Austin found a satisfying research career in the Bureau of Standards and McAdie pursued one less successfully in the Weather Bureau and later as Director of the underfinanced Blue Hill Meteorological Observatory. Slocum taught physics for forty years at the University of Vermont, but appears to have made no significant research contributions. The rest drifted off in other directions: Mayer to a distinguished career as a marine biologist; Hall as a physician and medical editor and teacher; Wadsworth and Warner into private business or governmental ventures requiring skill in engineering rather than in pure physics.

Of the ten men making up the Clark physics community in the Michelson era who were still living in 1906, seven are listed in the first edition of *American Men of Science*. Five of these (Michelson himself, Webster, Austin, Mayer, and Wadsworth) were starred; that is, ranked among the first 1,000 American scientists, although Mayer was starred in biology and Wadsworth in astronomy and astrophysics. Michelson clearly had attracted talented scientists around him, even though he had been unable to sustain a supply of research-minded Ph.D.'s in physics.

But the Michelson era at Clark is marked by significant contributions to physical measurement by Michelson and his associates with the resources Clark University had put at their disposal, particularly the laboratory phases of the meter bar experiments and the beginnings of the adaptation of interference methods to astronomical measurements. The road to Paris had begun in Cleveland, but its longest section ran through Worcester. It was that same road which, by 1907, had extended beyond Paris to Stockholm and terminated at that internationally visible milestone for American physics, Michelson's Nobel Prize.[58]

Table 1. The Physics Community at Clark, 1889-1892

NAMES/DATES	BACKGROUND	CLARK STATUS	CAREER
Austin, Louis Winslow (30 Oct. 1867-27 June 1932)	A.B. Middlebury, 1889; U. of Strasbourg, 1889-90	Scholar, Nov.-April 1890-91; Fellow, May-June 1891	U. of Strassburg 1891-93 (Ph.D., 1893); Physics, Wisconsin, 1893-1901; Physikalisch-Technische Reichsanstalt, 1902-04; Bureau of Standards, 1904-32
Brace, DeWitt Bristol (5 Jan. 1859-2 Oct. 1905)	A.B. Boston, 1881; A.M., 1882; M.I.T., 1879-81;Johns Hopkins, 1881-83; Berlin, 1883-85 (Ph.D., 1885); Physics, Michigan, 1886; Nebraska, 1887-	Docent-designate, 1890 (withdrew)	Physics and Astonomy, Nebraska, to 1905
Durand, William Frederick (5 March 1859-9 Aug. 1958)	Naval Academy, 1880; Ph.D. Lafayette 1881; Michigan State 1887-	Honorary Scholar, Nov.-Dec. 1889	Mechanical Engineering, Michigan State to 1891; Cornell, 1891-1904; Stanford, 1904-24
Hall, T. [Thomas] Proctor (7 Oct. 1858-25 March 1931)	A.B. Toronto, 1882; Fellow, Chemistry, 1882-84; A.M., Ph.D. Chemistry, Illinois Wesleyan, 1888; Science Master, Woodstock College (Ont.), 1885-90	Scholar, 1890-91; Fellow, 1891-93; Ph.D. 1893	Natural Sci., Tabor Coll., 1893-96; Physics, Kansas City U., 1897-1901; M.D. Nat. Medical U., 1900; Prof. Nat Med. U. & Editor, American X- Ray; Physician, Vancouver to 1931
McAdie, Alexander George (4 Aug. 1861-1 Nov. 1943)	A.B. Coll. City of N.Y., 1881; A.M., 1884; A.M. Harvard, 1885; Signal Service, 1882-88; Tutor, 1888-89	Fellow, 1889-90	Signal Service, 1890-91; Weather Bureau, 1891-1913; Blue Hill Met. Obs., 1913-1931
Mayer, Alfred Goldsborough (16 April 1868-24 June 1922)	M.E. Stevens, 1889	Assistant, 1889-90	Assistant, Kansas, 1890-92; Zoology, Harvard, 1892-95 (Sc.D., 1897); later career in biology
Michelson, Albert Abraham (19 Dec. 1852-9 May 1931)	Naval Academy, 1873; Instr. 1875-79; research, Berlin, Paris etc., 1880-82; Physics, Case, 1882-89; Ph.D. (hon.), Western Reserve, 1886; Stevens, 1887	Professor of Physics, 1889-92	Physics, Chicago, 1892-1929; Nobel laureate, 1907; Phys. D. (hon.) Clark, 1909
Ravené, Louis Gustav (-1900[?])	Columbia University (Senior, 1884-85)	Scholar, 1890-91 (Scholar-designate?)	?

Table 1 (continued)

NAMES/DATES	BACKGROUND	CLARK STATUS	CAREER
Slocum, Allison Wing (22 April 1866-15 Dec. 1933)	A.B. Haverford, 1888; A.M. (Math), 1889	Scholar, 1890-91 (Scholar-designate)	Physics, Harvard, 1890-93; A.M., 1891; Berlin, 1891-92(?); Math, West Chester, 1893-94; Physics, Vermont, 1894-1933
Wadsworth, Frank Lawton Olcutt (24 Dec. 1866-11 April 1936)	E.M. Ohio State, 1888; B.S., M.E., 1889; Assistant in Physics, 1880-89	Fellow, 1889-90; Assistant and Fellow, 1890-92	Simthsonian Astrophys. Obs., 1892-94; Physics and Astronomy, Chicago, 1894-98; Allegheny Obs., 1900-1904; various businesses, 1896-1936
Warner, Arthur Judson (22 Jan. 1870-24 March 1953)	A.B. Marietta, 1889; P.B.K., Valedictorian	Scholar, 1889-90; Fellow-designate, 1890-91; withdrew, July 1890	Johns Hopkins, 1890-92 (Cert.-E.E., 1892); A.M. Marietta, 1897; adv. engr. studies, Stanford, 1897-98; business, 1898-1906; City Engineer's off., Seattle, 1906-37
Webster, Arthur Gordon (28 Nov. 1863-15 May 1923)	A.B. Harvard s.c.l.,1885; Inst. Math, 1885-86; Berlin, Paris, Stockholm, 1886-90; Ph.D. Berlin, 1890	Docent, 1890-92	Physics, Clark, 1892-1923; Elihu Thomson Prize, Paris, 1895

NOTES

1. For biography, see Bernard Jaffe, *Michelson and the Speed of Light* (New York: Anchor Books, 1960); Dorothy Michelson Livingston, *The Master of Light: A Biography of Albert A. Michelson* (New York: Charles Scribner's Sons, 1973); Loyd S. Swenson, Jr., "Michelson, Albert Abraham," *Dictionary of Scientific Biography* (hereafter *DSB*), (New York: Charles Scribners Sons, 1974) Vol. 9, pp. 371–74 and references; and Albert E. Moyer, "Michelson in 1887," *Physics Today*, 1987, *40*, 5: 50–56 and references.

2. See Loyd S. Swenson, Jr., "Michelson and Measurement," *Phys. Today*, 1987, *40*, 5: 24–30. For the more famous work at Cleveland, see R. S. Shankland, "Michelson-Morley Experiment," *American Journal of Physics*, 1964, *32*: 16–35; and Swenson, *The Ethereal Aether: A History of the Michelson-Morley-Miller Aether-Drift Experiments, 1880–1930* (Austin: Univ. Texas Press, 1972), esp. chap. 4, notes, appendices, and bibliography. Jean M. Bennett et al., "Albert A. Michelson, Dean of American Optics," *Applied Optics*, 1973, *12*: 2253–79, is a clear analysis of Michelson's research contributions in scientific context.

3. For the institutional context, see William A. Koelsch, *Clark University, 1887–1987: A Narrative History* (Worcester: Clark Univ. Press, 1987), chap. 1.

4. G. Stanley Hall, *Life and Confessions of a Psychologist* (New York: D. Appleton and Company, 1923), p. 237. Hall's autobiography is self-serving and often factually unreliable; unfortunately it has often been used uncritically as a source for the interpretation of facts and events at Clark in the early years, as in Livingston, *The Master of Light*. Dorothy Ross, *G. Stanley Hall: The Psychologist as Prophet* (Chicago: Univ. Chicago Press, 1972) provides a balanced view of Hall's strengths and weaknesses. The Hopkins physics community of the 1880s is briefly described in Hugh Hawkins, *Pioneer: A History of the Johns Hopkins University, 1874–1889* (Ithaca: Cornell University Press, 1960), pp. 138–140.

5. G. Stanley Hall to Jonas G. Clark, 14 November, 22 November, 12 December 1888 and 18 February 1889, G. Stanley Hall Papers, Clark University Archives (hereafter CUA), reprinted in *Letters of G. Stanley Hall to Jonas Gilman Clark*, ed. N. Orwin Rush (Worcester: Clark Univ. Library, 1948): pp. 23, 25, 27, 32. See also Russell McCormmach, "Hertz, Heinrich," *DSB*, 6 (1972), p. 343. "Professor Gibbs'" plan for an ideal physics department does not survive either in the Willard Gibbs papers at Yale or in CUA.

6. Livingston, *The Master of Light*, p. 141; Hall, statement for Amy Tanner's manuscript history of Clark, p. 14, Hall Papers; Michelson, "A Plea for Light Waves," *Proceedings of the AAAS: Thirty-seventh Meeting Held at Cleveland, August, 1888* (Salem, Mass., 1889), pp. 67–78.

7. Hall to Clark, 10 September 1888, Hall Papers (also *Letters*, pp. 14–15); Hall, *Life and Confessions*, pp., 272, 277–78; Hall, statement for Amy Tanner, p. 13; *Clark University Minutes of the Board of Trustees*, Vol. 1, *1887–1901*, CUA, pp. 13–14, 16.

8. Michelson to Rowland, 6 November 1885, Rowland Papers, Johns Hopkins University; reprinted in Nathan Reingold, ed., *Science in Nineteenth Century America: A Documentary History* (London: Macmillan, 1966), pp. 311–12; Shankland, "Albert A. Michelson at Case," *A.J.P.*, 1949, *17*: 487–90.

9. Michelson to Alfred M. Mayer, April 23, 1886, Hyatt and Mayer Collection, Princeton University; C. H. Cramer, *Case Western Reserve: A History of the University, 1826-1976* (Boston: Little, Brown, 1976), pp. 62-66, 210-17. In an interview the day after Michelson's death, Eckstein Case, retired treasurer of the Case School, commented on his own difficulties with Michelson's handling of his accounts; *Cleveland Plain Dealer*, 10 May 1931, quoted in Cramer, p. 212, and Livingston, *The Master of Light*, p. 121.

10. Edward W. Morley to Sardis B. Morley, 2 June 1889, Morley Papers, Library of Congress; reprinted in Reingold, ed., *Science in Nineteenth Century America*, p. 313.

11. *Worcester Evening Gazette*, 11 July 1889; Michelson to Hall, 13 June 1889, Hall Papers. Rowland's nominee was offered the job but declined in favor of a business career. Michelson wanted an assistant "who will do just what he is told -- and who is not too ambitious;" Michelson to Rowland, 5 September 1889, Rowland Papers. The Lowell Lectures planned for 1890 were evidently not given that year; Michelson's Lowell Lectures of 1899 were published as *Light Waves and Their Uses* (Chicago: Univ. Chicago Press, 1903).

12. Michelson to Morley, 29 June 1889, Morley Papers; reprinted in Reingold, ed., *Science in Nineteenth Century America*, p. 314.

13. Hall's numbers in *Minutes of the Board of Trustees*, Vol. I, p. 37. Candidates in mathematical physics listed in Michelson to Hall, 20 August 1889, Hall Papers.

14. Kevles comment in Kevles, "The Study of Physics in America, 1865-1916" (Ph.D. diss., Princeton Univ., 1964), p. 136. Other commentators agree: see Jaffe, *Michelson and the Speed of Light*, p. 102. Michelson treated mathematics rather cavalierly: see Livingston, *The Master of Light*, index entries under "mathematics, AAM's use of." He once wrote to his friend Alfred M. Mayer that he made "no pretense of being astronomer or mathematician": Michelson to Mayer, 26 June 1880, Hiatt-Mayer Collection, Princeton Univ.; reprinted in Reingold, ed., *Science in Nineteenth Century America*, p. 286.

15. For Webster, see A. Wilmer Duff, "Arthur Gordon Webster: Physicist, Mathematician, Linguist, and Orator," *American Physics Teacher*, 1938, *6*: 181-94; Melba Phillips, "Arthur Gordon Webster, Founder of the APS," *Phys. Today*, 1987, *40*, 6: 48-52; Albert E. Moyer, "Webster, Arthur Gordon," *DSB*, Supplement II (New York: Charles Scribner's Sons, in press) and references; also Webster to Hall, 19 August 1890 and other materials in Webster files, Hall Papers. Webster's father was a businessman and Republican political leader who had earlier recommended his son for a Clark position to G. F. Hoar, U.S. Senator and Clark trustee: William Webster to Hoar, 29 January 1889, George Frisbie Hoar Papers, Massachusetts Historical Society

16. Notebooks, "Staff and Annual Appointments, Clark University, 1889," Jonas G. Clark Papers, CUA, and "Clark University: Consolidated Salary Records, 1889-1896," Louis N. Wilson Papers, CUA. Biographical data on students in Table I is compiled from these and other Clark records and from archives of the other institutions they attended.

17. Durand to Michelson, 15 July 1889 and other correspondence in William F. Durand file, Hall Papers. For biographical material, see Durand, *Adventures* (New York: McGraw-Hill, 1953) and Frederick E. Terman, "William Frederick Durand," National Academy of Sciences, *Biographical Memoirs*, 1976, *48*: 153-93.

18. Louis N. Wilson to Brace, 7, 23 November, 9, 16 December 1889; Hall to Brace, 17 December 1889, all in *Letterbook, 1889-90*, Hall Papers; Brace to Board of Regents, University of Nebraska, 17 December 1889, Records of the Board of Regents, University of Nebraska Archives; Alexander G. McAdie to Hall, 13 January 1890, McAdie file, Hall Papers; Michelson to Lord Rayleigh, 25 December 1889, Rayleigh Collection, Research Library of the Air Force Geophysics Laboratory, Hanscom Air Force Base, Mass. For Brace, see Eugene Frankel, "Brace, DeWitt Bristol," *DSB* 2 (1970): pp. 382-83, and Edward L. Nichols, "The Scientific Work of DeWitt Bristol Brace," *Physical Review*, 1907, *24*: 515-21.

19. Alfred Goldsborough Mayer, "Autobiographical Notes, January, 1917," ms., Archives, National Academy of Sciences, esp. pp. 8-12; see also Charles B. Davenport, "Alfred Goldsborough Mayor," National Academy of Sciences, *Memoirs*, 1926, *21*(8): 1-14. Mayer's name was legally changed to Mayor in August 1918.

20. William A. Koelsch, "Ben Franklin's Heir: Alexander McAdie and the Experimental Analysis and Forecasting of New England Storms, 1884-1892," *New England Quarterly*, 1986, *59*: 523-43 and references cited; Clark University, *Journal, May 1887-June 1891*, p. 46.

21. Warner to Wilson, 15 July 1890, Warner file, Hall Papers. Warner subsequently entered the electrical engineering program at Hopkins, at that time a subsection of Rowland's physics department.

22. Austin to G. Stanley Hall, 11 September, 10 October 1890; Hall to Austin, 25 September 1890, Austin file, Hall Papers. See also Charles Susskind, "Austin, Louis Winslow," *DSB* 1 (1970): p. 338 and standard references.

23. For T. P. Hall, see *Tabor College Monthly*, April 1894, *5*, 2: 64-66; also *Who Was Who in America* 1, p. 507 and scattered Clark sources; no T. P. Hall file survives in CUA.

24. Hall, statement for Amy Tanner manuscript, p. 20; Hall to Brashear, 16 October 1889, Wilson to Brashear, 8, 24 October, 11, 14 November, 4, 7, 12 December 1889, all in *Letterbook, 1889-90*: also accounts with Brashear in Clark University, *Journal, May 1887-June 1891*, pp. 59, 63 (apparatus and supplies) and *Journal, July 1891-August 1896*, pp. 31, 35 (gratings), CUA.

25. Michelson to Hall, 1 May 1890, Hall Papers.

26. Data on financing from "Consolidated Salary Records, 1889-1896," which contains summaries of annual expenditures by department and periodically cumulates them.

27. Hall to John H. Gray, Jr., 16 October 1889; Wilson to L. J. Stabler, 6 November 1889, *Letterbook, 1889-90*. Space allocations from Clark University, *Register and Second Official Announcement*, May 1890. Photos of interior spaces, taken for the World's Columbian Exposition in Chicago in 1893, are in CUA

28. Hall, *First Annual Report of the President*, 4 October 1890, pp. 14-15. Course offerings from Clark University *Registers*, 1890-93 and Arthur Gordon Webster, "Physics," in [William E. Story and Louis N. Wilson, comps.] *Clark University, 1889-1899: Decennial Volume* (Worcester: Clark Univ., 1899), pp. 91-92. The McAdie file, Hall papers, contains what appears to be a synopsis of McAdie's meteorology lectures.

29. Loeb was appointed docent in physical chemistry in 1889 and remained at Clark two years; T. P. Hall acknowledged his help in 1890-91. Biographical material and extracts from Loeb's Clark lectures in T. W. Richards, ed., *The Scientific Work of Morris Loeb* (Cambridge: Harvard Univ. Press, 1913), pp. xv-xxiii, 3-20. Mathematics attendees in Michelson's physics lectures noted in *Worcester Daily Spy*, 18 January 1890; for chemistry students in physics courses, see Hall to W. S. Myers, 13 November 1889, *Letterbook, 1889-90*.

30. Clark University, *First Official Announcement*, 23 May 1889, p. 17; reprinted in *Science*, 14 June 1889, *13*, 332:464.

31. See numerous letters from Wilson to Queen and other suppliers in *Letterbook, 1889-90*; quotation is from Wilson to Queen and Co., 29 November 1889.

32. Koelsch, "Ben Franklin's Heir," p. 533 and references; *Second Annual Report of the President*, 29 September 1891, p. 25

33. Michelson to Morley, 3 December 1889 and 6 March 1890, Morley Papers; also notes on tests of Brashear interferometer dated May 1890, Michelson notebooks, Mount Wilson Observatory Archives. The Brashear interferometer survives at Clark.

34. Michelson to Morley, 25 November, 3 December 1889, 6 March 1890, Morley Papers.

35. Koelsch, "Ben Franklin's Heir," p. 533; Hall to Secretary of the Navy, 23 November 1889, Hall to Simon Newcomb, 3 December 1889, Wilson to Brace, 7 November 1889, all in *Letterbook, 1889-90*.

36. Michelson, "Physics," in *First Annual Report of the President*, 4 October 1890, pp. 31-33; Michelson to Morley, 6 March 1890, Morley Papers; Michelson to J. Willard Gibbs, 24 April 1890, Gibbs Papers; E. S. Holden, article in *Christian Union*, 24 April 1890, quoted in *Worcester Evening Gazette* 28 April 1890. See also Michelson to E. C. Pickering, 6 July, 14 July, 12 August, 15 September 1890, Letters from Eminent American Astronomers files; Pickering to Michelson, 9 July 1890, *Letterbook B-4;* 9 August, 20 September, 9 October 1890, *Letterbook A-9*, all in Records of the Harvard College Observatory, Harvard University Archives. Also E. A. Robson, *Report of a Visit to American Educational Institutions* (London and Manchester: Sharrett and Hughes, 1905), p. 132, and David H. DeVorkin, "Michelson and the Problem of Stellar Diameters," *Journal of the History of Astronomy*, 1975, *6*: 1-5.

37. Michelson, "Physics," *Second Annual Report of the President*, 29 September 1891; *New York Tribune*, 26 January 1891; *Boston Transcript* 28 January 1891; unidentified and undated [Aug.-Sept. 1891] newspaper clippings, "An Astronomical Feat" and "Jupiter's Four Moons," Scrapbooks, CUA; correspondence between Michelson and Holden, 1891, Mary Lea Shane Archives of the Lick Observatory, University of California, Santa Cruz.

38. See Robert Shankland, "Michelson and His Interferometer," *Phys. Today*, April 1974, *27*(4): 37-43 and Swensen, "Michelson and Measurement," *passim*.

39. Albert A. Michelson and Edward W. Morley, "On a Method of Making the Wave Length of Sodium Light the Actual and Practical Standard of Length," *American Journal of Science*, 3d ser., 1887, *34*: 427-30; idem, "On the Feasibility of Establishing a Light Wave as the Ultimate Standard of Length." *A. J. Sci.*, 3d ser., 1889, *38*: 181-86; see also

Shankland, "Michelson and His Interferometer," p. 41.

40. Michelson to Morley, 3 December 1889, Morley Papers; Jaffe, *Michelson and the Speed of Light*, pp. 117-18; Livingston, *The Master of Light*, pp. 145-48.

41. Clark University, *Register and Third Official Announcement*, April 1891, pp., 31-32; Michelson, "Physics," in *Second Annual Report of the President*, 29 September 1891, pp. 22-23; Michelson to Morley, 1 May 1891, Morley Papers.

42. Michelson, "Autobiographical sketch, 1907," prepared for the Nobel Prize Committee; printed in Nobel Foundation, *Le Prix Nobel en 1907* (Stockholm: P. A. Norstedt and Sons, 1909), pp. 63-67, and reprinted in *The Albert A. Michelson Nobel Prize and Lecture* (China Lake, Calif.: Michelson Museum, 1966), pp. 21-24. See also Robert A. Millikan, "Albert A. Michelson, 1842-1931," National Academy of Sciences, *Biographical Memoirs*, 1938, *19*: 136.

43. Hall to Board of Trustees, 6 November 1891; draft telegram, Michelson to Benjamin A. Gould, 14 December 1891; Hall to Michelson, 12 February 1892; American Express receipt for "Drawings" consigned to Taylor and Francis, London, 5 July 1892, all in Michelson file, Hall Papers. See also Board Determination of Trustees, *Minutes*, vol. 1, pp. 68-69, and Michelson, *Détermination Expérimentale de la Valeur du Mètre en Longueurs d'Ondes Lumineuses* (Paris: Gauthier-Villars et Fils, 1894): pp. 3-6.

44. Wadsworth to Hall, 3 August 1892, w/encl. from G. Browne Goode to Wadsworth, 30 July 1892 (copy); 17 July, 30 August 1892; Hall to Wadsworth, 22 July 1893; Wadsworth to Wilson, 6 November 1892, 14 August 1893, all in Wadsworth file, Hall Papers; Jaffe, *Michelson and the Speed of Light*, pp. 118-20; Livingston, *The Master of Light*, pp. 171-78. A loan from Clark University paid for Wadsworth's Paris trip, subsequently to be reimbursed by the Smithsonian

45. See Koelsch, "Ben Franklin's Heir," pp. 533-34; Clark University, *Register and Second Official Announcement*, May 1890, p. 35.

46. *Second Annual Report of the President*, 29 September 1891, pp. 24-25; *Third Annual Report*, April 1893, pp. 70-71; Clark University, *Register and Third Official Announcement*, April 1891, pp. 32-33; Phillips, "Arthur Gordon Webster...," p. 49.

47. Clark University, *Register and Second Official Announcement*, May 1890, p. 35; *Second Annual Report of the President*, 29 September 1891, pp. 25-26; T. Proctor Hall, "New Methods of Measuring the Surface-Tension of Liquids," *Philosophical Magazine*, 1893, *36*: 385-413.

48. Clark University Faculty Records, *University Senate, Minutes, 25 November 1889-7 September 1891*; Clark University Faculty Records, *General Faculty, Minutes, 21 February 1890-2 February 1892*, CUA.

49. Ross, G. Stanley Hall, *passim*; Statement by Members of the Faculty Concerning Differences with President Hall, 1891-92 (typescript), Hall Papers.

50. Michelson to Harper, 11 August 1892, Harper Papers, University of Chicago Archives; Millikan, "Albert A. Michelson...," p. 123.

51. Michelson to E. S. Holden, 11 April, 1890 [1891]. Holden tried to get Michelson to move either to Stanford or to Berkeley during 1891; see Holden to Michelson, 4, 17 April, 7, 14 May 1891; Michelson to Holden, 25 May 1891, Mary Lea Shane Archives

52. Koelsch, *Clark University*, chap. 1; Ross, *G. Stanley Hall*, chap. 12; Michelson et al. to G. Stanley Hall for transmission to Board of Trustees, 21 January 1892, and "Statement by Members of the Faculty....," pp. 36-37, Hall Papers.

53. In addition to Koelsch, *Clark University*, and Ross, *G. Stanley Hall*, see Lincoln C. Blake, "The Concept and Development of Science at the University of Chicago, 1890-1905" (Ph.D. diss., Univ. Chicago, 1966), *passim*, and T. W. Goodspeed, *A History of the University of Chicago* (Chicago: Univ. Chicago Press, 1915), pp. 204-207, 211-12.

54. Michelson and Whitman, letter to the editor, *Worcester Evening Gazette*, 6 May 1892. See also other clippings, *Scrapbooks*, CUA; a detailed and balanced account is "Exodus From Clark," *Boston Globe*, 10 May 1892. An oft-quoted Hall phrase embodies his argument in a nutshell: "Thus Clark had served as a nursery, for most of our faculty were simply transplanted to a richer financial soil;" Hall, *Life and Confessions*, p. 297.

55. Michelson file, Hall Papers; see also copies of materials in the archives of the International Bureau of Weights and Measures, plus Michelson's original notebooks of his 1892-93 meter bar experiments, in the Michelson Collection, Nimitz Library, U.S. Naval Academy.

56. Austin to Wilson, 20 July 1892, Austin File, Hall Papers; Koelsch, "Ben Franklin's Heir," pp. 538-39.

57. Michelson and Hall, signed agreement, 14 April 1891; Wadsworth to Hall, 24 June, n.d. [June], 14 July, 1892; all in Wadsworth file, Hall Papers.

58. It seems clear that the Nobel Prize was awarded primarily for Michelson's work on the experimental determination of the length of the meter rather than on the bet-ter-known Michelson-Morley experiment; see Elisabeth Crawford and Robert M. Fried-man, "The Prizes in Physics and Chemistry in the Context of Swedish Science," in *Science, Technology, and Society in the Time of Alfred Nobel*, ed. Carl Gustav Bernhard (Oxford: Pergamon Press, 1982), pp. 320-21; Crawford, *The Beginnings of the Nobel Institution: The Science Prizes, 1901-1915* (Cambridge: Cambridge Univ. Press, 1982), pp. 173-74; and Moyer, "Michelson in 1887," p. 50. Of the three American universities contributing time, facilities and funds to that effort, Clark unquestionably made the greatest, though the least recognized, contribution.

THE CHICAGO CONNECTION:
MICHELSON AND MILLIKAN, 1894-1921

John L. Michel

For a quarter of a century, Albert A. Michelson (1852-1931) and Robert A. Millikan (1868-1953) were colleagues in the Physics Department of the University of Chicago. What was the nature of this long association? We will examine the relationship between two major occupations of each scientist's academic career at Chicago -- teaching and research. This study will also consider how the early institutional development of the University of Chicago influenced their endeavors in the classroom and the laboratory. Michelson and Millikan were from different generations and backgrounds. They never collaborated in their experimental pursuits. Nonetheless, their two academic lives became enmeshed in a complementary fashion, to their mutual benefit and to the educational advantage of their students.

MICHELSON AND MILLIKAN: CROSSING PATHS, 1894

When the University of Chicago officially opened in January of 1891, it had neither buildings nor faculty. Its young, energetic president, William R. Harper, scoured American and European universities for highly respected professors, with promises of superior laboratory facilities and generous salaries, to open its doors to students by the fall of 1892.[1] Harper took advantage of bitter discontent among the faculty at the young Clark University, and hired nearly half of its faculty, who had recently resigned. Among these was Albert A. Michelson, who was hired as the first "head professor" of physics at the University of Chicago.[2]

Michelson's immediate responsibilities were making recommendations to Harper for candidates to the physics department faculty and directing the design, construction, and furnishings for a building to house the department.[3] Martin A. Ryerson, the son of a wealthy lumber tycoon, and president of the board of trustees of the University of Chicago, had donated $150,000 to be used "as a Physical Laboratory building and to be known as the 'Ryerson Physical Laboratory' in memory of my father."[4] The facility was completed on New Year's Day, 1894, three months behind schedule and overbudget. This four-story structure, with a central square tower, was built of blue Bedford limestone in the mid-English Gothic style, to blend with the university architectural design. It contained "rooms for special purposes, small laboratories for work of investigation, large laboratories for general instruction, lecture rooms, class rooms, library and offices . . . Every laboratory [was] provided with gas for light or fuel, electricity for light and power, water, compressed air, and vacuum pipes." The heating system was "controlled automatically by the most improved form of temperature regulators . . ."[5] Similar attention was given to the undergraduate laboratories. The Ryerson Physical Laboratory rivaled any of its kind in America or Europe.[6]

The seventh University Convocation and formal dedication of the Ryerson Physical Laboratory in July of 1894 provided an opportunity for Michelson, fellow physicists, and university officials to speak on the importance of such facilities for the research and teaching of physics. Thomas C. Mendenhall noted the recent growing importance of original research in the education of undergraduate, as well as graduate, students.[7] Michelson sympathized with this trend in his convocation address on "Some of the Objects and Methods of Physical Science." Both the student and the investigator, Michelson believed, "must have at [their] command all the modern appliances and instruments of precision which constitute a well-equipped physical laboratory."[8]

These ceremonies also marked the beginning of the long relationship between Michelson and Millikan. Millikan was just beginning his doctoral research on polarization of incandescent light, under Ogden N. Rood at Columbia University. Rood had a reputation for making precision measurements and he greatly appreciated Michelson's experimental achievements.[9] Millikan came to Chicago to spend the summer with Michelson for some assistance on his doctoral research.[10] Millikan enrolled in Michelson's graduate lecture course in "Theoretical Physics on Hydrodynamics, Elasticity, Capillarity, Molecular Physics, Thermodynamics, Wave-Motion and Sound, Optical Theories, Electricity and Magnetism, History of Physics."[11] The course also included laboratory work devoted to the "repetition of classical experiments" on related topics. Millikan enrolled also in Michelson's Research Course, intended for "those graduate students who, having . . . shown aptitude for investigation, are prepared to undertake a special research."[12] Millikan told Michelson about his dissertation, and Michelson "showed the best understanding [Millikan] had yet found anywhere" on this optical topic.[13] Michelson outlined how he would study the problem, and thereafter left Millikan to himself, visiting him just a few times the rest of the summer.[14]

Despite these rather limited relations between the two, Michelson made a strong impression on Millikan, both physically and intellectually. Michelson's "jet black hair, his attractive hazel eyes, his faultless attire, and his elegant and dignified bearing . . . made [him] a striking figure, though his height was not over five feet seven or eight."[15] In his graduate lectures, Michelson impressed Millikan "with the fact that [he] was in the presence of one who had a far deeper understanding of optics than any one [he] had thus far met."[16] Millikan returned to Columbia "much more impressed by Michelson than anyone else [he] had thus far met."[17] At the time, Michelson appeared to be the personification of an experimental physicist that the young Millikan could admire: "elegance of observational technique, elegance of analysis, elegance of presentation."[18] In the next few years, Millikan would come to regard other physicists, such as Michael Pupin and Walther Nernst, as role models to be emulated like Michelson. It would be incorrect, as will become apparent in a later section, where the graduate teaching and research styles of the two physicists are compared, to maintain that Millikan attempted to emulate Michelson more than anyone else.[19]

Millikan completed his doctoral research under Rood's guidance in 1895. Another Columbia physicist in the Department of Mechanics, Michael Pupin, had taken a keen interest in Millikan's research. Millikan envied Pupin's analytical abilities, and his knowledge of the latest developments in electricity and magnetism had piqued Millikan's interest.[20] When Millikan failed to receive any job offers after graduation, Pupin recommended that Millikan spend a year in Europe, as Pupin had done twenty years before, to learn to become a first-rate experimental physicist.[21] After Pupin's third offer, Millikan accepted a loan of $300 (at 7 percent interest) from him, and sailed for Europe with first-class accommodations.[22]

After "cleaning up satisfactorily one problem, [Millikan] wanted time for orientation as to the whole field" of physics.[23] Consequently, Millikan attended lectures by a variety of famous physicists in Jena, Paris, and Berlin. In Berlin in January of 1896, he was among the first to see Röntgen's amazing X-ray pictures of the human skeleton. News of the X-ray pictures aroused the interest of many physicists, but Millikan decided to venture instead to Göttingen in the spring to study under the new rising star of German experimental science, Walther Nernst -- perhaps at Pupin's suggestion.[24] At Nernst's new institute for electrochemical research, Millikan learned two valuable lessons about the coordination of classroom and laboratory work. Nernst organized his students into groups to investigate different aspects of a particular research problem. For example, Millikan and three others carried out experiments on dielectric constants, based on Nernst's methods.[25] Nernst also coordinated classroom instruction and laboratory investigations in a complementary fashion. For example, Millikan attended Nernst's courses on the Physical Methods of Chemistry, and Electrical Methods of Measurements.[26] In both cases, Millikan learned about the latest experimental techniques and instruments and then used them in the laboratory to determine dielectric constants. Nernst apparently intended these courses to inform his graduate students about recent theoretical ideas and experimental methods concerning research topics in which Nernst was currently interested and that his students would be investigating. Within a few years, as a professor of physics, Millikan would recognize the same benefits of such an arrangement for himself and his students.

MICHELSON AND MILLIKAN AS HIGH SCHOOL
AND COLLEGE PHYSICS EDUCATORS

Just before he began to work with Nernst, Millikan had accepted an offer to teach physics at his alma mater, Oberlin, with the prospect of taking complete charge of its physics department in a year or two.[27] In August, Samuel W. Stratton cabled Millikan, on Michelson's behalf, with an offer to be an assistant in the Physics Department at the University of Chicago. While the Oberlin position had higher pay and better immediate prospects for advancement, Millikan wrote President Harper that he preferred the Chicago offer because he wanted "opportunities to do research work which would be denied [him], both on account [of Oberlin's] limited facilities and on account of the meager amount of time which [he] could get there to put into such work."[28] Unknown to Millikan, the Physics Department at Chicago had recently lost the undergraduate teaching services of an assistant and an assistant professor, due to illness and personality conflict.[29] Harper wanted replacements for these men "whose soul is teaching and who will make the introductory teaching in Physics a pleasure and a delight even to those who have no special interest in the subject."[30] Harper would have been impressed by Millikan's teaching credentials from Oberlin, where he had taught physics in the Preparatory Department as well as the college before going to Columbia to study for his Ph.D.[31] Apparently unaware of those underlying pedagogical considerations, Millikan obtained a release from his Oberlin obligation, and accepted Michelson's offer. So began a quarter century association between Michelson and Millikan as colleagues in the Physics Department of the University of Chicago (see Fig. 1).

Seldom has an American educational institution been molded so much by the personality and ideas of a single individual, as the University of Chicago was by William R. Harper. Harper's charismatic presence exemplified a new aggressive -- if not autocratic -- administrative style of American university presidents in the late nineteenth century.[32] He delighted in solving organizational problems. He compartmentalized the university into three divisions (The University, Extension, and Publication) and subdivided these into various academies, colleges, schools, affiliations, libraries, and a press. The administration of these subdivisions was regulated by a many-tiered hierarchy of different faculties, a university council and senate, which were continually restructured during his tenure. Just as the university was physically divided into quadrangles, so was its administration segmented into a multilayered system, by a man who believed that the "method and spirit of the work [of the university] are largely determined by these outside factors."[33]

Harper hoped that the men and women who staffed such a complex organization "would work together harmoniously and with one common spirit."[34] Being a brilliant Hebrew scholar in an age that identified original investigation with the university, Harper naturally wanted the research accomplishments of the graduate schools to be the "most prominent" part of that common spirit; its young scholars would be instilled with an

"overwhelming passion to discover new truth."[35] One organizational means to achieve
that end was his unusual division of the undergraduate program into two-year Junior and
Senior Colleges, to separate preparatory from advanced work. The Junior Colleges served
as "clearing houses" in which a student would undertake broad and balanced course work
to remove deficiencies in his or her secondary school education.[36] Harper's dream--
never fulfilled -- was to remove the Junior Colleges from the university campus, so that
the university "could devote its energies mainly to the University Colleges and to strictly
University work."[37] In the Senior Colleges, students had greater freedom to select
courses in a few subjects in preparation for advanced work in specialized graduate
studies.

Figure 1
University of Chicago Physics Department Professors Michelson, Millikan,
Gale, and Kinsley (Mann absent), 1908, in front of the Ryerson Physical
Laboratory. Photograph by Miss Crowe; courtesy of the Niels Bohr
Library, American Institute of Physics.

Although Harper emphasized research, he realized the danger that such an outlook
could deprecate teaching. He believed that "in general, [the] investigator will accomplish
most who is closely associated with a group of students."[38] One of the advantages of
his undergraduate program was that senior professors, like Michelson -- and Millikan,
later in his career there -- could devote a portion of their time to instruction in the
Junior Colleges.[39] This was not a requirement for a head professor, such as Michelson.
In fact, between 1896 and 1921, Michelson never was responsible for any Junior College
physics course; he did give an "occasional lecture" to these elementary courses for a few
years.[40] Before Millikan came, Michelson and Stratton taught a general physics lecture
course "intended for students [in the Senior College] who are prepared with the Mathematics necessary for an advanced course."[41] Michelson taught only two other Senior
College courses, Molecular Physics, and Physical Computations, and those only for two
years after Millikan came.[42] Throughout his long tenure at Chicago, undergraduate
instruction was a minor obligation for Michelson.

One of his important departmental responsibilities as head professor, which did
affect physics instruction in the colleges, was the arrangement of courses for each
quarter. Michelson believed that the Department of Physics offered courses that were
"sufficiently varied to suit the requirements of all classes, including those who desire
simply to complete a general education . . ."[43] Before 1900, freshmen and sophomore
students were offered a laboratory course that was not oriented simply toward the

156   The Michelson Era in American Science: 1870–1930

understanding of fundamental physical principles. Six years earlier, at a meeting of physicists from around the nation, gathered for the Ryerson dedication ceremony, Michelson had spoken of the greater importance to undergraduates of understanding fundamental principles, rather than improving accuracy.[44] But in the Junior College General Physics course the "object of the laboratory work [was] to teach the student how to handle apparatus, the degree of accuracy to be obtained with given conditions, and properly to arrange and tabulate the results of experimental work."[45] One reason these elementary courses were designed apparently for students interested in further work related to physics, rather than for students in general education, was that physics was only required for those seeking a bachelor of science degree; candidates for a bachelor of arts or philosophy were not required to take physics.

Michelson delegated the responsibilities for teaching the physics courses at the Junior and Senior College level to less senior professors, associates, assistants, and instructors -- such as Millikan. Millikan's first teaching responsibilities were the laboratory portion of the Junior College General Physics course, and a slightly advanced Experimental Physics course, for those who had the equivalent of General Physics.[46] Stratton conducted the lecture sections of General Physics. Stratton was "sympathetic" with Millikan's ideas and he was given "a perfectly free hand" to organize the laboratory class.[47] This gave Millikan the opportunity to begin to implement his notions about teaching physics in a laboratory setting, derived from his experiences as a student and teacher at Oberlin. That background had convinced him that he could introduce physics to high school and college students most effectively by using a problem-solving and laboratory approach, with a minimum of mathematics, to illuminate a limited number of physical principles.[48]

Based on their teaching experiences in those courses, Millikan and Stratton wrote *A College Course of Laboratory Experiments in General Physics*.[49] This textbook provides us with a view of how they organized laboratory instruction in their General Physics course in the late 1890s. Unlike some other popular laboratory manuals, such as Stewart and Gee's three-volume *Elementary Practical Physics*, Stratton and Millikan selected only fifty-four experiments, on mechanics, molecular physics, and physics of the ether.[50] Like many others, Stratton and Millikan designed their experiments to be confirmatory rather than exploratory; students were directed how to measure physical parameters or to verify well established relationships, rather than to test or to discover. These experiments were really didatic demonstrations conducted by the students. Their students worked on experiments Stratton and Millikan had selected from the standard repertoire found in many other similar texts.[51] The design and improvement of their apparatus was done mostly by Stratton. Only a few experiments, such as the more practically oriented ones on the LeClanche cell and the incandescent lamp, were unusual then. In summary, there was little that was novel or unique in Stratton and Millikan's joint approach to physics laboratory instruction for the college student.

In 1898, Stratton and Michelson helped to train troops in Chicago for the Volunteer Naval Militia, who later manned battleships in the brief Spanish-American War. After the war, Michelson resumed his duties at the university, while Stratton began to spend much of his time in Washington. Much to Michelson's shock, Stratton was hired as director of the new National Bureau of Standards in 1899.[52]

Stratton's departure from teaching responsibilities at the university opened up new opportunities for the young Millikan. He took over Stratton's lecture section for the Junior College General Physics course in the summer of 1899. Based on his teaching experiences in these courses, Millikan wrote *Mechanics, Molecular Physics, and Heat* in 1902. In several respects, this textbook represented a different approach to introductory college physics.[53] Millikan abandoned formal lectures and made it mainly an experimental course, "to establish an immediate and vital connection between theory and experiment."[54] He chose to treat only a few basic physical principles, which were the "most effectively presented in connection with laboratory demonstration."[55] He confined his experiments to mechanics, molecular physics, and heat; in some years he also included light and sound, but excluded heat or molecular physics. Although the students' time was divided almost equally between the laboratory and the classroom, their classroom work was devoted entirely to practical problem-solving related to the twenty-three principles that were demonstrated in the laboratory. Millikan intended the students to learn principles rather than methods of experimental physics, but he made "an especial effort . . . *to present Physics as a science of exact measurement*."[56] Apparatus was chosen and newly designed "with a special reference to its ability to yield accurate results in the hands of average students."[57] In this manner, Millikan hoped to give a college student "an insight into the real significance of physical things to introduce him to the very heart of the subject by putting him in touch with the methods and instruments of modern physical investigation . . ."[58]

This approach to the study of mechanics, molecular physics, and heat made up the first third of a year of the Junior College course of General Physics. The second third was devoted eventually to electricity, magnetism, light, and sound. Once again, Millikan's teaching experience became the source of a textbook, *A Short University Course In Electricity, Sound, and Light*.[59] Millikan maintained again that the methods and principles of these topics could not be "readily gained unless theory [was] presented in immediate connection with . . . concrete laboratory problems . . ."[60] The final third of this sequence was occupied with subjects that were not treated in the first two courses and that were more suitably presented by demonstration lectures rather than laboratory methods. These three course remained the basic format for Junior College physics at the University of Chicago over the next two decades.

Millikan had prepared the above triad of physics courses, and his accompanying textbooks, for freshmen and sophomores who had completed a year of high school physics. After 1898, the University of Chicago required physics for admission; those students (who numbered from twenty to forty) who had not fulfilled this requirement had to make up their deficiency.[61] After 1902, Millikan found his trio of General Physics courses unsuitable for them and he introduced a new two-quarter sequence of Elementary Physics. The character of this new course was developed over the next four years in conjunction with the University High School of the School of Education at Chicago, and with various affiliated secondary schools.

President Harper had devised a system of affiliations with other nearby colleges, high schools, and academies. He intended these associations to raise the standards of those institutions as well as to improve instruction at the university.[62] Millikan's teaching background, in the Preparatory Department at Oberlin, made him the most suitable choice in the physics department to carry out Harper's program in connection with the new Elementary Physics course. Besides working with the affiliated secondary schools in relation to his Elementary Physics courses, Millikan had the opportunity to present his ideas to secondary school physics teachers in a summer course, The Pedagogy of Physics, which he offered in the School of Education from 1904 to 1913.

In his course on physics pedagogy, Millikan lectured on the "choice of subject-matter and methods of presentation best suited to elementary courses in Physics."[63] Although Millikan's notes from this course have not been found, we can learn a great deal about what he probably lectured on from the textbooks he wrote with Henry G. Gale in 1906 for high school and college students in his Elementary Physics course. In *A First Course In Physics* and its companion *A Laboratory Course in Physics For Secondary Schools*, Millikan stated he believed that secondary school physics should not be simply a condensed version of the more technical and quantitative methods and instruments employed in a college physics curriculum. High school physics, he believed, should be presented in terms of the language and equipment that a high school student already understands to provide "a simple and immediate presentation . . . of the hows and whys of the physical world in which he lives."[64] For these students, Millikan still believed that this could be best accomplished by closely coordinating classroom and laboratory work. In many high schools inadequate laboratory facilities made this close correlation impossible. For those circumstances, he provided many substitute classroom demonstrations. When facilities permitted, *A First Course in Physics* and its complementary laboratory manual of fifty-one experiments were designed to be coordinated in a variety of ways.

These teaching responsibilities and publications naturally drew Millikan into the debates over the correlation of high school and college physics. By 1905 there was a growing criticism of the influence that colleges exerted on secondary schools, and on their physics instruction in particular. This authority was exerted through a variety of model syllabi, college entrance requirements, and the recently organized College Entrance Examination Board. The deleterious effect of these authorities, some maintained, had led to the drop in enrollment in secondary school physics courses.[65]

In a series of speeches and articles written between 1906 and 1911, Millikan adopted many of the positions of these critics. He believed that university professors had coerced high school teachers to place too much emphasis on exact measurement with complicated and expensive apparatus, and too much use of advanced mathematical manipulations to solve problems too narrowly technical to foster insight into the physical world known to high school students.[66] Rather than emphasizing precision measurement, Millikan believed that high school physics courses should expose students directly to a broad range of physical phenomena drawn from their everyday lives by means of an integrated program of lectures, demonstrations, and experiments with a minimum amount of mathematics.[67] The adoption of this more effective pedagogy, he thought, required the cessation of the "tyranny of the university over the high school."[68] He recommended instead that high school teachers were the most competent judges of what was the best way to teach physics to their students. This was a rather remarkable self-critical recommendation from a university professor who had devoted a decade of teaching and writing to re-fashioning the way physics was to be introduced to high school and college students.

These pedagogical endeavors were not without their rewards for Millikan. By 1907, at the age of thirty-nine, he had risen steadily up the many-tiered academic ladder at the University of Chicago to hold the tenured rank of associate professor. This had been achieved largely on the basis of his successful reorganization of the Junior College physics curriculum, which closely coordinated classroom and laboratory activities, and the publication of four popular high school and college physics textbooks and laboratory manuals. His reputation as a physics educator brought him job offers from other universities, but Michelson persuaded him to stay at Chicago.[69] Millikan's pedagogical pursuits had been personally and financially rewarding; he was building a home for his growing family with money from his book royalties.

He was aware, however, that these successes as a physics educator could promote him no further at the University of Chicago, or in the eyes of his fellow physicists. President Harper made it clear that he regarded excellence in original research -- not teaching and textbook writing -- as the paramount mark of a university professor; promotions to full professor were judged accordingly.[70] Harper's successor, Henry P. Judson, quickly established the same priorities for his own administration. "Teaching [is] an essential part of university work," Judson declared, but "it should not be given a place prior to investigation . . . [The] essential underlying thought of the university [is] research."[71]

Millikan's professional peers likewise lavished their highest praise and honors on those physicists who had excelled in research. At this juncture in his career, Millikan was made keenly aware of this outlook when Michelson was awarded the 1907 Nobel Prize in Physics "for his optical precision instruments and the spectroscopic and metrological investigation carried out with their aid."[72] Millikan knew the path to greatness as an academic physicist. How could he realize his ambition? We will see later how he attained a reputation as an experimentalist the equal of Michelson.

MICHELSON AND MILLIKAN AS GRADUATE PHYSICS TEACHERS IN THE CLASSROOM

Instructor Millikan taught his first graduate physics course in the summer quarter of 1899.[73] As his graduate teaching responsibilities expanded, there emerged a distinct pattern in his pedagogical style compared with that of his senior colleague. In this section we will examine, by means of their students' recollections, the differences between these two physicists' approaches to graduate physics instruction in the classroom. In the next part we shall compare their dissimilar procedures in the laboratory. These two analyses will combine to show how their distinct styles complemented each other to the benefit of their careers and of their students' education.

As a teacher, Michelson was known to most of the physics graduate students by his three- or four-quarter sequence of courses on theoretical physics. Michelson lectured to his students in a formal, dignified manner that suited his personality and military background.[74] Michelson could be polite and affable in other settings, but in the classroom he was aloof and commanded his students' total and uninterrupted attention. No questions or comments from the students were permitted, so a wide range of topics could be covered in one class period. He spoke in a carefully worded manner, seldom repeating himself. Students were encouraged just to listen and to write up their notes after class; few found this procedure suitable. Michelson earned the respect of his brighter students for his precise and lucid lectures. He often left out intermediary steps in derivations of equations, which took even the best students hours to write out in detail after class. Less capable students often found his methods too demanding. His difficult weekly oral examinations could embarrass many students. Those students who successfully completed these courses were well prepared for the questions Michelson posed at their four-hour oral doctoral examination. In this manner, Michelson set an example of discipline and incisiveness as a physics teacher that only the most able graduate students could emulate.[75]

Michelson and Millikan taught in the same classroom, but their graduate students were exposed to very different pedagogical practices. Millikan spoke to his students in an informal manner, which invited questions and comments. His enthusiasm for physics spilled into the classroom, and he often kept talking after the scheduled period. Millikan enjoyed the company of his students in the classroom, his office or home. His agreeable personality and vitality inspired students of different abilities and ambitions. He usually went over a topic three times; first writing it out in long hand on the blackboard, then talking about it, and finally reviewing what he had said. Thereby, most students could take notes in class that accurately reflected what he had presented. He often put his remarks in historical context and included information about the most recent developments.[76]

Michelson was more strict and more distant from his students, and yet he gave them more responsibility for understanding what he had presented. In contrast, Millikan established a close rapport with his students and took more responsibility for making certain they understood what he had said. Each teaching style satisfied a different sense of obligation: Michelson wanted to present the material as he understood it, while Millikan wanted to be certain his students understood the subject. Each physicist's approach had its appeal to the different needs and abilities of his students.

The content of Michelson's and Millikan's graduate physics courses offered a similar complementary mix of topics. After 1902, Michelson's graduate teaching responsibilities were confined primarily to his three-quarter sequence of lectures on theoretical physics. These courses took up a broad range of traditional mathematical-physical topics in mechanics, electricity and magnetism, optics, molecular physics, thermodynamics, and sound.[77] Michelson taught virtually this same sequence of courses in the next two decades. Graduate students enrolled in these courses were given a comprehensive review of the fundamental and well-established principles of late nineteenth-century physics. This was the backbone for subsequent more advanced and specialized studies, as well as preparation for the final doctoral examination. Michelson's courses were often taken in conjunction with another sequence entitled Experimental Physics, which comprised three quarters of laboratory courses taught by Michelson, Millikan, and other physics faculty members. These laboratory courses were "devoted to the repetition of classical experiments, such as: Determination of the Mechanical Equivalent of Heat, Maxwell's "V," Hertzian Oscillation, Relative and Absolute Wave-lengths, etc."[78]

As the graduate students looked to Michelson for their education in classical physics, so they turned to Millikan to learn about the new physics of X rays, radioactivity, electrons, and quantum theories of matter and radiation.[79] Millikan offered a graduate course first in the summer of 1904 that covered "the work of the last ten years on the electrical properties of gases, the electron theory, and radioactivity."[80] In 1909, Millikan expanded this course, now entitled Electron Theory, over two quarters to include topics "dealing with application of the electron theory to metallic conduction, to the Seebeck, Peltier, Thomson, and Hall effects, to optical phenomena in magnetic fields, to the subject of electromagnetic mass, etc."[81] Millikan offered one of the first courses in America on quantum theory in the fall of 1910, entitled Heat Radiation.[82] This combined lecture and seminar course dealt with the thermodynamics of radiation, as treated in Max Planck's *Theorie der Wärmestrahlung* of 1906. Two years later, Millikan extended his treatment of quantum theory of radiation to include the work of J. J. Thomson, William Bragg, and Albert Einstein.[83] In subsequent years, Millikan continued to expand his teachings on quantum theory with new courses on the Atomic Theories of Radiation and their Experimental Basis, and X-Rays and Theories of Atomic Structure.[84]

With these course offerings, Michelson and Millikan were able to provide their graduate students with a complementary selection of advanced physics instruction in classical and the most recent topics.

MICHELSON, MILLIKAN, AND THEIR GRADUATE STUDENTS IN THE RESEARCH LABORATORY

Michelson's reputation for precision measurement had been established in the international physics community by 1896 with his determination of the speed of light, the null detection of ether drift and the measurement of the standard meter bar with his interferometer. Rather than venture into investigations of the mysterious new invisible radiations, he preferred to continue pursuing new applications of his interferometer (e.g., measuring the diameter of Jupiter's moons) and his echelon spectroscope.[85] He made many frustrating attempts over the next two decades to build engines to rule large diffraction gratings. In 1917, Michelson told a graduate student who asked why he had no research interests in the new physics that "the new work fascinated and appealed to him; however, to prepare to change his field of work implied a two-year period of orientation and study with no productive work. He felt that since he was master of the fields of activity for which he was known and there was still far too many problems to do, it was more effective and beneficial to science to devote his full energies in his later years to his present field of competence."[86]

Throughout his tenure at Chicago, Michelson had an uneasy relationship with his graduate students in the classroom and the laboratory. His formal lecture style, discussed above, engendered distance as well as respect from his students. Between 1896 and 1902, Michelson taught three graduate courses on Spectral Analysis, Applications of Interference Methods, and Velocity of Light, which might have stimulated students' interests in his research fields.[87] Michelson's optical investigations could have readily provided excellent topics for doctoral research. But by 1921 only two of sixty-five graduate students had selected topics related to Michelson's optical studies for their doctoral dissertations.[88] Why did Michelson not foster a greater following among the doctoral candidates in the laboratory?

Between 1902 and 1908, Michelson became increasingly frustrated with the responsibility of supervising the research of doctoral candidates. Millikan had begun to help oversee some of these projects in the summer of 1903.[89] He recalled that after a number of "unfortunate experiences" with the research problems of graduate students, Michelson called Millikan into his office and said,

> If you can find some other way to handle it I don't want to bother any more with this thesis business. What these graduate students always do with my problems, if I turn them over to them, is either to spoil the problem for me because they haven't the capacity to handle it as I want it handled, and yet they make it impossible for me to discharge them and do the problem myself; or else, on the other hand, they get good results and at once begin to think the problem is theirs instead of mine, when in fact the knowing of what kind of a problem it is worth while to attack is in general more important than mere carrying out of the necessary steps. So I prefer not to bother with graduate students' theses any longer. I will hire my own assistant by the month, a man who will not think I owe him anything further than to see that he gets his monthly check. You take care of the graduate students in any way you see fit and I'll be your debtor forever.[90]

After this frank admission about his own temperament and needs, Michelson entrusted Millikan with this duty. Michelson had successfully collaborated with fellow professors before (with Edward Morley at Case in the 1880s and William Stratton at Chicago in the late 1890s) and he would thereafter (with Henry Gale at Chicago in the late 1910s). But he recognized that his drive for self-control and perfection were seriously impaired in the laboratory by cooperative efforts with students. Fortunately for him, there was someone in the department whom he respected and to whom he could turn to relieve himself of this burden. Michelson's reputation attracted a growing number of graduate students to Chicago. Once there, they most often became Millikan's doctoral candidates (see Fig. 2).

Millikan pursued his early (1896-1902) research projects alone. They were not successful efforts. He had to abandon a theoretical treatment of anomalous dispersion when he learned of Paul Drude's more complete treatment in 1897. His first two graduate courses in thermodynamics (1900) and molecular physics (1899) stimulated his research interest in the measurement of specific heats.[91] Within a year or two, his experiments proved disappointing and he ceased work on this topic. At this same time, a fortuitous assignment from Michelson turned Millikan's attention toward more recent discoveries.

Michelson asked Millikan in 1899 "to take off his hands" the Friday afternoon

Physics Club.[92] Faculty and students presented brief reports on their own research or on other recently published investigations at these weekly meetings. Millikan began to read articles on electron theory of matter, cathode rays, canal rays, radioactivity, X rays, and other new developments.[93] Before long, he presented reviews of diffraction experiments with X rays and J. J. Thomson's experiments on cathode rays.[94] Students

Figure 2

University of Chicago Physics Department faculty, staff, and graduate students, 1916, by the new annex to the Ryerson Physical Laboratory. Photography from the Leonard Loeb Photography File; courtesy of the Niels Bohr Library, American Institute of Physics.

soon began to report on these exciting new discoveries. Millikan started to build apparatus to study electron emission from metal surfaces in high vacuua with large electric fields in 1899. This first exploration in the new physics would continue unsuccessfully for another eight years. He incorporated discussions of his own and others' investigations of cathode rays, spark discharges, various e/m measurements, Becquerel rays, X rays, and electron theory in his courses as early as the summer of 1901.[95]

This early series of Millikan's teaching and research activities related to the new discoveries illustrates a strategy that he adopted increasingly over the next two decades. The weekly physics club and courses stimulated Millikan's interest in new research topics. Those educational duties also gave him an opportunity to read, think and discuss about his own and others' investigations. Those occasions introduced his students to the new problems Millikan thought were worthwhile and exciting to solve. This suggested to the students possible subjects for their own doctoral research. So when Michelson began to turn the responsibility for supervising doctoral dissertations over to Millikan, he had already accumulated a number of possibilities. These proposals often concerned the new field of electron physics, which was becoming the focus of Millikan's own research efforts.

Millikan conducted three different research projects during the next five years (1902-1906), but none of them proved successful. He abandoned the studies of electrical discharges in high vacuua by large electric fields when they failed to yield any reproducible results for him.[96] (One of his students, Glenn Hobbs, did successfully complete research on this phenomenon in 1905.[97])

Millikan's second research program was very short lived. After attending lectures by Henri Becquerel and Pierre Curie on radioactivity at the Exposition Universelle

Internationale at Paris in 1900, Millikan became interested in radioactive substances. Like the study of the electron, radioactivity posed fascinating new questions about the structure of matter. In 1903, Austria placed an embargo on the export of pitchblende ore, the primary source of radioactive materials. This cutoff precipitated a worldwide search for other sources of uranium and radium. It also probably prompted Millikan's examination of the relation between radioactivity and the uranium content of some American and English minerals.[98] When he learned similar findings had been achieved by others, he abandoned forever any other investigations of radioactivity. The subject did become the focus of several popular addresses and articles in 1904, and a topic in his undergraduate and graduate courses for many years. Despite this attention, his lack of research interest in radioactivity accounts for the absence of any doctoral research in this field by Chicago students.

Millikan embarked on a third research venture into the discharge of electricity from illuminated metal surfaces -- the photoelectric effect -- with another doctoral candidate, George Winchester, from 1903 to 1906. Together they studied the influence of temperature on photoemission. Winchester's contributions became the subject of his doctoral dissertation.[99] Millikan turned his attention toward determining what properties of the light had a fundamental effect on the photoelectric effect. These studies initially yielded puzzling results, which would not be resolved for several more years.

By 1907, Millikan felt that he had little to show for his research efforts, in sharp contrast to his many successful pedagogical achievements. This inequality brought Millikan to a critical juncture in his career at Chicago, as mentioned above. How was he to surpass his modest research accomplishments, and achieve the distinction as an experimental physicist and the further advancement at the university that he so ardently sought? Contrary to his autobiographic accounts, Millikan did not dramatically turn away from his successful teaching efforts and "gamble" on achieving fame in research.[100] He learned instead how to promote his experimental program in a manner quite different from Michelson's. Millikan had already established the two major components of his effective research style, which would become the hallmarks of his subsequent research endeavors. (1) From his previous teaching experiences, he had learned the advantages of the integration of classroom and laboratory studies in the new physics. (2) From Nernst and from his own earlier research efforts, he had learned the benefits of cooperative work with graduate students. This proved to be not only the best way to advance his own career, but "the best way for both training and selecting of the oncoming group of research men -- an obvious part of [his] university job."[101]

Millikan decided to concentrate his research efforts on two subjects -- the photoelectric effect and the measurement of the elementary electrical charge -- by the end of 1907. Both of these topics had drawn the attention of many distinguished physicists around the world. But these projects had different objectives. His photoelectric experiments were explorations at the frontier of physicists' knowledge of the fundamental nature of light and electricity. His measurements of the elementary electrical charge were attempts to determine more precisely the value of a recognized fundamental physical constant. In both cases, Millikan's integration of classroom and laboratory and his team approach to research would contribute to his success. We shall examine hereafter only one research program, his photoelectric studies, to demonstrate why Millikan's dualistic research style worked so effectively for him and his students.[102]

The findings of Millikan's photoelectric research teams led to a quandary by the spring of 1912. His initial investigations with George Winchester of the temperature dependence of the photoelectric current were followed by attempts with Winchester to obtain absolute measurements of the fundamental influence of light on the photocurrent, between 1907 and 1909. Their experiments showed no fatigue effects (the diminution of the photocurrent over long periods of illumination).[103] Millikan's experiments had found increased values of the positive potentials acquired by non-alkali metals illuminated by ultra-violet light over a period of three years (1906-1909).[104] These unusual findings led Millikan and a new graduate student, James Wright, to begin in 1910 a three-year investigation of the possible influence of different properties of their light sources.[105] They found that light from spark sources produced curiously high values of the positive potentials compared with those produced by light from arc sources. This disparity seemed to be due to the light, rather than secondary effects.[106] This presented Millikan with a dilemma: either he must make the radical conclusion that some property of the light other than frequency or intensity was responsible for the disparity, or he must reject the well-established fact of the independence of the positive potential with the illumination intensity.[107]

At this juncture, Millikan mentioned for the first time in print the relation between his own photoelectric studies and the "Planck-Einstein light-unit theory."[108] He noted, as a sidelight, that his values of the positive potentials were "completely at variance" with the productions of the Planck-Einstein theory. Up to this point, Millikan's photo-electric research program had been disinterested in Einstein's light quanta hypothesis and his photoelectric equation. Within a year, this would change dramatically.

Unlike his photoelectric investigations, Millikan's measurements of the elementary electric charge had progressed quite successfully by the spring of 1912. In need of uninterrupted time away from his many departmental responsibilities to compile his findings on the charge, he decided to take his family (his wife and two boys) to Europe for six months. After meeting with German scientists engaged in photoelectric experi-ments, Millikan became aware of their interest in Einstein's photoelectric equation, the validity of which had not been established experimentally with any certainty.

Millikan managed to quickly resolve his puzzling data on the disparity of positive potentials, upon returning to Chicago, by properly screening inductive effects from the light sources. He now realized that he and his students had perfected an experimental method for precisely determining the relationship between the frequency of light and the positive potentials of the illuminated surfaces emitting electricity. It was just this relationship that Einstein had predicted on the basis of his postulated light quanta. Millikan knew that he must turn his attention there now.

At this important turning point in his photoelectric studies, Millikan undertook an extensive critical review, from an experimentalist's point of view, of the quantum theories of Max Planck, Walther Nernst, William H. Bragg, J. J. Thomson, and Albert Einstein. This analysis in 1912 persuaded Millikan that "we are forced to conclude that an atomistic structure of some sort must be applied to radiant energy."[109] He then decided that "however radical it may be," Einstein's quantum theory "most completely interpreted" the experimental data.[110] This conclusion, however qualified, strengthened Millikan's decision to experimentally test Einstein's prediction for the photoelectric effect.

Millikan struggled with his assessment of Einstein's light quanta hypothesis over the next four years (1912-1916). His remarks in public are well known to historians of science.[111] Much less known, though equally revealing about his thinking on Einstein, are his classroom discussions on Einstein's theory of the photoelectric effect, during these four years. These remarks to his students disclose more about his scientific judgment of this controversial theory, as well as how Millikan integrated his laboratory and classroom work.

We do not know how Millikan presented Einstein's ideas, if at all, in his first course on quantum theory in 1910. After 1912, Einstein's treatment of the photoelectric effect was dealt with more and more in his graduate courses on Heat Radiation, Electron Theory and Atomic Theories of Radiation and Their Theoretical Basis. Millikan told his students in the fall of 1912 that just as there was a modern tendency toward "atomism" (i.e. discontinuity) in new theories of matter and electricity, so there was in recent theories of radiation.[112] Millikan usually characterized Einstein's light quantum hypothesis and explanation of the photoelectric effect as an amalgam of special unitary theories (like Bragg's and Thomson's) and Planck's temporal unitary theory. Millikan repeated this erroneous description of Einstein's ideas in print and class for the next eleven years.[113] Millikan misunderstood Einstein's original arguments from thermodynamics and statistical mechanics, which did not build upon either Planck's or Thomson's theories.[114] This misconception led Millikan to raise objections to Einstein's theory that were not relevant.[115] Millikan also told his students and colleagues on numerous occasions that he thought that Einstein had abandoned his quantum theory; this was untrue.[116] This incorrect reasoning contributed significantly to Millikan's unfavorable opinion of Einstein's interpretation of the photoelectric effect for many years.

When Millikan compared Einstein's quantum theory to other "atomic" theories of radiation in class and in print, he was most favorably impressed with the ability of Einstein's theory to account so well for the experimental findings.[117] But Einstein's theory faced one major obstacle, which made it impossible for Millikan to adopt it: "*no one [had] thus far seen any way of reconciling such a theory with the facts of diffraction and interference so completely in harmony in every particular with the old theory of ether waves.*"[118] This objection remained a major impediment to Millikan's adoption of Einstein's quantum theory after 1912.[119] This apparent irreconcilability of the particle and wave conception of light was equally troublesome to many of Millikan's colleagues at this time.[120]

Millikan saw no a priori reason why some quantum theory of radiation could not be reconciled with the electromagnetic wave theory; it was just that no one had been able to do it yet. Millikan realized that some discontinuous theory of light "had to come."[121] He attempted, however, to reconcile his findings on the photoelectric effect with the wave theory of light by proposing an unusual atomic resonance mechanism of continuous absorption, as Planck had suggested.[122] Despite the implausibility of such a mechanism in the atom, Millikan continued to favor some variation of this explanation into 1917.[123]

Throughout the period 1912-1916, when Millikan was devoting ever more of his time in the laboratory to his photoelectric investigations, his discussions of this phenomenon in the classroom also expanded. In his 1912 course on quantum theories of radiation, he only briefly mentioned the photoelectric effect.[124] A year later, he devoted much more attention to the most recently published experimental determinations of the relationship between the maximum kinetic energy of the ejected photoelectrons and the frequency of the light.[125] It was this relationship, predicted by Einstein to be linear, that Millikan was just beginning to measure in his laboratory. The next year, the photoelectric effect was the subject of the first twelve of thirty-two discussions on electron theory.[126] Millikan presented a detailed history of the experimental findings to date, and used those facts to assess different theoretical interpretations (Lenard's trigger theory, J. J. Thomson's "ether-string" theory, and Einstein's quantum theory). He included information about the apparatus and methods currently being used by him and his students to test Einstein's photoelectric equation. He pointed out why his approach was superior to others. In this manner, Millikan treated his graduate courses as research seminars in the new physics of the electron and quantum theory. These classes informed his graduate students about the pros and cons of the latest theories and experiments, and in particular about his own research programs. Millikan even shared correspondence from other physicists with his students.[127] These courses became the training ground for his students' doctoral research in related subjects. For Millikan, his instruction stimulated critical thinking about his and others' experiments and interpretations, which he carried back and forth from the laboratory.

Millikan's inclusion of his graduate students in various aspects of his own research programs took place in the laboratory as well as the classroom. He invited graduate students to participate in his most important research problems, "first, just as assistants, and later if they developed originality and initiative, as collaborators."[128] In some cases he "turned over the field completely to them when [he] thought they had become sufficiently resourceful and effective."[129] Millikan encouraged his doctoral candidates to work on problems related to his research interests. But he was not dictatorial about their choice of research, and he supported those who wanted to work on a different problem of their choosing.[130]

Four graduate students (William Kadesch, Wilmer Souder, Leopold Lassalle, and A. Melcher) made significant contributions to Millikan's photoelectric studies between 1912 and 1916, as assistants and collaborators.[131]   Portions of these studies became the subjects of doctoral research for three students -- Albert Hennings, Kadesch, and Souder.[132] These three men then conducted post-doctoral research on the photoelectric effect, in close conjunction with Millikan's own work.[133] Therefore, a total of eight-- including the early contributors Winchester and Wright -- constituted Millikan's photo-electric research team. After 1916, Millikan persuaded six more graduate students to do their doctoral research on related topics.[134]   In a similar fashion, at least eleven graduate students conducted research related directly or indirectly to Millikan's measure-ment of the electron charge.[135] Of the approximately fifty graduate students Millikan supervised between 1905 and 1921, twenty-one conducted their doctoral research on topics related to Millikan's two major research interests. In comparison, only two doctoral students supervised by Michelson carried out experiments associated with Michelson's primary research interest in light.

There were many benefits of this cooperative approach for Millikan and his graduate students. There was a noticeable upturn in the progress of Millikan's research efforts after his graduate students joined him in the laboratory. Their contributions of time and skills as research assistants enabled Millikan to tackle a broader range of problems in a shorter span of time. Five of his graduate students and postdoctoral collaborators conducted experiments that corroborated Millikan's findings on the photoelectric ef-fect.[136] This enabled Millikan to defend himself better against physicists, such as Ramsauer, with conflicting data.[137] The many classroom discussions on topics related to Millikan's research interests also added, rather than detracted, from the time Millikan could devote to keeping up with the recent publications and to critically evaluating what he and others were doing.

This partnership in the classroom and the laboratory provided the underlying organizational structure that enabled Millikan to successfully complete his many research objectives. Those achievements soon earned Millikan from his peers and the university the high recognition as an experimental physicist that he had so desired. His measure-ment of the electron charge was rewarded with a promotion to full professor in 1910, a star in the 1910 edition of *American Men of Science*, and the $1,000 Comstock Prize of the National Academy of Sciences in 1913. He was soon elected to prestigious profes-sional societies: The American Philosophical Society (1914), the American Academy of Arts and Sciences (1914), the National Academy of Sciences (1915). He was elected president of the American Physical Society (1916), and vice-chairman of the National Research Council (1916). Millikan achieved the greatest international recognition in 1923 with the award of the Hughes Medal of the Royal Society of London and the Nobel Prize in Physics for his precision measurements of the electron charge and the photoelectric effect.

There were also rewards for his students. In the classroom they were introduced to the latest theoretical and experimental developments in the new physics by an enthusias-tic instructor who specialized in those topics. In the well-equipped Ryerson Laboratory, they were guided by the expertise and insights of a man who was becoming one of the best experimental physicists in the world. As members of his research team, they were able to accomplish their own research objectives more efficiently than if they had had to proceed on their own. In time, their efforts earned them a doctoral degree from a university with a growing reputation for excellence in research.

There was also a price to pay for these benefits. The students who followed Millikan's lead relinquished some independence in the pursuit of their research goals. A few, such as Harvey Fletcher, were even deprived of credit for some of their important contributions to Millikan's investigations.[138] In many other cases, Millikan openly acknowledged in print the assistance his students provided him. It was a team effort, but Millikan was the master and the students were the disciples.

MICHELSON AND MILLIKAN: TWO STYLES OF ACADEMIC PHYSICS

New American universities, like the University of Chicago, were creating a new dualistic role for academic scientists by 1900. The professor's once primary responsibility to be an educator was being redefined to require that he perform also as a researcher. His students were seen as novice investigators as well as the audience of his instruction. While administrators, like Harper and Judson, extolled a new common university spirit, there developed a troublesome conflict between their professors' obligations to teaching and research. The organizational diversity and complexity of their universities tended to intermingle inharmoniously those two primary functions.[139] Presidents Harper and Judson created a many-layered hierarchy of academic programs and positions, gave speeches, allocated resources (e.g., for laboratories and libraries), and applied promotion criteria that promoted excellence in research; their system of affiliations with other educational institutions and their statements encouraged improved teaching. There was little doubt which purpose they believed was primary. Somehow they hoped research could thrive, but not at the expense of teaching. It was left to their professors to struggle with this dualism.

There was no one widely accepted way for professors to balance their duties to promote the dissemination of established knowledge to students as well as to foster their discovery of new information. As a head professor with an established reputation for precision measurements, Michelson was more insulated from university pressures and more secure in his preferences as a physicist. Millikan, as an ambitious young professor, had to struggle more with the duality of his academic obligations and with building his reputation in the physics community.

The careers of Michelson and Millikan at the University of Chicago between 1896 and 1921 exemplified two different approaches to the resolution of this antagonism between teaching and research, which benefited their own personalities and ambitions and satisfied the different educational needs of their students. The tension between investigation and instruction was most acute with their obligations to their doctoral candidates. Both physicists chose to exercise control over these duties in the classroom and the laboratory, but in dissimilar manners. Michelson took command of the content and conduct of his lectures in classical physics; Millikan exercised his dominance in the laboratory by directing a team approach to solving experimental problems in the new physics of the electron. Michelson gradually separated his responsibilities in the laboratory from those in the classroom; Millikan increasingly learned to effectively intermingle his duties in those two academic work places. Each style enabled Michelson and Millikan to win the esteem of their colleagues and students and to provide a complementary educational experience for their students.

NOTES

1. Richard J. Storr, *Harper's University* (Chicago: Univ. Chicago Press, 1966), pp. 65–86.

2. Albert A. Michelson to William R. Harper, 30 April and 2 May 1892, William R. Harper Papers, Box 14, Folder 12, Univ. of Chicago Special Collections (hereafter cited as UCSC); Dorothy Michelson Livingston, *The Master of Light* (New York: Scribner's, 1973), pp. 157–170.

3. Michelson to Harper, 9 May, 10 June, and 17 June 1892, Harper Papers, Box 14, Folder 12; Michelson to Harper, 28 May 1892, 3 July 1892, 11 August 1892, 17 August 1892, and 25 February 1893, President's Papers, 1889–1925, Box 46, Folder 7, UCSC. While Michelson corresponded with Harper from Paris, Assistant Professor Samuel W. Stratton oversaw the construction of the new facility.

4. Martin A. Ryerson to Board of Trustees of the University of Chicago, 7 November 1892, President's Papers, 1889–1925, Box 68, Folder 22, UCSC.

5. Harper, "Response by President Harper [to the Formal Gift of the Ryerson Physical Laboratory]," *The Quarterly Calendar of the University of Chicago*, 1894, *3*: 32–33.

6. Paul Forman, John L. Heilbron, and Spencer Weart, "Physics circa 1900," *Historical Studies in the Physical Sciences*, 1975, *5*: 104–114; David Cahan, "The Institutional Revolution in German Physics, 1865–1914," *HSPS* 1925, *15*, Part 2: 15–37.

7. Thomas C. Mendenhall, "The Evolution and Influence of Experimental Physics," *Quarterly Calendar*, 1894, *3*: 3–11. Michelson delivered this speech in Mendenhall's absence, Livingston, *Master of Light*, p. 183.

8. Michelson, "Some of the Objects and Methods of Physical Science," *Quarterly Calendar*, 1894, *3*: 15.

9. Edward L. Nichols, "Biographical Memoir of Ogden Nicholas Rood, 1831–1902," *National Academy of Sciences Biographical Memoirs*, 1909, *6*: 461, 468; Ogden R. Rood to A. A. Michelson, 16 October 1894, File X-49, A. A. Michelson Collection, Nimitz Library, U.S. Naval Academy.

10. Millikan may have thought about transferring to the University of Chicago, following the loss of his $500 University Fellowship at Columbia. Robert A. Millikan, holographic note (n.d.), pp. 6–7, Robert A. Millikan Collection (hereafter cited as RAM), Folder 64.6, California Institute of Technology Archives.

11. "The Department of Physics," *The Annual Register* [Univ. Chicago], 1894, p. 117.

12. "Physics," *Annual Register*, 1894, p. 117.

13. Millikan, "Some Episodes in the Scientific Work of Robert A. Millikan," TS, 17 January 1944, p. 4, RAM, Folder 67.8.

14. Millikan, "Scientific Recollections of R. A. Millikan," 1939, p. 20, RAM, Folder 67.7; Millikan, "Albert A. Michelson, 1852-1931," *National Academy of Sciences Biographical Memoirs*, 1929, *19*: 125.

15. Millikan, "Michelson," p. 124. Millikan was of similar stature and both men were physically fit.

16. Ibid., p. 125.

17. Millikan, "Scientific Recollections," p. 20.

18. Millikan, "Michelson," p. 125.

19. Robert H. Kargon, *The Rise of Robert Millikan* (Ithaca: Cornell Univ. Press, 1982), pp. 36, 59.

20. Millikan, "Scientific Recollections," pp. 14-15; "Episodes," pp. 2-3.

21. Millikan, "Scientific Recollections," pp. 18, 21.

22. Ibid., p. 22.

23. Ibid., p. 24.

24. Pupin wrote to Millikan during his year in Europe (none of this correspondence has been found) "suggesting special matters to be taken up." Millikan, "A Minister's Son Knocks Out Old Prejudice," *Kansas City Star Magazine*, 14 March 1926, p. 13, RAM, Folder 68.1.

25. Walther Nernst, "Methode zur Bestinnung von Dielektrizitätskonstanten," *Zeitschrift für Physikalische Chemie*, 1894, *14*: 622-663; James C. Philip, "Das dielektrische Verhalten flüssinger Mischung, besonders verdümter Lösungen," *Z. Physik. Chemie*, 1897, *24*: 18-38; Richard W. H. Abegg, "Dielektricitätsconstanten bei tiefen Temperaturen," *Annalen der Physik und Chemie*, 1897: *60*: 54-60; Florian Ratz, "Ueber die Dielektricitätsconstant von Flüssigkeiten in ihrer Abhangigkeit von Temperatur und Druck," *Z. Physik. Chemie*, 1896, *19*: 94-112; in Millikan, "Scientific Recollections," p. 26.

26. Millikan, "Nernst [on] The Physical Methods of Chemistry," holographic notes, 24 April 1896 to 3 July 1896, pp. 1-25, and "Nernst Electrische Messmethode," holographic notes, 29 April 1896 to 29 July 1896, pp. 48-91, RAM, Folder 4.2.

27. Millikan to Harper, 22 September 1896, p. 2, President's Papers, 1889-1925, Box 46, Folder 9, UCSC.

28. Millikan to Harper, 1896, p. 4.

29. Correspondence of the Secretary of the Board of Trustees of the University of Chicago, 11 August 1896, Box 1, Folder 9, UCSC; Michelson to Harper, 16 May 1895, President's Papers, Box 46, Folder 7, UCSC.

30. Harper to Michelson, 22 February 1896, Harper Papers, Box 2, Folder 22, UCSC.

31. John L. Michel, "Millikan's Physics Education at Oberlin." This unpublished address was delivered at the 1978 annual meeting of the History of Science Society.

32. Lincoln C. Blake, "The Concept and Development of Science at the University of Chicago" (Ph.D. diss., Univ. Chicago, 1966), p. 18; Lawrence R. Veysey, *The Emergence of the American University* (Chicago: Univ. Chicago Press, 1965), pp. 360-380.

33. Harper, *The Trend in Higher Education* (Chicago: Univ. Chicago Press, 1905), pp. 27-28.

34. Harper's letter transmitting *The President's Report*, 1897-98, quoted in *The University of Chicago Survey* (Chicago: Univ. Chicago Press, 1933), III, p. 4.

35. Harper, "The Statement by the President of the University for the Quarter Ending July 1, 1893," *Quarterly Calendar*, 1893, 2: 9; "President Harper's Baccalaureate Address," *The University of Chicago Weekly*, 1894, 2: 3.

36. Edward Capps, "The Junior Colleges," *The President's Report, 1897-98*, p. 85.

37. *Official Bulletin* [Univ. Chicago], 1891, No. 2, p. 3. The University Colleges were renamed Senior Colleges in 1896.

38. Harper, "Quarterly Statement of President Harper," *The University [of Chicago] Record*, 1897, 2: 11, "President's Forty-First Quarterly Statement," *The University Record* 1902, 6: 387.

39. Capps, "Junior Colleges," p. 93.

40. "Physics," *Annual Register*, 1894, p. 119; "Physics," *Annual Register* 1901, p. 26; "Physics," *Annual Register*, 1902, p. 285; "Physics," *Annual Register*, 1907, p. 241.

41. "Physics," *Annual Register*, 1895, p. 151.

42. "Physics," *Annual Register*, 1897, pp. 285-86; "Physics," *Annual Register*, 1898, p. 294.

43. "Physics," *Annual Register*, 1893, p. 82. Michelson repeated this same statement until 1907.

44. Michelson, "Objects and Methods," p. 15.

45. "Physics," *Annual Register*, 1898, p. 293.

46. "Physics," *Annual Register*, 1897, p. 284.

47. Millikan, *The Autobiography of Robert A. Millikan* (New York: Prentice-Hall, 1950), p. 40.

48. Michel, "Millikan's Physics Education," p. 4.

49. Samuel W. Stratton and Robert A. Millikan, *A College Course of Laboratory Experiments in General Physics* (Chicago: Univ. Chicago Press, 1898).

50. B. Stewart and W. W. H. Gee, *Elementary Practical Physics* (New York: MacMillan, 1893, 1896, 1897), I. *General Processes*; II. *Electricity and Magnetism*; III. *Practical Acoustics.*

51. See, for example, Joseph S. Ames and William J. A. Bliss, *A Manual of Experimental Physics* (New York: Harper, 1898). Millikan often referred to this textbook.

52. Livingston, *Master of Light*, pp. 208-209. Michelson told Stratton privately that he was interested in the position. Stratton continued to be listed on the Chicago faculty until 1905, although he no longer taught there.

53. The most similar approach perhaps was taken by Edwin H. Hall and Joseph Y. Bergen, in *A Text-Book of Physics: Largely Experimental* (New York: Holt, 1891). Hall and Bergen covered many more topics and recommended that classroom and laboratory work be given equal time.

54. Millikan, *Mechanics, Molecular Physics, and Heat* (Boston: Ginn, 1902), p. 3. See also his lecture notes, "Heat," 1899, RAM, Folder 1.1. Millikan realized a few years later that he had "gone too far" and reintroduced lectures once a week. Millikan, *Autobiography*, p. 41.

55. Millikan, *Mechanics*, p. 4.

56. Ibid., p. 5. His emphasis.

57. Ibid., p. 5.

58. Ibid., p. 6.

59. Millikan and John Mills, *A Short University Course in Electricity, Sound, and Light* (Boston: Ginn, 1908). John Mills was a physics instructor from Western Reserve University who taught this course in the summer of 1908.

60. Ibid., p. iii.

61. Capps, "Junior Colleges," p. 80.

62. Storr, *Harper's University*, pp. 211-222.

63. "Physics," *Annual Register*, 1904, p. 325.

64. Millikan and Henry G. Gale, *A First Course in Physics* (Boston: Ginn, 1906), p. iii; Millikan and Gale, *A Laboratory Course in Physics* (Boston: Ginn, 1906). An analysis of this and other early Millikan textbooks is given by Alfred Romer in his "Robert A. Millikan, Physics Teacher," *The Physics Teacher*, 1978, *16*: 82-83.

65. Charles R. Mann, *The Teaching of Physics for Purposes of General Education* (New York: MacMillan, 1912), pp. 9-22. Mann joined the Chicago faculty the same year as Millikan. Sidney Rosen, "A History of the Physics Laboratory in the American Public High School," *American Journal of Physics*, 1954, *22*: 200-203; Edward A. Krug, *The Shaping of the American High School* (New York: Harper, 1964), pp. 369-370; Albert E. Moyer, "Edwin Hall and the Emergence of the Laboratory in Teaching Physics," *Phys.*

Teacher, 1976, *14*: 101-102.

66. Millikan's untitled discussion of George Mead's "Science in the High School," *The School Review*, 1906, *14*: 251. Michelson concurred with this assessment two years later, at a Symposium on the Purpose and Organization of Physics Teaching in Secondary School, *School Science and Mathematics*, 1909, *9*: 3.

67. Millikan, discussion of Mead's "Science in the High School," pp. 251-252.

68. Ibid., p. 251.

69. Millikan to Harper, 12 April 1898, President's Papers, 1889-1925, Box 46, Folder 9, UCSC.

70. Thomas W. Goodspeed, *William Rainey Harper* (Chicago: Univ. Chicago Press, 1928), pp. 123-124, 157-158.

71. Harry P. Judson, "President's Quarterly Statement," *University [of Chicago] Record*, 1906, *10*: 160; see also his "The Idea of Research," *The University of Chicago Magazine*, 1911, *4*: 14-18.

72. *Nobel Lectures: Physics, 1901-1921* (Amsterdam: Elsevier, 1967), p. 157. For more about this decision, see Elisabeth Crawford, *The Beginnings of the Nobel Institution* (Cambridge: Cambridge Univ. Press, 1984), pp. 173-174. The high regard for research among physicists is discussed in Daniel Kevles, *The Physicists* (New York: Knopf, 1978), pp. 45-90; Kevles, "The Physics, Mathematics, and Chemistry Communities: A Comparative Analysis," in *The Organization of Knowledge in Modern America, 1860-1920*, ed., Alexandra Oleson and John Voss (Baltimore: Johns Hopkins Univ. Press, 1979), pp. 150-154.

73. "Physics," *Annual Register*, 1900, p. 301. The lecture notes for this course, entitled "Molecular Physics," are labeled "Kinetic Theory," RAM, Folder 1.3.

74. Michelson was a graduate of the U.S. Naval Academy and taught physics there from 1876 to 1880. Livingston, *Master of Light*, pp. 42-75.

75. Transcripts of taped interviews, and autobiographical writings of Millikan's graduate students Karl K. Darrow, Harvey Fletcher, Vern O. Knudsen, Leonard B. Loeb, and Ralph A. Sawyer are deposited at the Center for History of Physics, American Institute of Physics (hereafter cited as CHP). See also an interview with Earnest C. Watson by A. B. Christman of the Naval Weapons Center, China Lake, California.

76. Transcripts of taped interviews, and autobiographical writings of Millikan's graduate students Fred Allison, Karl K. Darrow, Harvey Fletcher, Katherine Kelly (wife of Mervin Joe Kelly), Vern O. Knudsen, Leonard B. Loeb, Ralph A. Sawyer, CHP; Earnest C. Watson interview, NWC.

77. "Physics," *Annual Register*, 1903, p. 321. See also Ralph A. Sawyer's outline and lecture notes for these courses in 1916, CHP.

78. "Physics," *Annual Register*, 1905, p. 286.

79. Only one other faculty member, Carl Kinsley, taught a course on these topics. One member of the Mathematics Department, Arthur C. Lunn, taught graduate courses on advanced mathematical physics and relativity.

80. "Physics," *Annual Register*, 1904, p. 326; Millikan lecture notes, "Recent Advances in Kinetic Theory," RAM, Folder 1.10.

81. "Physics," *Annual Register*, 1909, pp. 347-348; Millikan lecture notes, "Electron Theory," RAM, Folder 1.15.

82. "Physics," *Circular of Information* [Univ. Chicago], 1910, *10*: 23. Millikan's courses on quantum theory will be discussed later in this article; see also Gerald Holton, "The Hesitant Rise of Quantum Physics Research in the United States," in this volume.

83. Millikan, lecture notes, "Unitary Theories," RAM, Folder 1.16.

84. "Physics," *Annual Register*, 1913, p. 248; "Physics," *Annual Register*, 1915, p. 285.

85. Michelson briefly speculated on X rays in "A Theory of X-Rays," *American Journal of Science*, 1896, *1*: 312-313, and with Stratton, "Source of X-Rays," *Science*, 1896, *3*: 694-696.

86. Loeb, "Autobiography of Leonard B. Loeb," TS, 1962, pp. 26-27, quoted by permission of the Niels Bohr Library, American Insititute of Physics.

87. Michelson taught a graduate course, "Light Waves and Their Uses," from 1909 to 1913, but he was supervising only one doctoral candidate then.

88. Gordon F. Hull did his doctoral research "On the Use of the Interferometer in the Study of Electric Waves" (1897), and Thomas E. Doubt on "The Effect of Intensity upon the Velocity of Light" (1904). "Physics," *Announcements* [Univ. Chicago], 1931, *31*: 114.

89. Millikan, *Autobiography*, p. 60.

90. Millikan, "Michelson," p. 126.

91. Millikan, *Autobiography*, p. 59.

92. Ibid., p. 58. The Physics Club was organized in 1898.

93. Millikan kept a notebook, "References to Important Articles," which listed articles on the new physics published from 1897 to 1914. RAM, Folder 3.8.

94. "Physics Club," *The University [of Chicago] Record*, 1899, *4*: 174; Millikan, *Autobiography*, p. 58.

95. Millikan lecture notes, "Electricity and Magnetism," RAM, Folder 1.5. This was an undergraduate course.

96. Millikan, *Autobiography*, p. 61. Michelson took an interest in the spectra of these sparks. After 1919, Millikan returned to more fruitful studies of this phenomenon.

97. "Physics," *Announcements*, 1931, p. 114.

98. Millikan, "The Relation Between the Radioactivity and the Uranium Content of Certain Minerals," in *Congress of Arts and Science: Universal Exposition, St. Louis, 1904*, ed. Howard J. Rogers (Boston: Houghton, Mifflin, 1906), IV, p. 187. See also the transcript of the interview of Millikan by Watson Davis, for the CBS radio program "Adventures in Science," 24 April 1948, pp. 7-8, RAM, Folder 61.29.

99. Millikan and George Winchester, "The Influence of Temperature upon Photo-electric Effects in a Very High Vacuum, and the Order of Photoelectric Sensitiveness of the Metals," *Philosophical Magazine*, 1907, *14*: 188-210; Millikan and Winchester, "Upon the Discharge of Electrons from Ordinary Metals under the Influence of Ultra-Violet Light," *Physical Review*, 1907, *24*: 116; "Physics," *Announcements*, 1931, p. 115.

100. Millikan, *Autobiography*, pp. 68-69.

101. Ibid., p. 88. There were usually several women graduate students in physics at Chicago every year, but only three earned their doctorate by 1921; none of them were supervised by Millikan.

102. Millikan's measurement of the electron charge has been treated extensively elsewhere; see Millikan, *The Electron* (Chicago: Univ. Chicago Press, 1917); Paul Epstein, "Robert Andrews Millikan as Physicist and Teacher," *Reviews of Modern Physics*, 1948, *20*: 10-25; Gerald Holton, "Subelectrons, Presuppositions, and the Millikan-Ehrenhaft Dispute," *HSPS*, 1978, *9*: 161-224; Kargon, *Millikan*, pp. 58-66; Alan Franklin, *The Neglect of Experiment* (Cambridge: Cambridge Univ. Press, 1986), pp. 140-162, 215-225, 229-232. None of these works takes a serious look at the contributions of his students; see Harvey Fletcher, "My Work with Millikan on the Oil-Drop Experiment," *Physics Today*, 1982, *35*: 43-47.

103. Millikan and George Winchester, "The Absence of Photoelectric Fatigue in a Very High Vacuum," *Phys. Rev.*, 1909, *29*: 85.

104. Millikan, "Some New Values of the Positive Potentials Assumed by Metals in a Very High Vacuum under the Influence of Ultra-Violet Light," *Phys. Rev.*, 1910, *30*: 287-288.

105. Millikan's Scientific Notebook, "Photo Phenomena, 1911," RAM, Folder 3.6; James R. Wright, "The Positive Potential of Aluminum as a Function of the Wavelength of the Incident Light," *Phys. Rev.*, 1911, *33*: 4; Millikan and Wright, "The Effect of Prolonged Illumination on Photo Electric Discharge in a High Vacuum," *Phys. Rev.*, 1912, *34*:68-70.

106. Millikan, "The Effect of the Character of the Light upon the Velocities of Electrons Liberated by Ultra-Violet Light," *Phys. Rev.*, 1912, *35*: 76.

107. Millikan, "Über hohe Anfangsgeschwindigkeiten durch ultraviolettes Licht ausgelöster Elektronen," *Verhandlungen der Deutsche Physikalische Gesellschaft*, 1912, *14*: 712-726. Milikan reluctantly favored the latter possibility.

108. Millikan, "Effect of the Character of the Light," p. 76.

109. Millikan, "Atomic Theories of Radiation," *Science*, 1913, *37*: 122.

110. Ibid., p. 132.

111. Michel, "Controversy and Contribution: Einstein's Light Quanta Hypothesis and Millikan's Photoelectric Research Program" (M. A. Thesis, Univ. Wisconsin, 1975); Roger H. Stuewer, *The Compton Effect* (New York: Science History, 1975), pp. 72-75; Kargon, *Millikan*, pp. 71-72; John Hendry, "The Development of Attitudes to the Wave-Particle Duality of Light and Qauntum Theory, 1900-1920," *Annals of Science*, 1980, *37*: 67-68.

112. Loeb, student notebook, "Millikan's Quantum Theory Course," 1912, Lecture I, CHP; Millikan, lecture notes, "Unitary Theories," 1912, RAM, Folder 1.16.

113. Millikan, "Atomic Theories," p. 130; Millikan, "A Direct Determination of Planck's 'h'," *Phys. Rev.*, 1916, *7*: 383-384; *Nobel Lectures: 1922-1941*, p. 64; Loeb, student notebooks for "Electron Theory III, Prof. Millikan," 1914, Lecture 6, CHP, and "Millikan's Qauntum Theory Course," Lecture IV.

114. Martin Klein, "Einstein's First Paper on Quanta," *The Natural Philosopher*, 1963, *2*: 80-85.

115. Millikan, "Determination of Planck's 'h'," pp. 383-384; Millikan, *Electron*, pp. 218-219.

116. Loeb, "Electron Theory," Lecture 6, and "Quantum Theory," Lecture 35; Millikan, "Atomic Theories," p. 133; "A Direct Determination," p. 384; *Electron*, p. 230.

117. Millikan, "Atomic Theories," p. 132; lecture notes, "Atomic Theories of Radiation," 1913, RAM, Folder 1.17, cards 112-113.

118. Millikan, "Atomic Theories," p. 132. Millikan's emphasis.

119. Loeb, "Quantum Theory," Lecture 35, and "Electron Theory," TS summary [p. 6]; Millikan, "Determination of Planck's 'h'," p. 355.

120. Hendry, "Attitudes to the Wave-Particle Duality," pp. 73-75; Stuewer, *Compton Effect*, pp. 23-31; Martin Klein, "Einstein and the Wave-Particle Duality," *The Natural Philosopher*, 1964, *3*: 3-49.

121. Millikan, "Atomic Theories," p. 133.

122. Ibid., p. 144.

123. Millikan, *Electron*, pp. 231-238.

124. Loeb, "Quantum Theory," Lecture IV.

125. Millikan, lecture notes, "Atomic Theories," cards 106-109.

126. Loeb, "Electron Theory."

127. Vern O. Knudsen, "Interview History," 1966-1969, p. 150, CHP.

128. Millikan, *Autobiography*, p. 88.

129. Ibid.

130. Loeb, "Autobiography," p. 17; Sawyer interview by Charles Weiner, 1967, p. 6, CHP.

131. Millikan, "Effect of the Character of Light," p. 76; Millikan, "On the Cause of the Apparent Differences Between Spark and Arc Sources in the Imparting of Initial Speeds to Photo-Electrons," *Phys. Rev.*, 1913, *1*: 74; Millikan and Wilmer Souder, "Effect of Residual Gases on Contact E. M. F.'s and Photo-Currents," *Phys. Rev.* 1914, *4*: 73; Millikan and Souder, "Experimental Evidence for the Essential Identity of the Selective and Normal Photoelectric Effects," *Proceedings of the National Academy of Sciences*, 1916, *2*: 19-24.

132. "Physics," *Announcements*, 1931, p. 116.

133. Millikan, "A Direct Photoelectric Determination of Planck's 'h'," *Phys. Rev.*, 1916, *7*: 361, 380, 388; William Kadesch and Albert Hennings, "The Value of 'h' Determined Photoelectrically from the Ordinary Metals," and "The Relation of the Photo Potentials Assumed by Different Metals when Stimulated by Light of a Given Frequency," *Phys. Rev.*, 1916, *7*:147-148; Hennings and Kadesch, "Photo-Potentials due to Light of Different Frequencies," *Phys. Rev.* 1916, *8*: 209-220; Kadesch and Hennings, "The Value of 'h' Determined Photo-Electrically from the Ordinary Metals," *Phys. Rev.*, 1916,*8*:221-226; Souder, "Photoelectric Effect in Lithium, Sodium, and Potassium," *Phys. Rev.*, 1916, *8*:310-319.

134. "Physics," *Announcements*, 1931, p. 117.

135. Ibid, pp. 115-117.

136. Besides Kadesch, Hennings, and Souder, Winchester and Wright also did ancillary investigations on the photoelectric effect.

137. Carl Ramsauer, "Über eine direkte magnetische Methode zur Bestimmung der lichtelektrischen Geschwindigkeitsverteilung," *Ann. Physik.*, 1914, *45*: 961-1002; "Über die lichtelektrische Geschwindigkeitsverteilung und ihre Abhängigkeit von der Wellenlange," ibid., 1121-1159.

138. Fletcher, "My Work with Millikan," pp. 46-47. Some denounced openly "the use of graduate students to promote the interest of the teacher [as] simply a type of scholarly graft." G. A. Miller, "Ideals Relating to Scientific Research," *Science*, 1914, *39*: 812.

139. Kevles, "The Study of Physics in America, 1865-1916" (Ph.D. diss., Princeton Univ., 1964), pp. 80-87; Charles Weiner, "Science and Higher Education," in *Science and Society in the United States*, ed. David D. Van Tassel and Michael G. Hall (Homewood: Dorsey Press, 1966), pp. 83-84; Hamilton Cravens, "American Science Comes of Age," *American Studies*, 1976, *17*: 61-62; Kevles, "Physics, Mathematics, and Chemistry Communities," pp. 144-145; Hugh Hawkins, "University Identity: The Teaching and Research Function," in *The Organization of Knowledge in Modern America*, 1860-1920, ed. Alexandra Oleson and John Voss (Baltimore: Johns Hopkins Univ. Press, 1979), pp. 285-312.

ON THE HESITANT RISE OF QUANTUM PHYSICS RESEARCH IN THE UNITED STATES[*]

Gerald Holton

A. A. Michelson accused himself of having been responsible, in part through the ether-drift experiment performed at the Case School of Applied Science, for spawning the "monster" of Einstein's relativity theory, the devourer of his beloved ether. Michelson has been acquitted of that charge.[1] But if he also regarded the launching of quantum physics research in the United States as a monstrosity, he might have been a little more justified to feel implicated -- if only quite indirectly, through a response one of his other experiments evoked, and through actions of two of his close followers, D. C. Miller and R. A. Millikan. This paper will touch on this unwitting connection while pursuing its main aim: to determine when, and under what circumstances, original research based on a belief in quantum physics began in the United States.

The usual answer to this question is that whereas in Europe quantum physics had become a fruitful research field by the time of the Solvay Conference of 1911, in America this became the case only with A. H. Compton's announcement in 1922 of his unanticipated and reluctant discovery of the effect bearing his name.[2] But while that was indeed a turning point for physics everywhere, it was not the result of quantum-theory-based research. Moreover, the evidence to be presented here is that the successful beginning of this type of investigation occurred years earlier, long before American physicists, according to the traditional view, were considered ready for it.

Our study will also contrast the two main types of responses in the United States to the new physics during the second decade of this century, with special attention to the underlying thematic differences. It will use as exemplars the contributions of Edwin C. Kemble and Robert A. Millikan. Kemble, who had been a student of D. C. Miller at the Case School of Applied Sciences, was an isolated beginner in his twenties, but eager to accept quantum ideas as the starting point of his theoretical researches -- and a harbinger of the new group of self-confident young theorists that would soon come to the fore. Millikan, twenty-one years older than Kemble, was then a well-established colleague of A. A. Michelson at the University of Chicago, and like him a superb experimental physicist, typical of the best of the older style of American physicists, yet also typical of the community at the time in resisting as long as he could the basic quantum conceptions which his own experiments were supporting.

The main action in this case, therefore, takes place about 1912-1917 and involves only a handful of American physicists. Ten years later there would be in place a sufficient number of quantum theoreticians of as high a caliber as was associated with the earlier tradition of American experimentalists. Not everyone saw at once that it was happening,[3] but by then the fundamental change had occurred.[4] And in their different ways, Kemble and Millikan had a large share of responsibility for making the transition possible.

THE UNINTENDED PREPARATION OF A QUANTUM PHYSICIST

In 1916-17, in now largely forgotten publications, Kemble became arguably the first American to use quantum-theoretical considerations to predict results that were not yet

---

[*] Copyright © 1988 by Gerald Holton. This essay is substantially Chapter 5 of the second (1988) edition of *Thematic Origins* (Ref. 1).

in the domain of experimental knowledge, to obtain experimental verification for them, to become thereby one of the first Americans who would be noted for work in this field by elite European scientists, and to start on a career that enabled him to train a considerable fraction of the new generation of American theoretical physicists.[5]

Unlike Millikan's biography, details of which have been amply documented and are widely available, Kemble's personal background has been little known. Some of the main facts will illuminate an understanding both of his own career and of the profession he was entering. He was born in a Methodist parsonage in 1889 in Delaware, Ohio, the son of a minister (as were so many scientists of his time). His parents had first met in a small frontier school, where his mother was a student and his father a temporary teacher on a year's leave from college; they were married immediately after her freshman year and spent some years in Mexico as Methodist missionaries. As a boy of eleven, Kemble reported later, he had a conversion experience during a Methodist Episcopal church revival meeting, and in his old age he still looked back on it as "the most important act of my life."[6] His earliest ambition was to be a missionary, and in that expectation he enrolled in the denominational college Ohio Wesleyan, in the town of his birth. But he soon became convinced that he was unsuited for such a calling because he found he "did not believe the religious theories were true," and thus discovered himself unable to accept the "personal divinity of Christ." This revelation, in his late teens, appears to have been a crucial event in his life. Those who knew him well felt that it had left a sense of guilt which Kemble never quite overcame and which may have been a factor in his habitual demand to exact of himself the highest standards and most intense service.

Apparently in reaction to this "unconversion" experience, he turned to the study of science, although a concern with religion and active church membership remained ever present in his life. His father had been an inventor on the side and had introduced him to scientific ideas. In pursuit of a new career, Kemble transferred to the Case School of Applied Science in Cleveland, Ohio (later named Case Institute of Technology). He had put himself through college, and he first intended to go into mechanical engineering. But he soon discovered his strength and interest in physics as well as in mathematics, taking courses in the Mathematics Department in, for example, partial differential equations, Fourier Series, and spherical harmonics. This was unusual for a college physics student at that time in the United States. Indeed, as John W. Servos has shown, the whole pattern of preeminence of the experimental rather than the theoretical aspects in most sciences in the United States until the 1920s is correlated with the limited mathematical training future scientists received in schools and colleges.[7]

Kemble also received a thorough introduction to the subject of waves, vibrations, and acoustics generally. That was not an accident. His main instructor throughout was Michelson's successor at Case, Dayton C. Miller (1866-1941), one of the leading experimentalists in acoustics. Kemble was the only physics student in his college class, with only one other in the class above and the class below his, and it appears that Miller, having come to see the quality of this student, worked closely with him. Kemble flourished under his care, and continued to have such high regard for Miller that he later dedicated his major book, *The Fundamental Principles of Quantum Mechanics* (1937), to him.

It is worth recalling that in Kemble's student years acoustics was a natural bridge to optics and hence to physics of the atom. In the last decade of the nineteenth century, mechanical-acoustical analogies were still the reigning ones, not least for explaining how spectral lines resulted from the emission or absorption of the atom. During the first decade of this century, the analogy had been modified to an electrical-acoustical one, with spectral lines considered to arise from the vibrations of electrons in the atom.[8]

Although his encounters with mathematics and acoustics would be an important preparation for the future quantum theorist, another occurrence in Kemble's senior year was more fortuitous, and concentrated his determination to become a research scientist. It came about while he was trying to repair a Phonodeik, a sound-pressure recording device Miller had invented. For this purpose Kemble decided to fashion a general mathematical theory -- which became part of his senior thesis --of how viscosity dampens a mechanical resonator. As he put it in an interview later, he had a sudden "state of intense alertness," a burst of bright ideas. This exalted state lasted about a week. It left him with a "vivid sense of the way in which mental activity propagates itself" and the feeling that such an experience produces a "break . . . in the picture." After Kemble graduated from Case in 1911, however, further study toward a higher degree seemed financially out of reach. To make ends meet, he joined the recently founded Carnegie Institute of Technology in Pittsburgh, where he taught elementary physics as an assistant instructor for two years.

MILLIKAN REPORTS ON QUANTUM PHYSICS, 1912

During this period there seems to have occurred an event that would be important in Kemble's career. To one of his interviewers, Kemble mentioned later that on a visit to Cleveland in December 1912 he had heard a lecture by R. A. Millikan. Millikan did give an address at that time at Case (as vice president of Section B of the American Association for the Advancement of Science, in a joint meeting with the American Physical Society), with the announced title "Unitary Theories of Radiation."[9] Just back from a six-month excursion to Europe, Millikan appears to have given at Case the earliest analysis by an American physicist in a scientific society meeting of the new science that was taking shape abroad. This was very likely the first time Kemble had heard about it, so it will be useful to set the stage for Millikan's first-hand report.

As Millikan recounted later in his *Autobiography*,[10] his trip had been an auspicious one at a time of rapid development. He had met Rutherford, Moseley, Darwin, Geiger, and Marsden at Manchester, Lord Rayleigh at Dundee, W. H. Bragg at Leeds, and Planck, Rubens, Nernst, Warburg, and others in Berlin, as well as visitors there such as von Laue. He had heard Franck and Pohl's analysis of current research on the photoelectric effect, which left it still unclear whether the available data could be uniquely related to Einstein's predictions. Above all, Millikan had attended Planck's lectures in Berlin on *Wärmestrahlung*,[11] and in this way, as well as socially and presumably in the weekly colloquia, he "saw much of Planck at this time."[12]

It deserves to be noted what a fortunate symbiotic relationship existed between these two, the world-renowned German theoretical physicist at one of the most distinguished universities and the younger American visitor from one of the newest universities, ready to present his first major experimental results. Millikan was just then obtaining, from the laboratory notebooks he had brought along for this purpose, the best value then available from his oil drop experiments for the charge on the electron ($e = 4.806 \times 10^{-10} \pm 0.005$ esu); he reported in Berlin on his investigations, later recalling: "If my memory serves me rightly, both Planck and Einstein were present at this meeting."[13]

Millikan's finding was just what Planck needed. In the first edition of his book *Wärmestrahlung* (1906),[14] Planck had derived the value of what he called the "*Elementarquantum der Elektrizität*" from the constants in his formula for radiation, obtaining $e = 4.69 \times 10^{-10}$ esu, whereas the best experimental value then available (J. J. Thomson's) was 6.5×10^{-10} esu, that is, 38 percent higher. Planck drew prominent attention to what had been a serious discrepancy only in the preface of the second (1913) edition of his book; by then he could add immediately, with evident pleasure, that "more recently, through the researches of E. Rutherford, E. Regener, J. Perrin, R. A. Millikan, T. Svedberg, and others, admirably more refined methods of measurement have decided, without exception, in favor of the value derived from the theory of radiation, falling between the values of Perrin and Millikan."[15]

Satisfying as this finding must have been for Planck, there was also another, more indirect connection with Millikan's work. If, despite the continuing attacks by Felix Ehrenhaft and his collaborators, one took Millikan's demonstration of the atomistic nature of the fundamental electric charge to be the final and sufficient evidence for it, as Planck evidently did,[16] one might come nearer to an understanding of the chief and troubling puzzle at the heart of quantum physics itself. In 1906 Planck had put the challenge in these words: it was the "disclosure [*Enthüllung*] of the full physical significance of the elementary quantum action of *h*."[17] How to reach this aim was a task for the future. But a cautious hint had appeared in the last sentence of his Conclusion, in the final paragraph of that book; there Planck had returned to the thought that one might expect a future stage of the theory to reveal "a more faithful explanation of the physical significance" of his universal quantum of action, one which will "certainly not take second place to that of the electrical elementary quantum."[18]

A suggestive coupling of this sort had been plausible enough from the beginning, not only because the electric charge appeared in the radiation formula, but because the discovery of two separate universal constants for atomic phenomena, *h* and *e*, raised quite naturally the tantalizing hope that some underlying connection would be found. It is therefore significant that six years later, in the second (1913) edition, Planck used those same words in the corresponding passage of his new book -- but now, with the new status of the electron, he was able to make an insertion in his final paragraph that signaled more visibly his long-standing hope of finding a bridge between *h* and *e*, a connection between the abstract oscillators on which his quantum theory of radiation was based and the now more firmly grounded conception of the elementary quantum of electric charge: Planck added the notion that electrons would be moving among, and interacting with, the oscillators.[19]

Clearly, Millikan had good reason to feel elated by the recognition being given to his first serious research results. On his return from Berlin in the fall of 1912, he evidently regarded himself as a proper presenter of Planck's theory of radiation-- although Planck's own views were about to be seen, at least in Germany, as among the most conservative and hesitant of the interpretations of quantum theory. But because of the even greater skepticism and conservatism of most American scientists at this time, including Millikan himself, there was a good fit between Millikan and his audience at Case as he gave his December 1912 talk.

Millikan began by dismissing the opinion, current two decades earlier, that physics in the main was complete, "and that future advances were to be looked for in the sixth place of decimals." On the contrary, in part through his own cited researches, physics had recently been fundamentally transformed through what he termed "the Triumphs of an Atomistic Physics," with "the last of its enemies" now silenced. The reign of the thema of atomicity or discreteness had failed to penetrate only one area, the "lost domain of *radiant energy*." But even there, Millikan said, claims were now being made for atomicity, that is, for the conception that there, too, "atoms of energy" or units of energy are involved (hence the original title of his talk, "Unitary Theories of Radiation").[20]

Millikan, always a good speaker, now surveyed in easily accessible terms the experimental basis of the successes and failures of the various opposing theories concerning both the "nature of radiant energy" and "the conditions under which such energy is absorbed or emitted by atomic or sub-atomic oscillators." But instead of dwelling on a division between the role of energy quanta in matter and of energy quanta in the ether, or presenting the historical flow from Planck's quantization (1900) to Einstein's light particles (1905) and specific heats (1907) to Planck's recent publications, he put forward for examination an array of "five distinct brands" of what he called "'quantum' theory of various degrees of concentration," ranging from the "least concentrated," namely that of Planck, to Einstein's 1905 theory, the "most concentrated form of quantum hypothesis." Millikan confessed that he found Einstein's theory, "however radical it may be," to interpret most fully the recent experimental facts. Nevertheless, he said, he could not recommend its adoption, above all because it could not be reconciled with "the facts of diffraction and interference, so completely in harmony in every particular with the old theory of either waves";[21] in any case, as he would indicate later, the abandonment of the wave theory of light was clearly impossible, particularly for a physicist from Chicago's Ryerson Laboratory, Michelson's home.[22] Indeed, a corpuscular theory of light was for Millikan "quite unthinkable"[23] and, as we shall see, so it remained for years.

Reporting on Planck's theory, Millikan noted that the original version had been modified after Poincaré's objections; in 1912 Planck had changed to the view that only emission can occur discontinuously, while the absorption process was continuous (although Millikan indicated in passing that Planck's earlier theory had accounted better for the observations by Nernst and Eucken on the anomalous specific heat of gases).[24] Where to place the mechanisms of discontinuity was still very puzzling. Perhaps, Millikan speculated, there was merit in J. J. Thomson's old theory, according to which units of light energy traveled as transverse tremors along tightly stretched Faraday tubes that acted as discrete threads embedded in the continuous ether, giving it a fibrous structure.[25] At any rate, Millikan was forced to this conclusion: "We have at present no quantum theory which has thus far been shown to be self-consistent or consistent with even the most important of the facts at hand."[26]

But while the acceptance of the thema of discreteness throughout physics was unthinkable for Millikan in his December 1912 speech, another frontier was opening before his eyes. One could, he said, now try to assign "a structure to the ether," to account for the apparent localization of radiant energy in space. Thus, he said, we seem to be "on the eve" of at last learning more about the ether. Never had physics offered greater tasks and required better brains than now. And waxing more fervent still -- as well as reaching back to his own roots as a son of a minister -- Millikan ended his talk with this exhortation: "It may be that 'THOU art come to the Kingdom for such a time as this.'"[27]

Kemble had of course heard a command like this many years earlier, in quite another context. If he was now listening to these words, they must have left a very complex impression on him. Those who knew him well were always puzzled why anyone's mention of Millikan's name would quite uncharacteristically produce a cloud over his face, and a quick change of subject.

THE EXCITEMENT OF *h*

At about that time there occurred yet another crucial turn in Kemble's career, also owing to D. C. Miller. On the basis of Miller's recommendation, Wallace Sabine at Harvard, himself distinguished both as an acoustician and as a discoverer of young talent -- he brought to Harvard several promising young scientists, including Theodore Lyman, P. W. Bridgman, and Francis Wheeler Loomis -- offered Kemble a fellowship for graduate study. It was discovered later that Sabine was in fact quietly financing it out of his own pocket.

Kemble arrived in Cambridge in 1913, and in his first graduate-student year took Sabine's course on optics, a side interest of Sabine's. Kemble liked it exceedingly. Unlike most physics students, he also enrolled in several advanced mathematics courses,[28] mostly in the Department of Mathematics, intent on studying a combination of physics and mathematics that today would be called theoretical physics but that was then still a rare pursuit for an American.[29] As almost everywhere else in America at that time, at Harvard, physics research was primarily experimental. The department was located in the Jefferson Physical Laboratory, itself designed to make a statement as a veritable temple of experimentation, with an unusual concentration of investigators and apparatus.[30] Bridgman was well started on his high-pressure researches, Edwin H. Hall was chiefly concerned with what one would call now solid state physics, and Sabine worked in acoustics. The rest of the Physics Department had divided the electromagnetic spectrum among themselves, so that the physics faculty was referred to as "the Spectroscopy Department." Theodore Lyman worked on the spectroscopy of the ultraviolet region, William Duane in X-ray spectroscopy, Emory L. Chaffee and George Washington Pierce on electrical oscillations associated with radiowaves, and (from 1919) Frederick A. Saunders on series spectra in the visible region.[31] As in most American physics departments, a graduate student was expected to become proficient in machine-shop work soon after arrival and to begin building the equipment after having chosen an experimental thesis topic in the first year.

Jefferson Laboratory was thus in many ways the least likely site for a young person to launch a career in quantum theory. The tradition and the investment of energy, personnel, and resources were all in the opposite direction. Not only had nobody in America done a Ph.D. thesis in quantum-theoretical physics, nobody at Harvard had written a theoretical thesis on any topic in physics. David L. Webster, who received his Ph.D. the year of Kemble's arrival, had tried but had not been given permission to do so. As Lyman was fond of saying, "Physics is an experimental subject."

Yet another determining event for Kemble occurred in the spring of 1915, toward the end of a course recently retitled "Radiation and the Quantum Theory," given by G. W. Pierce.[32] The course used Morton Masius's 1914 translation of the second edition of Max Planck's *Wärmestrahlung*.[33] That book -- derived from the lectures Millikan had attended in 1912 -- presented Planck's "second theory," in which absorption of radiation was held to be continuous. Like other Americans who taught this topic at the time, Pierce himself, according to Kemble,[34] was bothered by "the inconsistency between it and Maxwell's theory, [and Pierce] continued for years to search for a purely classical explanation of the phenomena on which quantum theory was based." It was in this setting that Kemble began to study the new physics. Once he had shaken off the initial doubts that he may have picked up from his instructor, he was ready for these ideas. As Kemble recalled later: "Anything with quantum in it, with h in it, was exciting."[35] He had hit on the area in which he would do his doctoral thesis research.

A look at Planck's book in Masius's translation shows how quantum theory may have appeared to the young student at the time. It is a strange book, much of it a classical survey, with the introduction of the "hypotheses of quanta" delayed to well past the halfway point in the book. But Masius's preface starts with a bugle call: "The profoundly original ideas introduced by Planck in the endeavor to reconcile the electromagnetic theory of radiation with experimental facts have proven of the greatest importance in many parts of physics. Probably no single book since the appearance of Clerk Maxwell's *Electricity and Magnetism* has had a deeper influence on the development of physical theories."[36]

In contrast, Planck's own, much more modest preface stressed that the greatest challenges lay still in the future. His theory, he warned, called "for improvements in many respects, as regards both its internal structure and its external form . . .[It] will require painstaking experimental and theoretical work for many years to come to make gradual advances in the new field . . .[The] fruits of the labor spent will probably be gathered by a future generation."[37] His caution, evident throughout his text, signaled both his realization of the incomplete state of the theory and his deep-seated ambivalence about what these strange ideas were doing to his beloved science. Planck's motivation was in a direction orthogonal to the one in which his findings were taking him. He hinted at this when he wrote that he had begun to feel early that the law governing black body radiation, depending not on the nature of the emitting bodies but only on temperature, "represented something absolute, and since I had always regarded the search for the absolute as the loftiest goal of all scientific activity, I eagerly set to work."[38] In 1899, just before he understood that the two constants a and b turning up in his equation for the entropy of a resonator in a radiation-filled cavity would reveal themselves as closely equivalent to h/k and h, and began to recognize the disruptive potential of what he was doing, he confided: "[It] may not be without interest to note that the use of the two constants a and b . . . offers the possibility of establishing units for length, mass, time and temperature which necessarily maintain their meaning for all time, and for all civilizations, even extraterrestrial and non-human ones, constants which therefore can be designated as 'natural units of measurement.'"[39]

"IS DISCONTINUITY DESTINED TO REIGN?"

Planck saw himself chiefly as taking part in the perfection of the physics of Newton and Maxwell, and even used his Nobel Lecture of 1920 in good part to express his concern, and virtually his apology, that his quantum hypothesis, which he called there "der Eindringling" [the intruder], must strike a theoretician brought up in the classical school as "a monstrous, and for the imagination almost unbearable, arrogation."[40] As late as 1926 he said that despite the usefulness of the concept of quanta, "the difficulties encountered in fitting the quantum hypothesis into the system of theoretical physics" were formidable and threatened what for him was still an overriding goal and undefiled thema, namely that "the picture of the physical universe . . . regain the complete unity of classical physics,"[41] -- a scientific *Weltbild* in which the "principle of determinism" and the "strictly causal outlook" were preserved also and were "strictly valid." Planck had been repelled from the beginning by any need for discontinuity, for remodeling "basically the physical outlook and thinking of man which . . . were founded on the assumption that all causal interactions are continuous."[42] In his publications and letters, as he developed his theory of radiation through its various stages, one can almost see him squirming under the stress of having to accommodate somehow a notion that was thematically unappealing to him. Planck later confessed that his introduction of what he usually pointedly called "the hypothesis of energy quanta" had been "a purely formal assumption," to which he felt driven because "no matter what the circumstances, may it cost what it will, I had to bring about a positive result."[43]

Just where he would place the discontinuity in the interaction between matter and radiation changed with time. Thus in a letter to H. A. Lorentz on 7 October 1908, it was the "resonator" (later called oscillator) that "does not respond at all to very small excitations," and it was therefore absorption which was not classical. Fifteen months later, writing to Lorentz on 7 January 1910, Planck said frankly: "The discontinuity [*Unstetigkeit*] must enter somehow . . . therefore I have located the discontinuity at the point where it can do the least harm, at the excitation of the oscillators. The decay can then occur continuously with constant damping." During 1911 and 1912, fashioning his "second theory," Planck again changed his mind. While an asymmetry between emission and absorption was preserved, the discontinuity was relocated. As Kemble would read in Planck's treatise, Planck now insisted that although energy is lost in quanta during emission, "in the law of absorption . . . the hypothesis of quanta has as yet found no room." "The absorption of radiation by an oscillator takes place in a perfectly continuous way."[44]

As an aside, let us note that the switch did not make Planck feel better about the importation of this unappealing conception. He wrote to Paul Ehrenfest on 23 May 1915: "For my part, I hate discontinuity of energy even more than discontinuity of emission."[45]

Planck's own reactions to the demands of the new physics serve to remind us that, as has been amply documented, the majority of European physicists were puzzled by the fundamental assumptions of the quantum theory, even while the successful applications to spectra and specific heats were beginning to be obtained. Thus Einstein tortured himself both in private and in public about his inability to find either an interpretation of Planck's constant in visualizable form or a theory built from fundamentals which would yield both the energy quanta and the electric charge quantum as consequences.[46] And Poincaré, in one of his last and most eloquent essays, written on his return from the summit meeting on quantum physics at the 1911 Solvay Conference, spoke for many when he concluded: "The old theories, which seemed until recently able to account for all known phenomena, have recently met with an unexpected check . . . A hypothesis has been suggested by M. Planck, but so strange a hypothesis that every possible means was sought for escaping it. The search has revealed no escape so far . . . Is discontinuity destined to reign over the physical universe, and will its triumph be final?" Only James Jeans, responding to this query in 1914, was unequivocal, and ready to hail the new thema: "The keynote of the old mechanics was continuity, *natura non facit saltus*. The keynote of the new mechanics is discontinuity."[47]

Nevertheless, like Planck and Einstein, a growing number of European physicists were eager to explore the potential of quantum physics despite their various doubts, accepting the theory in some version at least as the basis of vigorous theoretical or experimental research programs on their part (even while perhaps agreeing with W. Nernst's remark of January 1911 that quantum theory was at the bottom still only "a very odd rule, one might even say a grotesque one"). As we shall soon see, during the same decade such an approach was the exception in the United States. There our future quantum theorist would be surrounded by a community which, if it was attentive to quantum theory at all, used it chiefly for pedagogic purposes or as a challenge to demonstrate its implausibility, or -- more rarely -- to test experimentally one of its predictions.[48]

"A LONE WOLF, SMALL SIZE"

As far as one can tell from his writings, letters, and interviews, none of this widespread thematic anguish infected young Kemble. He was ready to be a discontinuist if it had pragmatic value. In any case, he had greater problems to attend to first. Excited by *h* and hoping to do his doctoral research on quantum theory, he faced the obstacle that the Physics Department at Harvard was not ready to accept a theoretical thesis. But worse, nobody on its faculty was sufficiently familiar with quantum ideas to guide a budding theoretician. It is therefore a remarkable fact, and perhaps an illustration of the greater willingness in America than in other parts of the world at that time to let sufficiently bright, insistent, and self-confident young people challenge the institutional traditions and take the risk of breaking into new fields, that the department gave in to Kemble's plea and accepted his plan for a purely theoretical thesis. (As it turned out, in the late stages of the work he added an experimental section to examine whether his theoretical predictions were reasonably closely fulfilled.) Initially Kemble had chosen Sabine to be his formal supervisor. But Sabine left for war work in 1916, so Bridgman agreed to take on this function, being perhaps more ready than his colleagues on the faculty to tolerate daring and, from all the evidence, less hostile to the quantum.

Thanks to Kemble's consistent writings, notes, and interviews on this subject, one can establish with fair confidence what his own general attitude to quantum physics was at that time, and one can also pinpoint when he found the specific problem to work on. On 26 February 1916, a fellow student of Kemble's in the department, James B. Brinsmade, gave a talk at the weekly colloquium on his own doctoral thesis research then in progress in experimental spectroscopy, the department's chief research interest.[49] That subject, however, contained a well-known, serious fundamental puzzle. As Kemble put it later:

> If the distribution of angular velocities of gas molecules followed the Maxwell-Boltzmann law, and if radiation were emitted in accordance with the classical electromagnetic theory, each emission and absorption frequency in a multi-atomic gas would be spread out into a continuous band whose width would depend on the [rotational] inertia, but would always be large compared with the width of a normal spectrum line. The absence of such continuous bands in the spectra of gases is one of the most incontrovertible evidences known that there is something radically wrong with the classical mechanics, or the classical electrodynamics, or both.[50]

Band spectra might thus be a key to the new physics.

The significant portion of Brinsmade's talk was his discussion of publications on the theory of the band structure of infrared absorption spectra by the young Danish physicist Niels Bjerrum.[51] Bjerrum had been dealing with the absorption spectra of gas molecules such as HCl, CO_2, and H_2O. Until recently, none of these spectra had been resolved; they had appeared merely as smears. But now, with better experimental techniques, they came to be seen as double bands, and there were even suggestions of line-like fine structure in some bands. Bjerrum had thought the width of these bands was of the order of magnitude to be expected from a superposition of molecular rotation on molecular vibration, the frequency center of each band being equal to the longitudinal oscillation frequency of the corresponding molecule.

That idea went back to Lord Rayleigh's proposal of 1892, according to which, on entirely classical grounds, an oscillator that would be emitting or absorbing light of frequency ν_0, if at rest, will absorb and emit in about equal proportions at frequencies $\nu_0 + \nu_r$ and $\nu_0 - \nu_r$ when also in rotation with frequency ν_r around an axis perpendicular to the direction of vibration.[52] As it happened, Lord Rayleigh's paper, on which much of the research depended that led later to the quantum theory of absorption spectra, had been triggered by one of Michelson's experiments; it was in fact a specific response to Michelson's finding that in applying interferometry to spectroscopy -- an advance on which Michelson's Nobel Prize award was based in part -- spectral lines became narrower as the pressure of the gas became lower.[53]

Bjerrum's role was a familiar one in the transition from one stage of a theory to another stage that challenges the earlier one fundamentally: he grafted a new mechanism onto the old one to see how well the recombinant form would do. Bjerrum's mentor, Walther Nernst, had recently turned his attention to the application of quantum theory to explain the specific heat of gases. Thus encouraged, Bjerrum proposed to adopt Nernst's suggestion that the rotational energy of the molecule is quantized;[54] hence the separation between the peaks in the absorption band should be given $\Delta\nu = h/(2\pi^2 I)$ (where I = rotational inertia). A corresponding doublet structure in absorption spectra had in fact been found experimentally in 1913 by W. Burmeister[55] and by Eva von Bahr.[56]

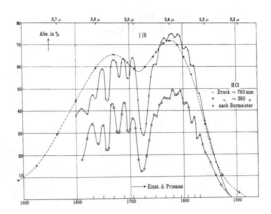

Figure 1

Absorption spectrum of HCl gas, as determined by Eva von Bahr (*Über die ultrarote Absorption der Gase*, p. 1153 [see note 56]), for two pressures, plotted against W. Burmeister's unresolved double band (*Untersuchungen über die ultraroten Absorptionsspektra einiger Gase*, p. 595 [see note 55]).

From the observed spacing of the doublets, one could calculate I and hence what the nuclear separation of the molecules should be. The plausible value of about 1.89×10^{-8} cm resulted from HCl. Moreover, $\Delta\nu$ was found to vary experimentally at least approximately with $\sqrt{T}$ as expected. And when von Bahr used Rubens's quartz prism instead of fluorite, the earlier hints of discontinuities in the absorption bands were confirmed as the bands resolved themselves into definite, nearly equispaced maxima and minima (Fig. 1), which Kemble soon came to call "quantum lines." Those were of course not plausible on the basis of classical theory; on the contrary, evidence of the quantization of the motions of the absorbing molecules now seemed to stand before one's very eyes.[57] As Kemble was to emphasize less than six months after Brinsmade's talk, "*Great as are difficulties to which we are led by that form of the quantum theory which assumes* [contrary to Planck's second theory] *that both the absorption and emission take place by quanta, it would appear that there is little hope of escape from this essentially kinematic confirmation of that theory*" (emphasis in original).[58]

We note that this quantization was still simply a superposition on the classical Rayleigh mechanism, the mechanical frequency of the molecule being equal to the frequency of the absorbed light. Also, the experimental data raised certain puzzles, such as the reason for the asymmetry found in the intensity in the two wings of the double band. But evidently Kemble's interest, as he listened to Brinsmade's account of the results, focused chiefly on two findings: the existence of quantum lines and the discovery of a mysterious, faint absorption band for CO, found at approximately half the wavelength of the main absorption band.[59] That showed Bjerrum's theory to be too limited, and cried out for a new idea. And Kemble was ideally prepared to propose one at this point; it would constitute a chief accomplishment of his Ph.D. thesis work during the fourteen months that elapsed between Brinsmade's talk and Kemble's submission, 1 May 1917, of his three-part dissertation.[60]

To summarize briefly, Kemble considered it axiomatic that contrary to Planck's authority,[61] absorption was a discontinuous process, with the molecular motions being responsible for its quantization. He saw that in a diatomic gas molecule, the lowest states of vibration would have to involve excursions that are an appreciable fraction of the mean distance between the atomic centers. Indeed, the amplitude of vibration of the molecule associated with absorption would typically be of the order of 10 percent of the normal distance between nuclei -- much too large for the force law to remain linear, as anyone trained in acoustics would immediately suspect. Therefore, vibrations of such an anharmonic oscillator, and the corresponding absorption spectrum, would contain overtones of the fundamental.[62]

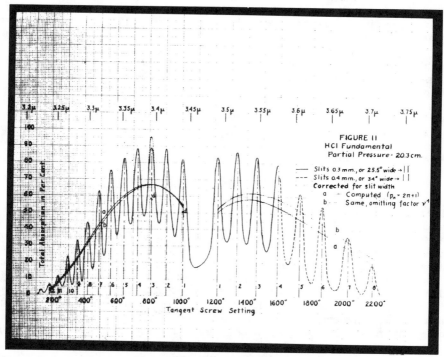

Figure 2

Absorption spectrum for HCl gas, from Figure 11 of Kemble's thesis essay (see note 60), p. 77, showing the well-resolved "quantum lines."

Thus Kemble was ready not only to apply quantum ideas but to modify significantly Planck's theory in each of the forms available to him. Contrary to Planck's second theory, Kemble considered absorption to be a fundamentally discontinuous process, and contrary to Planck's first theory he considered the oscillators to be anharmonic. His own version of the quantum theory not only explained the faint second absorption band in the spectrum of CO, with his calculations fitting the available observations quite well, he also went on to predict the wavelengths at which similar "first overtones" were to be sought in the spectra of HCl and HBr. In collaboration with Brinsmade,[63] whose quartz-prism spectrometer had been set up for a quite different study, Kemble found the expected absorptions quite close to (typically within 2 percent of) the predicted values.

The principal spectra he obtained also showed far better resolution (see, for example, Figure 2) than had been previously available, and he concluded that the observed "sharpness of the maxima of double absorption bands in the spectra of CO and HBr is incompatible with the newer form of the quantum theory [Planck's 1913 version] in which absorption is supposed to take place gradually and not by quanta." For as Kemble explained, according to Planck's second theory, the absorbing molecules should be able to have "all vibrational energies between zero and one quantum. In fact they are assumed to be uniformly distributed over the 'region elements' between these extreme values." Therefore, they should be able to absorb equally all corresponding wavelengths and so give rise to a "flat-topped elementary absorption band," one much wider than the separation of the doublet maxima that had been observed.[64]

His first results were reported at the December 1916 meeting of the American Physical Society.[65] When Kemble's thesis work is read today, two points are particularly arresting, and should have been when it was submitted to the Harvard Physics Department. First is its boldness under the circumstances. Kemble had taken on a difficult problem; three years later, in his Nobel address, Max Planck still had to admit that so far the "strange rotation spectra" had not yielded all their secrets to anyone. But despite the homemade form of the theory and its other inadequacies, Kemble challenged the current theory as too timid, proposed a version of his own, and obtained support for it from experiments he had designed in order to test it. There is no evidence that he was able to discuss his ideas with any sympathetic, informed person above the level of his fellow students. He had been trying to work it out on his own, a beginner surrounded by professors whose ideas and orientations were quite different from his own. As he put it in an interview later: "I was a lone wolf, small size."

THE RESISTANCE TO "RECKLESS" HYPOTHESES

The second surprising point is that none of the three parts of Kemble's dissertation made reference to energy levels or stationary states; they did not even carry a footnote to Niels Bohr's 1913 conception of the atom. The absorption frequencies were still considered to be the mechanical frequencies of rotation-vibration. Even the ideas discussed in James H. Jean's book,[66] or in the proceedings of the first (1911) Solvay Conference, to both of which Kemble made occasional reference in his work, were used so sparingly that one wonders if Kemble really had extensive access to these publications.

Toward the end of his life, Kemble, as always painfully concerned about any shortcomings in his work, felt so keenly the conceptual gaps which had become evident in his dissertation that he found it necessary to make a confession. He asked the librarian to attach a note to his thesis, then still on the shelves of the Physics Department library: "It would appear that during the period when the theoretical work here described was in progress, 1916-17, I had not heard of the Bohr theory."

Taken at face value, it is a baffling statement because on the published program of the American Physical Society meeting of 26-29 December 1916,[67] in New York, at which Kemble spoke on his thesis research, the very next paper to be delivered was listed as "A Criticism of the Rutherford-Bohr Atomic Hypothesis. Based upon a Theorem of Phase Equilibrium of Two Electrons," by Albert C. Crehore. Moreover, the high point of the same meeting was an elaborate symposium, "The Structure of Matter," led by an address by Millikan, now as president of the 600-member American Physical Society, titled "Radiation and Atomic Structure," in the course of which he summarized and endorsed Bohr's theory (as a theory of atomic structure, although not as a theory of radiation, since "the mechanism of radiation was still illusive").[68]

But other references to Bohr's theory in American science up to that time -- in journals, for example -- were few in number and usually critical in tone. This was true even of the other seven speakers at the December 1916 symposium. For example, G. N. Lewis voiced his "fundamental objection to the theory of the revolving electron," and spoke on the advantages of his "static atom"; W. J. Humphreys, on "The Magnetic Field of an Atom," managed to avoid any mention of Bohr; so did William Duane in his lecture "Radiation and Matter," warning also "that we are not compelled to believe that because Planck's radiation law fits the fact of black body radiation more or less closely, therefore energy must be radiated in quanta $h\nu$." Lauder William Jones, speaking on the distribution of valence electrons, relegated Bohr to one sentence in a footnote, and William D. Harkin of the University of Chicago, in "The Structure of Atoms," mentioned Bohr only once, in passing, in order to warn that despite the successes which his colleague Millikan had just reported, Bohr's idea of a planar atom was not in accord with the apparently three-dimensional arrangements of the outer electrons of organic molecules.[69]

Therefore Kemble's retrospective note on not having heard of Bohr's theory becomes plausible if one interprets it as an indication of what was being taught at Harvard, of his isolation, and of the delay in the acceptance in the United States of Bohr's theory, as of quantum physics generally. Kemble himself, when still a student in Pierce's course in 1915 and before he fell under the spell of h, had begun one of his reports to Pierce with this paragraph: "I presume that all of us would agree that the quantum theory is quite distasteful. In working with the theory we have no definite mechanical picture to guide us nor have we any definite clear-cut principle as a basis of operations--physicists everywhere have been making strenuous efforts to find a method of escape from the theory. If such a method could be found I presume that we should all breathe a sigh of relief and sleep better thereafter."[70]

While Kemble had soon thereafter abandoned the escape attempt, the more typical reaction of his fellow countrymen was to regard the early forms of the quantum theory not as a challenge to improve them but -- aside from pedagogic use -- as an argument not to accept them as guides to research, either as a whole or in large portions.[71] To cite here only two more examples: A. L. Parson, from 1915 on, elaborated a widely cited theory of atomic structure entirely opposite to Bohr's; in it, each electron was a thin ring of approximately atomic size, with negative charge "distributed continuously on it," and which rotated on its axis "with a peripheral velocity of the order of that of light."[72] A. H. Compton published a theory of the electron in 1919,[73] partly based on Parson's, and proposed, on the basis of the data for the scattering of hard X rays and gamma rays, that there were just three possible models for the electron: a thin and flexible ring, a rigid spherical shell incapable of rotation, or a flexible spherical shell. In 1921 he modified that model; instead of a ring of electrons, there was now to be a more nearly isotropic distribution, with strong concentration of charge near the center.[74]

During his thesis work, Kemble must have kept a sharp eye especially on the positions of three of the more established physicists with respect to quantum ideas. Two were in the same laboratory building, and he seemed destined to cross the path of the third repeatedly. They were David L. Webster, already mentioned, who had obtained his doctorate in the same department three years earlier and who had stayed on as an instructor until 1917; William Duane, who had joined the Harvard Faculty in 1913 and had remained there; and Millikan, whose publications of 1913 and 1917 were cited in Kemble's thesis as giving "the latest values for the constants h and k."[75] But as a brief glance at each shows, none of these three was comfortable with the new physics. Thus Kemble found it necessary, in his thesis, to dismiss explicitly Webster's "suggestion . . . that the absorption and emission of radiation may both take place in accordance with the law of the classical electrodynamics, the quantum effect being really only a collision effect."[76] Nor did Webster soon change his mind. Writing with Leigh Page, he objected to Bohr's theory as late as 1921: "In the form in which the theory has usually been stated, the orbits in an atom are all supposed to lie in one plane. In this form it seems impossible to reconcile the theory with the chemical evidence, because a set of orbits in a single plane are so different from the positions where the chemist would put the electron."[77] The very successes of Bohr's theory were held against it:

The point that is unexpected then is that plane orbits can give such exact quantitative agreements where they do. This, in fact, tends to cast some suspicion on quantitative agreements as a sure proof of real truth in a theory. Indeed they are not such sure proof. There have been many mechanical ether theories that have been just as good quantitatively for a certain distance, but no more. And Fourier's theory of heat conduction in solids gave remarkable quantitative agreements with the facts of temperature distributions within the conducting bodies, although it was based entirely on the assumption that heat was a material substance, "caloric." But this assumption was not necessary. It was merely convenient, in the absence of anything better. And so it may be with Bohr's orbits.[78]

Professor Duane also would not have given Kemble much support while he was working on his thesis in a room nearby. Duane's experimental researches (with his student F. L. Hunt) had yielded the law for the X-ray wavelength limit. Although Duane and Hunt made it clear that their research had not been intentionally based on quantum theory, they at first could not deny that "our results furnish strong evidence in favor of the fundamental principle of the quantum hypothesis."[79] But Duane quickly retreated from the implications of his findings and proposed a derivation of Planck's radiation formula that avoided "any law according to which radiant energy can be produced or exist only in quanta $h\nu$."[80]

Millikan's stance may have been the most difficult one for Kemble. We recall that Millikan had stated in December 1912, presumably in Kemble's presence, that "no quantum theory was yet self-consistent," and that he held a corpuscular theory of light to be "quite unthinkable." Indeed, he repeatedly testified later on how great a struggle it was for him eventually to have to leave behind the continuum and accept the discreteness of light energy. Thus, even in 1916, after his exhaustive work had forced him to accept his own experimental proof of the exactness of Einstein's *equation*, Millikan did not also accept the underlying theory. Rather, like so many others, he opposed its implication, making the oft-quoted remark that the "theory by which Einstein arrived at his equation seems at present to be wholly untenable," and referring to Einstein's proposal of 1905 as "the bold, not to say the reckless, hypothesis of an electromagnetic corpuscle of energy $h\nu$."[81] Even in his acceptance speech of 1923 on receiving the Nobel Prize "for his researches upon the elementary charge of electricity and the photoelectric effect," Millikan confessed that "after ten years of testing and changing and learning and sometimes blundering, all efforts being directed from the first toward the accurate experimental measurement of the energies of emission of photoelectrons, now as a function of temperature, now of wavelengths, now of material (contact e.m.f. relations), this work resulted, contrary to my own expectation, in the first direct experimental proof in 1914 of the exact validity, within narrow limits of experimental error, of the Einstein equation, and the first direct photoelectric determination of 'Planck's constant h.'"[82]

At a time when Kemble was well advanced in his thesis work, Millikan, in his presidential address on 27 December 1916, to the American Physical Society meeting in New York,[83] cautiously expressed the thought that at least in some cases the emission of electromagnetic radiation by an electronic constituent of an atom must be "a sudden or explosive" process. But he was still embracing a classical theory of radiation; in 1917, in the conclusion of his widely read book *The Electron*,[84] Millikan expressed his strong misgivings about any theory except an ether-based one, coupled with what he called the "hazy" idea that "within the atom there exists some mechanism which will permit a corpuscle continually to absorb and load itself up with energy of a given frequency until a value at least as large as $h\nu$ is reached" -- a mechanism harking back to Planck's "second theory" which Kemble, in his dissertation research, had explicitly rejected, in order to forge ahead under the standard of discontinuity.

EPILOGUE

Kemble himself later expressed the reasonable speculation that the relative backwardness of the responses to the new quantum theories in the United States at that late date may have been caused in part by the onset of the war in Europe in 1914, having interrupted the flow of scientific books and journals. In particular he acknowledged in 1920 that until then he had no access to Sommerfeld's important work that improved on Planck's and Bohr's ideas and made them so convincing.[85] At any rate, in pursuing his research in 1916-17, Kemble was evidently deflected neither by his own ignorance of the latest advances abroad nor by the general disbelief and confusion in the United States.

If one tries to identify the idiosyncratic characteristics that most separated Kemble from other Americans, one must note two: his skill in mathematical methods within physics (in part reinforced by his self-study of Gibbs's writings) and his acceptance of discontinuity in physical processes at the atomic level. But in two other significant respects Kemble's approach in 1916-17 had, despite all the differences, also recognizable similarities with that of his less venturesome countrymen. First, he still had the penchant for visualizable processes and for what Sopka has properly identified as "the strong American interest in a variety of mechanically formulated models of atomic structure, which interest would serve to inhibit American participation in developing a quantum-theoretical approach to atomic structure."[86]

Moreover, he was at the time as handicapped as his compatriots in this continent-wide country by the absence of the more concentrated, continuous, lively intellectual interaction among bright fellow pioneers Millikan had glimpsed on his visit to a few research centers in Europe. That lack in turn was of course a function, despite some signs of improvement, of a whole set of continuing institutional weaknesses that fed on one another at the time: the absence of enough centers of productive faculty research; the ambivalence of university administrators about supporting research; the lack of a national program of subventions for physics (with fellowships for promising physicists and chemists not available until the postwar National Research Council program); the small size and relative passivity of the National Academy of Sciences in the prewar years; the inadequacy of the American Physical Society, under the sway of an elite that had most of its training and rewards in nineteenth-century science; the difficulty for U.S. journals, such as the *Physical Review*, of setting new standards through informed peer review.

Therefore something more was needed than the steadily improving mathematical training of science students, and the eventual growth to a kind of critical size of the groups of competent theoreticians. At least three other factors played a role in the rise of quantum theory in the United States after Kemble's first foray. One was the multiple effects of World War I on the scientific community, for example, in bringing together working teams, in enlarging the support structure, and in demonstrating the need for research both in the physical sciences and in applied mathematics.

The second major factor was the cumulative impact in the United States, from 1920 on, of the persuasive, experimentally certified successes of the Bohr-Sommerfeld atom and of Einstein's general relativity, added to the triumphs of the theories of black body radiation and specific heats. These drove home the lesson throughout the community in the United States that a career in theoretical physics was in every way as serious and fruitful an endeavor as had been the pursuit of experimental phenomena in the laboratory -- in short, that "science walks forward on two feet, namely theory and experiment," as Millikan put it in 1924 in the first sentence of his Nobel lecture.[87]

A third essential factor was an act of will among the better-trained young Americans to take on leadership roles in their profession in order to help one another raise the level of research in theoretical as well as experimental physics. An example of this spirit comes through in I. I. Rabi's recollections contained in the recent biography by John S. Rigden. While on a fellowship in Europe in the mid-1920s, Rabi "chafed under the general contempt toward American physics," particularly the scorn of Arnold Sommerfeld. Rabi saw that at the level below a Bohr, a Pauli, a Stern, or a Heisenberg, Americans were "equal to or better than the best," and he concluded: "What we needed were the leaders." Right then and there, Rabi, Edward Condon, and Robert Robertson "promised one another that they would put an end to the second-class status of American physics."[88] The European experience at the time seems to have had a similar effect on other Americans, such as Oppenheimer.

But all this lay in the future. During the transition years, from the middle of 1917 to the very early 1920s, this is how matters stood: the puzzles concerning the dualistic nature of matter and radiation, as well as their interaction, were not resolved. The likelihood -- so ominous to many -- was increasing that essential discontinuity and inherent probabilism would have to stay on as fundamental themata, threatening the ancient, rational assumption that all causal interactions are continuous. Indeed, these matters were getting more intractable, waiting for the new approaches that would soon culminate in the rise of quantum mechanics.[89]

Planck, who had helped give Millikan early prominence and whose book had introduced Kemble to the quantum, was yielding intellectual dominance to Einstein, Bohr, and other, younger theoreticians. Millikan, having beaten back all challenges to his work on the photoelectric effect, but not yet being fully reconciled to his unintended verification of the quantum theory of light, was about to consolidate his entry into the pantheon of great American experimenters. The new Institute in Pasadena, under his leadership, would become one of the greatest breeding grounds for fundamental experimental advances to be found anywhere, but would also support high theory. And by a curious symmetry, Kemble, becoming Harvard's first faculty appointee in the field, moved into position on the opposite coast, eager to build in Cambridge, right in the stronghold of experimentation, one of the major schools of theoretical physics. In the department's letter of invitation, Bridgman (16 March 1919) called on Kemble's "old-time idealistic ambition" to accept (despite a low starting salary) a clear local and national mandate: "If we can get the courses [two and a half advanced courses on theoretical physics, including quantum theory] well given, it ought to put Harvard pretty near the top in this country. What is more, it is a good beginning to putting this country on the map in theoretical physics."

That programmatic phrase,"a good beginning to putting the country on the map in theoretical physics," also summarizes well the outcome of Kemble's solitary labors of 1916-17. Taken by themselves they had not been in a class with the groundbreaking advances being made in Europe at the time. But his work demonstrated the emergence, on this side of the ocean, of the venturesome spirit that was prerequisite for making such advances in quantum theory. It countered the reigning hesitancy about embracing quantum ideas and put Kemble in a position of taking a leadership role during the transition period. He threw himself into the new missionizing task laid out by Bridgman with all his energy through his research,[90] through his meticulously prepared courses, and above all through his thesis supervision. Between 1922 and 1935, the twenty-six dissertations by Kemble's students, and by *their* students, represented about one-third of all theoretical physics dissertations completed during that period in all United States institutions.[91] He also taught a number of advanced courses to undergraduates who were to become influential in physics, including Robert Oppenheimer.

Everywhere in the country, the momentum was also gathering, with more than a dozen physics departments building up centers for theoretical research in the 1920s.[92] Some Europeans were not aware of the activity or had written off this part of the world as being averse to theory. But when the time came for quantum mechanics, from the mid-1920s on, to transform physics once more, the ground had been laid for the participation of the United States as a full member. As John H. Van Vleck, who had been Kemble's first Ph.D. student, put it so graphically,[93] "Although we did not start the orgy of quantum mechanics, our young theorists joined it promptly."

NOTES

1. See Holton, *Thematic Origins of Scientific Thought: Kepler to Einstein* (Cambridge: Harvard Univ. Press, 1973), first edition, pp. 317, 350, footnotes 160-162, for reports of Michelson's self-indictment. Émile Picard's quotation of Michelson's remark to him on relativity, "Je n'ai pas voulu cela," is referred to in the biography of Michelson by his daughter; see Ref. 53, p. 335. See *Thematic Origins*, Chapter 8, for the evidence, based on all documents that were available by 1973, that the influence of Michelson's ether-drift experiments on the genesis of Einstein's relativity theory was at best "indirect," and that Einstein guessed he "just took it for granted that it was true," as he explained later (ibid., pp. 283-285). That account has been supported by all additional first-hand documents that have come to light since. Chief among them are the following:

(a) Einstein's indication in a newly found letter of 1899 (in Document 57 of *The Collected Papers of Albert Einstein*, v. I, Princeton Univ. Press, 1987) that he read Wilhelm Wien's paper, "Ueber die Fragen, welche die translatorische Bewegung des Lichtäthers betreffen," *Annalen der Physik und Chemie 65, Beilage* (1898), pp. i-xvii; in it Einstein would have seen a discussion of ten "experiments with negative result" on the supposed existence of a fixed ether, the Michelson-Morley experiment being the last one on Wien's list, with Wien's clear acknowledgment of the need to adopt the "hypothesis" of a compensatory shrinking of the length dimensions of rigid bodies in order to rescue the interpretation of the experiment. See also John Stachel, "Einstein and Ether Drift Experiments," *Physics Today*, 1987, *40*: 45-47. Stachel concludes that from the documents available for the period 1899-1902, "the new evidence thus serves to confirm . . . [the] conclusion that the experiment [of Michelson-Morley] did not play a significant role in Einstein's work."

(b) Einstein's remark in a letter of 19 January 1909 to Arnold Sommerfeld (see M. Eckert and W. Pricha, "Die ersten Briefe Albert Einsteins an Arnold Sommerfeld," *Physikalische Blätter 40* (1984), p. 32: "If Fizeau's Experiment and the measurements concerning the velocity of light in vacuum had not been at hand, the material for establishing the Relativity Theory would have been lacking." The existing velocity-of-light values then available had come, first of all, from the well-established stellar aberration measurements, to which Einstein often referred later (e.g., as R. S. Shankland reported, Einstein on 4 February 1940 "continued to say that the experimental results which had influenced him most were the observations on stellar aberration and Fizeau's measurements on the speed of light in moving water. 'They were enough,' he said." Quoted in *Thematic Origins*, Ref. 1, p. 282). Among other well-established velocity-of-light experiments was Foucault's. On the other hand, Einstein accurately saw the use of the Michelson-Morley experiment as a force for gaining credibility for the relativity theory in the community of physicists, and he wrote to Sommerfeld on 14 January 1908 (Eckert and Pricha, ibid., p.30): "If the Michelson-Morley experiment had not placed us in greatest embarrassment, nobody would have perceived the relativity theory as a (half) salvation." The same reference to the justificatory and pedagogic usefulness of the Michelson-Morley experiment was later made by Einstein repeatedly; see the quotation and discussion in *Thematic Origins* Ref. 1, pp. 272-275.

For discussion of passages in a third-hand document consisting of a debatable English translation of a Japanese scientist's account (in Japanese), purporting to contain "the gist" of an impromptu talk which Einstein had given at Kyoto in 1922 (in German), see Arthur I. Miller, "Einstein and Michelson-Morley," *Phys. Today*, 1987, *40*: 8-13.

2. A. H. Compton, "A Quantum Theory of the Scattering of X-rays of Light Elements," *Physical Review*, 1923, *21*: 297 (abstract of talk given on 1 Dec. 1922, at the American Physical Society), and ibid., 483-502. "By the time Compton finally enunciated his quantum theory of scattering, he had explored every modification of *classical* electrodynamics known to man at the time," according to Roger H. Stuewer, *The Compton Effect: Turning Point in Physics* (New York: Science History Publications, 1975), p. 223. Emphasis in original.

3. Thus, as reported by S. S. Schweber, "The Empiricist Temper Regnant: Theoretical Physics in the United States, 1920-1950," *Historical Studies in the Physical Sciences*, 1986, *17*: 55-56, as late as the spring of 1927, when E. U. Condon visited P. A. M. Dirac in Göttingen, he asked Dirac if he would like to visit America, and received this reply: "There are no physicists in America." R. T. Birge, commenting on that response at the time, added: "That is worse than Pauli, who I understand credits America with only two." The work of Josiah Willard Gibbs was then widely appreciated, but Gibbs had died twenty-four years earlier.

4. For an exhaustive study of the relative standing of various Western countries with respect to personnel, funding, and productivity of the academic establishment, see Paul Forman, John L. Heilbron, and Spencer Weart, "Physics circa 1900," *HSPS*, 1975, *5*: 1-185, and a summary in John L. Heilbron, "Lectures on the History of Atomic Physics, 1900-1922," in *History of Twentieth Century Physics*, ed. Charles Weiner (New York: Academic Press, 1977), pp. 48-52. The academic physics profession in the United States in 1900 and 1910 was by many *quantitative* indicators not only not "backward" but was on a par with, say, Britain and Germany, and in some respects was better off and growing more rapidly. For example: Academic Physicists circa 1900 (all ranks, from professor to assistant) per 10^6 of population: Britain 1.7, Germany 2.9, United States 2.9. Annual investment in academic physics circa 1900, total per academic physicist: about the same in Britain and the United States, rather less in Germany. Posts at physics institutes (professors and junior faculty): at 25 United Kingdom institutions: 68 in 1900, 80 in 1910; at 21 German universities: 52 in 1900, 59 in 1910; at 21 United States institutions: 68 in 1900, 103 in 1910 (with an estimated 150 advanced students in 1900).

However, the one important difference, of special relevance to this discussion, is indicated by the number of posts in theoretical and mathematical physics (senior and junior faculty combined) for 1910: Austria-Hungary 9, Britain 2, France 4, German universities (i.e., excluding Technische Hochschulen) 12, Italy 12, *United States* 1.

For an account of the institutional developments, research, and teaching in theoretical physics after 1900 in Germany, see Christa Jungnickel and Russell McCormmach, *Intellectual Mastery of Nature: Theoretical Physics From Ohm to Einstein*, vol. 2 (Chicago: Univ. Chicago Press, 1986), chap. 26.

5. The analysis is based in part on interviews in 1962 and 1963 with E. C. Kemble by the project "Sources for History of Quantum Physics" in "Archives for the History of Quantum Physics," at the American Institute of Physics and other depositories; discussions with Kemble by K. R. Sopka, P. Galison, and myself; interviews transcribed and deposited at the American Institute of Physics Center for the History of Physics with a number of Kemble's contemporaries, including E. Feenberg, S. Goudsmit, J. R. Oppenheimer, I. I. Rabi, R. S. Shankland, J. C. Slater, H. C. Urey, and J. H. Van Vleck; publications by C. Fujisaki, J. L. Heilbron, M. Jammer, D. Kevles, M. J. Klein, T. S. Kuhn, A. I. Miller, A. Moyer, A. Pais, S. S. Schweber, R. H. Stuewer, and others; and the letters and manuscripts in Kemble's *Nachlass*, now deposited, together with the transcription of the discussions referred to, at the Harvard University Archives. I have also profited from discussions with Kemble's colleagues and members of his family; with A. J. Assmus; and above all with Katherine R. Sopka, on whose exhaustive treatise, *Quantum Physics in America, 1920-1935* (New York: Arno Press, 1980, and American Institute of Physics, 1988), I have drawn extensively. An early version of the material in this paper was presented at the History of Science Society's annual meeting, 24 October 1986.

6. When not otherwise identified, quotations of Kemble's are taken from one of the interviews, discussions, or papers listed in note 5.

7. John W. Servos, "Mathematics and the Physical Sciences in America, 1880-1930," *Isis*, 1986, *77*: 611-629. For example, in 1910 "a one-year course in elementary calculus was sufficient to satisfy the degree requirements for a physics major at Yale, Harvard, Stanford, California, and Michigan" (p. 616). (Of these, only California required that much for a chemistry major.) Moreoever, the deficiencies were rarely made up fully during postcollege studies. Servos points to Michelson and Millikan as eminent scientists whose careers were influenced by such a preparation in mathematics.

8. See C. L. Maier, *The Role of Spectroscopy in the Acceptance of the Internally Structured Atom, 1860-1920* (New York: Arno Press, 1981), p. 531.

9. Soon thereafter published as "Atomic Theories of Radiation," *Science*, 24 Jan. 1913, *37*: 119-133.

10. R. A. Millikan, *The Autobiography of Robert A. Millikan* (New York: Prentice-Hall, 1950).

11. Ibid., p. 94. Millikan reports there that he had been impressed by Planck's rejection "as completely untenable the idea that radiation itself could be corpuscular (photonic) in nature, although the emission of radiant energy by the atom had to take place discontinuously or quantum-wise."

12. Ibid., p. 95.

13. Ibid., p. 96. The *Berichte der Deutschen Physikalischen Gesellschaft* for 1912 (vol. 10) show that Millikan gave at least two talks on his researches, on the charge e and on the photoelectric effect.

14. M. Planck, *Vorlesungen über die Theorie der Wärmestrahlung* (Leipzig: Barth, 1906), p. 163.

15. M. Planck, *Vorlesungen über die Theorie der Wärmestrahlung*, 2nd ed., rev. (Leipzig: Barth, 1913), p. vii. Planck's *Vorwort* is dated November 1912. (Unless indicated otherwise, all translations are mine.) Fritz Reiche, in his *Quantentheorie: Ihr Ursprung und ihre Entwicklung* (Berlin: Springer, 1921), p. 16, was brought, by the near coincidence of Planck's and Millikan's value for *e*, to exclaim: "Ein wahrhaft erstaunliches Resultat!"

Planck also mentioned Millikan (twice) in his Nobel Lecture of 1920, *Die Entstehung und bisherige Entwicklung der Quantentheorie* (Leipzig: Barth, 1920).

16. For an account of the controversy between Millikan and Ehrenhaft, see G. Holton, *The Scientific Imagination: Case Studies* (New York: Cambridge Univ. Press, 1978), chap. 2. The battle, by no means over in 1912, dragged on for some years more. For example, the recently opened Nobel Prize Archives show that although Millikan was nominated for the prize in physics regularly from 1916 on, Svante Arrhenius, in the report he prepared on Millikan's work for the deliberations of the committee, noted as late as 1920 that even though most physicists had come to agree with Millikan in the dispute with Ehrenhaft, the matter was not yet resolved, and that Millikan should therefore not then be recommended for the prize. (I thank Robert Marc Friedman for providing this information.)

17. Planck, *Wärmestrahlung* (1906 ed.), p. 162.

18. Ibid., p. 221.

19. In the 1906 edition Planck had written that "an investigation would be required concerning the influence which a movement of the oscillators exerts on the processes of radiation" (pp. 220–221). In the 1913 version, he added, after the word "oscillators," the phrase "and of the electrons that fly back and forth between them" (p. 204). This idea may have been the forerunner of Planck's ill-fated and short-lived "third theory" of 1914, intended to remove the quantum altogether from the radiation process of both emission and absorption and to relegate it to the interaction between particles of matter (M. Planck,"Eine veränderte Formulierung der Quantenhypothese," *Sitzungsberichte der Königlichen Preussichen Akademie der Wissenschaften*, 28 (1914), I: 918–923; II: 330–335.)

M. J. Klein, "The Beginnings of the Quantum Theory," in Weiner, *Twentieth Century Physics*, discussed Planck's desire to find a connection between *h* and other basic natural constants, without which the quantum might remain only an ad hoc hypothesis. Klein first published there Planck's revealing remarks in a letter to F. Ehrenfest of 6 July 1905, that there may be "a bridge" from what then was still an "assumption" of the existence of an elementary charge quantum *e*, "to the existence of an elementary quantum of energy *h*, especially since *h* has the same dimensions and also the same order of magnitude as e^2/c" (ibid., pp. 15–16).

20. Millikan, "Atomic Theories of Radiation," pp. 119, 120, 122.

21. Ibid., pp. 120–121, 132.

22. R. A. Millikan, "Scientific Recollections," pp. 56–57, as cited in Robert H. Kargon, "The Convervative Mode: Robert A. Millikan and the Twentieth-Century Revolution in Physics," *Isis*, 1977, *68*: 520.

23. Millikan, "Atomic Theories of Radiation, p. 133.

24. M. Planck, "Über die Begründung des Gesetzes der schwarzen Strahlung," *Ann. Physik*, 1912, *37*: 642-656. Planck's 1913 book was of course not yet available, but Millikan may well have had a preview of it in Planck's lecture course in 1912.

25. J. J. Thomson, *Electricity and Matter* (New Haven: Yale Univ. Press, 1904), pp. 62-63. Millikan must have been much interested in Thomson's ideas, first given as the Silliman Lectures during Thomson's visit to America, if only because of chapter 4, "The Atomic Structure of Electricity."

26. Millikan, "Atomic Theories of Radiation," p. 133.

27. Ibid.

28. Kemble recorded that he took these courses, listed as mathematics courses and given by the Mathematics Department: Classical Mechanics (W. F. Osgood), Newtonian Potential Functions (B. O. Pierce), Functions of a Complex Variable (Osgood, Maxime Bôcher), Ordinary Differential Equations (Charles L. Bouton), Linear Partial Differential Equations of the Second Order ("Bouton?"). In addition, he enrolled in eight physics courses: Optics, Electric Oscillations and Waves, Kinetic Theory, Thermal Properties, Electron Theory (seminar), Electron Theory and Relativity (under Bridgman), Electricity, and the course of G. W. Pierce, to be mentioned shortly. On his own, Kemble read Gibbs, Christiansen's *Theoretische Physik*, Jean's *Dynamics*, Pell's *Rotational Mechanics*, Föppl's *Introduction to Maxwell's Theory of Electricity*, and Voigt's *Kristallphysik*.

29. Practitioners in theoretical physics, such as H. A. Bumstead, Max Mason, and Leigh Page, came somewhat later and were mostly concerned with electromagnetism. Apart from Gibbs, who had died in 1903, one of the few exceptions to this pattern in the early years of the twentieth century was G. N. Lewis at MIT; although primarily a physical chemist, he produced the first substantial work on relativity in the United States in 1908, with his student, R. C. Tolman: see S. Goldberg, "The Early Responses to Einstein's Special Theory of Relativity, 1905-1911" (doctoral thesis, Harvard Univ., 1968).

30. See Gerald Holton, "How the Jefferson Physical Laboratory Came to Be," *Phys. Today*, December 1984, *37*: 32-37.

31. Benjamin O. Peirce, a member of the Physics Department, taught mathematics, but he died in 1914.

32. As K. R. Sopka has found, similarly titled courses had begun to be given earlier at other institutions, such as MIT (1908-09) and Chicago (1909-10, under Millikan); by 1915 there were such courses also at Johns Hopkins, Princeton, and Yale. Sopka, *Quantum Physics in America*, p. 1.94.

33. M. Planck, *The Theory of Heat Radiation* (Philadelphia: Blakiston, 1914), "authorized translation" by M. Masius of the second edition (1913) of Planck's *Wärmestrahlung*. An important difference between the theory there presented and the earlier version was that the development now proceeded no longer from the statistics of energy elements but from the theory of the elementary domains in "state space." For a discussion of the development of Planck's ideas, see M. J. Klein, "Planck, Entropy, and Quanta, 1901-1906," *The Natural Philosopher*, 1963, *1*: 83-108; H. Kangro, *Early History of Planck's Radiation Law* (New York: Carne, Russak, 1976); T. S. Kuhn, *Black-Body Theory and the Quantum Discontinuity, 1894-1912* (Oxford Univ. Press, 1978), particularly pp. 197-222 and 235-

254, including the controversial interpretation that not until his "second theory" did Planck "allow any place at all for discontinuity" (p. 244); Kuhn's defense of this view in "Revisiting Planck," *HSPS*, 1984, *14*: 231-252; Abraham Pais, *"Subtle Is the Lord"* ... *The Science and the Life of Albert Einstein* (New York: Oxford Univ. Press, 1982), chap. 19; Armin Hermann, *The Genesis of Quantum Theory (1899-1913)* (Cambridge, Mass.: MIT Press, 1971); and Jungnickel and McCormmach, *Intellectual Mastery of Nature*, pp. 213-216, 252-268, 304-319.

In the United States, very few reviews had noted Planck's books of 1906 or 1913, or its 1914 translation, and these reviews had been mixed. For example, C. E. M. (probably C. E. Mendenhall, as a member of the editorial board of *Physical Review*), in the review of the 1913 edition in *Physical Review*, 1914, *4*: 76-77, revealed his reaction as follows: "Planck was led to the change in hypotheses which we have just outlined because the hypothesis of discontinuous absorption seemed unreasonable. A skeptic might question whether discontinuous absorption is any more unreasonable than discontinuous emission --or indeed whether 'reasonable' and 'unreasonable' have any meaning whatever in connection with much of the present day discussion."

34. Sopka, *Quantum Physics in America*, p. 1.58.

35. Interview of 1 Oct. 1963, in "Sources for History of Quantum Physics," p. 2.

36. Masius, in Planck, *Theory of Heat Radiation*, p. v.

37. Ibid., p. ix. See also p. 154, bottom.

38. M. Planck, *Scientific Autobiography and Other Papers* (New York: Philosophical Library, 1949), pp. 34-35. See also p. 13: "The outside world is something independent from man, something absolute, and the quest for the laws which apply to this absolute appeared to me as the most sublime pursuit in life."

39. M. Planck, *Physikalische Abhandlungen und Vorträge*, vol. I (Braunschweig: F. Vieweg, 1958), reprint of M. Planck, "Über irreversible Strahlungsvorgänge," *Ann. Physik*, 1900, *1*: 121.

40. Planck, Nobel Lecture (see note 15). His language evidences his painful state of mind, concerning, e.g., the idea that the light quantum would require "sacrificing" the chief triumph of Maxwell's theory, "for today's theorist a most unpleasant consequence"; the quantum of action announces "something unheard of," which "fundamentally overturns" our physical thinking based on continuity.

41. M. Planck, "Foreword," in N. M. Bligh, *The Evolution and Development of the Quantum Theory* (New York: Longmans, Green, 1926), p. 9.

42. Planck, *Scientific Autobiography*, pp. 136, 143-144, 44.

43. M. Planck, letter to R. W. Wood, 7 Oct. 1931, in "Sources for History of Quantum Physics," Microfilm 66; the original is reprinted in Armin Hermann, *Frühgeschichte der Quantentheorie* (1899-1913) (Mosbach in Baden: Physik Verlag), pp. 31-32.

A related notion, which also clearly gave Planck concern, was what he termed "the hypothesis of elemental chaos [*Hypothese der elementaren Unordnung*]." On first mentioning it, he added that he is not implying that this hypothesis really applies everywhere in nature (*Wärmestrahlung* [1913 ed.], p. 115); but it leads him directly and "necessarily to the 'hypothesis of quanta'" (Ibid., p. 118).

On the delay in full acceptance of the themata of intrinsic discontinuity and intrinsic probabilism in the atomic realm, see G. Holton, *The Advancement of Science, and Its Burdens: The Jefferson Lecture and Other Essays* (New York: Cambridge Univ. Press, 1986), chap. 7.

44. Planck, *Theory of Heat Radiation*, pp. 153, 161.

45. The original wordings of Planck's letters to Lorentz and to Ehrenfest appear in Kuhn, *Black-Body Theory*, pp. 303, 317, and 322.

46. E.g., in Einstein's letter to Sommerfeld, in M. Eckert and W. Pricha (note 1), and A. Einstein, "Zum gegenwärtigen Stand des Strahlungsproblems," *Physikalische Zeitschrift*, 1909, *10*: 185-193.

47. H. Poincaré, "L'hypothese des Quanta," *Dernières Pensées* (Paris: Flammarion, 1913), in the translation by J. H. Jeans with which Jeans ended his book *Report on Radiation and the Quantum Theory* (London: The Electrician, 1914), p. 90. I use this version because Kemble indicated in his thesis that he had access to Jeans's book. Jeans's opinion is given on p. 89.

48. One such relatively rare case was the work of Margaret Calderwood Shields, a student of Millikan's, who published her Ph.D. thesis research in the same year as Kemble, entitled "A Determination of the Ratio of the Specific Heats of Hydrogen at 18°C and - 190°C," *Phys. Rev.*, 1917, *10*: 525-540. It is purely experimental, but ends with the sentence: "For hydrogen at - 191°C, γ becomes 1.592, in general accordance with the quantum theory of specific heats."

Another case is Karl T. Compton's experimental work, with his professor at Princeton, the distinguished English physicist O. W. Richardson, on the photoelectric effect, starting in 1912, specifically to test the theories of this effect by Einstein and by Richardson (cf. *Philosophical Magazine*, 1912, *24*: 575-594. See also A. W. Hull, "The Maximum Frequency of X-rays," *Phys. Rev.*, 1916, *7*: 156-158, finding that "the maximum frequency is given by the quantum relation," contrary to the statement in a recent publication by E. Rutherford.

49. Brinsmade's thesis, entitled "Studies in the Absorption of Light by Gases and Crystals," was completed and accepted in 1917.

50. E. C. Kemble, R. T. Birge, W. F. Colby, F. Wheeler Loomis, and Leigh Page, *Molecular Spectra in Gases*, National Research Council Bulletin, 57 (Washington, D.C., 1926), p. 10, in the introductory chapter by Kemble.

51. N. Bjerrum, "Über die ultraroten Absorptionsspektren der Gase," in *Festschrift W. Nernst* (Halle: Knapp, 1912), pp. 90-98; "Über die ultraroten Spektren II. Eine direkte Messung der Grösse von Energiequanten," *Verhandlungen der Deutschen Physikalischen Gesellschaft*, 1914, *16*: 640-642, and "Über die ultraroten Spektren der Gase III," *Verh. Dtsch. Phys. Ges.*, 1914, *16*: 737-753.

52. Lord Rayleigh, "On the Interference Bands of Approximately Homogeneous Light; in a Letter to Prof. A. Michelson," *Philos. Mag.*, 5th. ser., 1892, *34*: 410.

53. A. A. Michelson, "On the Application of Interference Methods to Spectroscopic Measurements, II," *Philos. Mag.*, 5th ser., 1892, *34*: 280-299.

Because Lord Rayleigh's view on the radiation mechanism continued to be embraced by several of the actors in our story, its genetic connection with Michelson adds to the earlier suggestion that Michelson may be counted among the spirits that attended the birth of quantum physics -- if only reluctantly, and without ever endorsing an idea apparently so hostile to the conception of light waves. See Dorothy Michelson Livingston, *The Master of Light: A Biography of Albert A. Michelson* (New York: Charles Scribner's Sons, 1973), pp. 301-302, for the account of how Niels Bohr, "well aware of Michelson's devotion to the wave theory in spite of the popularity of the newer discoveries" in quantum physics, used his "tact, not to say cunning" in 1924 to obtain a grating from Michelson by implying that its use might help "harmonize the contradictory conceptions of light."

54. Probably Bjerrum proceeded independently of similar conceptions advanced by Pierre Weiss, H. A. Lorentz, and J. W. Nicholson. See Kuhn, *Black-Body Theory*, pp. 219-220.

55. Wilhelm Burmeister, "Untersuchungen über die ultraroten Absorptionsspektra einiger Gase," *Verh. Dtsch. Phys. Ges.*, 1913, *15*: 598-612.

56. Eva von Bahr, "Über die ultrarote Absorption der Gase," *Verh. Dtsch. Phys. Ges.*, 1913, *15*: 710-730, 1150-1158. See also pp. 731-737. Neither von Bahr nor Burmeister referred to Bohr's recent publication.

57. The resolution was still fairly primitive, but advances in technique much beyond what theory can handle are not always a service. I tend to agree with Fujisaki's remark, in an essay that contains a brief survey of the work of Bjerrum and von Bahr: "It was historically fortunate that vibration-rotation bands occurred in double bands. If fine-structure had occurred in advance of the development of the quantum theory, it would have been very difficult to speculate as to its physical meaning." Chiyoko Fujisaki, "From Deslandres to Kratzer, I: Development of the Understanding of the Origin of Infrared Band Spectra (1880-1913)," *Historia Scientiarum*, 1983, *24*: 74.

58. Kemble, "The Distribution of Angular Velocities among Diatomic Gas Molecules," *Phys. Rev.*, 1916, *8*: 689-700. The paper is dated 7 Aug. 1916, and was immediately followed, on Aug. 9, by his publication, "On the Occurrence of Harmonics in Infra-red Absorption Spectra of Gases," *Phys. Rev.*, 1916, *8*: 701-714. These would count later as the first two of three parts of his Ph.D. dissertation. The main results were undoubtedly presented by Kemble in his talk, "The Infra-red Absorption Bands of Gases and the Application of the Quantum Theory to Molecular Rotation," listed in the "Minutes of the New York Meeting" of the American Physical Society for 26-29 Dec. 1916, as printed in *Phys. Rev.*, 1917, *9*: 170.

59. Burmeister, "Untersuchungen über die ultraroten Absorptionspektra," p. 609.

60. Parts 1 and 2 are given in note 58. The third part was his typed essay, "Studies in the Application of Quantum Hypothesis to the Kinetic Theory of Gases and to the Theory of Their Infra-red Absorption Bands," dated 1 May 1917. In its Introduction, Kemble wrote: "With this paper and as a part of the dissertation submitted by the writer with his application for the degree of Doctor of Philosophy, are reprints of two articles [see note 58] which were recently published by the writer in the *Physical Review*." He then summarizes the nonexperimental portion of the subsequent material as follows: "In the

theoretical part of this essay the writer proposes (a) to discuss Planck's recent extension of the quantum theory to systems having more than one degree of freedom, pointing out the simple modification required to bring it into conformity with von Bahr's observation on the structure of the infra-red absorption bands of gases; (b) to discuss several reasons which indicate that the hypothesis of a zero-point energy must be retained though the second form of the Planck quantum theory is abandoned, and in particular, to treat the relation of the specific heat of hydrogen to this question; (c) to present certain considerations tending to show that the behavior of the more complicated molecular systems cannot be accounted for by the Planck theory even when modified in the manner indicated; (d) to formulate a theory of the asymmetry of the infra-red absorption bands of gases."

61. Kemble referred to Planck's second theory as given in the Masius translation of the 1913 book (see note 33) and to its elaboration in M. Planck, "Die Quantenhypothese für Molekeln mit mehreren Freiheitsgraden," *Verh. Dtsch. Phys. Ges.*, 1915, *17*: 407-418.

62. Kemble reported later in an interview of 1 Nov. 1981, that he felt there had been reasons for not bringing in a photon theory of light at the time, one argument being that the observed behavior of the dispersion curve in the neighborhood of a line fitted the classical theory, and he thought that this would be unlikely to be the case also under the photon theory.

As it turned out, Kemble's theory of the relation between the strong and weak absorption bands did not fit exactly the observations he made subsequently. At the time that difference seemed to him small and tolerable, but later it became a serious preoccupation. When he returned to these ideas after the war, having now studied Bohr's theory, he devoted much of his energy to developing his theory further, as shown in his detailed "Lecture Notes" for the first course he gave in quantum theory in the fall term of 1919-20 (Physics 16a). Kemble's first paper using the Bohr theory to explain the small deviations of the approximate harmonics as well as the asymmetries in the infrared spectra of diatomic gases was published early in 1920 ("The Bohr Theory and the Approximate Harmonics in the Infra-red Spectra of Diatomic Gases," *Phys. Rev.*, 1920, *15*: 95-109).

63. The collaboration also resulted in J. B. Brinsmade and E. C. Kemble, "Harmonics in the Infra-red Absorption Spectra of Diatomic Gases," *National Academy of Sciences, Proceedings*, June 1917, *8*: 420-425.

64. Kemble, "On the Occurrence of Harmonics," p. 714. Ibid., pp. 713-714. Kemble tried also to explain the asymmetrical shape of the observed principal absorption band spectra, where the high-frequency components were narrower and more intense, as being caused by the lowering of the frequency of atomic vibration by the expansion of molecules with larger angular velocities. Among several other significant contributions in Kemble's thesis are discussions of zero-point energy, the specific heats of gases, and the absorption spectra of poly-atomic molecules. It can also be read as a case study of the relation between instrumentation and theory construction.

A succinct and more up-to-date treatment of the chief problems on which Kemble worked is provided in Arnold Sommerfeld, *Atombau und Spektrallinien*, 3rd ed. (Braunschweig: F. Vieweg, 1922), pp. 505-520, where Kemble is cited. The English translation appeared as A. Sommerfeld, *Atomic Structure and Spectral Lines* (New York: E. P. Dutton, 1923), pp. 413-425. Kemble himself summarized and updated his findings in the monograph with R. T. Birge et al., *Molecular Spectra in Gases*, chap. 1.

65. "Minutes of the New York Meeting" (see note 58).

66. Jeans, *Report on Radiation*.

67. "Minutes of the New York Meeting" (see note 58), pp. 169-173.

68. R. A. Millikan, "Radiation and Atomic Structure," *Phys. Rev.*, 1917, *10*: 194-205. Millikan's greatest praise for Bohr's theory was occasioned by the fact that Millikan's calculation of the value of the Rydberg constant, using his own measurements of e and h, agreed to within 0.1 percent of its observed value: "This agreement constitutes most extraordinary justification of the theory of non-radiating electronic orbits" (ibid., p. 202). He may also have noticed that Bohr had adopted Millikan's value for e in his first paper on the model of the hydrogen atom.

69. These papers were published in *Science*, vol. 46, between 21 Sept. and 23 Nov. 1917. An even more negative picture results from reading issues of *Physical Review* between 1913 and mid-1917, the period within which Kemble's thesis was in progress. For further documentation of the uncertain progress of Bohr's as well as Planck's ideas in the United States, see Sopka, *Quantum Physics in America*, particularly pp. 1.49-1.62; and Daniel J. Kevles, *The Physicists* (New York: Alfred A. Knopf, 1978), chaps. 6 and 7.

70. Quoted in Sopka, *Quantum Physics in America*, p. 1.56. For an analysis of the general bewilderment upon the breakdown of mechanical pictures and visualizable processes, see Arthur I. Miller, "Visualization Lost and Regained: The Genesis of the Quantum Theory in the Period 1913-1927," in *On Aesthetics in Science*, ed. Judith Wechsler (Cambridge, Mass.: MIT Press, 1978).

71. The reactions between about 1915 and 1920 of G. S. Fulcher, G. N. Lewis, I. Langmuir, A. L. Parson, R. A. Millikan, E. Q. Adams, R. C. Tolman, A. H. Compton, and William Duane are summarized in Sopka, *Quantum Physics in America*, pt. I. A possible exception was R. T. Birge.

72. A. L. Parson, "A Magneton Theory of the Structure of the Atom," *Smithsonian Miscellaneous Collection*, 1915, *65*: 1-80.

73. A. H. Compton, "The Size and Shape of the Electron," *Phys. Rev.*, 1919, *14*: 20-43.

74. Compton, "The Magnetic Electron," *Journal of the Franklin Institute*, 1921, *92*: 145-155. Bohr's conception is mentioned in passing, only to be rejected.

75. Kemble, "Studies in the Application of the Quantum Hypothesis," p. 19. Incidentally, it is interesting to note that like Millikan and Kemble, Duane and the two Compton brothers cited above were all sons of ministers.

76. Ibid., p. 90. He added: "In view of the fact that the form in which Dr. Webster's suggestion would throw the quantum theory gives no explanation of the phenomena of photoelectricity, the writer is not greatly surprised at its failure when applied to infrared absorption bands" (ibid., p. 91). Kemble did not say where Webster had made his suggestions, but he could well have referred to D. L. Webster, "Experiments on the Emission Quanta of Characteristic X-rays," *Phys. Rev.*, 1916, *7*: 599-613, which opts for "the laws of classical electromagnetics" to explain X rays and mentions Bohr's atom only to indicate that Parson's theory is more likely.

77. D. L. Webster and L. Page, *A General Survey of the Present Status of the Atomic Structure Problem*, Bulletin of the National Research Council, 14 (July 1921): 345.

78. Ibid., pp. 345-346. Emphasis in original.

79. W. Duane and F. L. Hunt, "On X-ray Wave-Length," *Phys. Rev.*, 1915, 6: 169.

80. W. Duane, "Planck's Radiation Formula Deduced from Hypotheses Suggested by X-ray Phenomena," *Phys. Rev.*, 1916, 7: 143-146.

81. R. A. Millikan, "A Direct Photoelectric Determination of Planck's *h*," *Phys. Rev.*, 1916, 7: 355, 383. See also Millikan's first extensive proof of the validity of Einstein's equation for the photoelectric effect, published on 1 June 1916, in the *Physikalische Zeitschrift*, 1916, 7: 220, where he confessed -- just as he had indicated four years earlier in his talk at Case -- that although the result of his experiment undoubtedly corresponded exactly with the equation, "nevertheless, the physical theory behind it appears to be completely invalid." In his 1917 book *The Electron* (Chicago: Univ. Chicago Press), p. 230, Millikan reasserted his rejection. (This contradicts his later reminiscence, in his *Autobiography*, pp. 101-102, that he accepted Einstein's theory as early as April 1915, a statement Roger H. Stuewer, *The Compton Effect*, p. 88, rightly calls "rather shocking." But the policy of conducting experiments that tested Einstein's *equation* without accepting the underlying theory was then not unusual.)

82. Millikan, *Autobiography*, pp. 102-103. Still later he went so far as to stress that all of the "most notable discoveries" in physics between 1895 and 1905 -- including X rays, radioactivity, the electron, "Planck's Quanta," and Einstein's relativity and energy-mass relation -- merely "supplemented the discoveries of the preceding centuries without notably clashing with or contradicting any of them." The only exception to this was that the "photoelectric equation had been set up on the definite assumption of an essentially corpuscular theory of light" (ibid., p. 106) -- an assumption to which Millikan had been unable to give his assent for years.

83. Millikan, "Radiation and Atomic Structure." As noted, Kemble may well have heard this talk by Millikan.

84. Millikan, *The Electron*, pp. 234, 237-238.

85. A. Sommerfeld's articles of 1916 in the *Annalen der Physik* and the great book *Atombau und Spektrallinien* (Braunschweig: F. Vieweg, 1919). The volume quickly became the "bible of the modern physicist" (as noted in Max Born's letter to A. Sommerfeld, 13 May 1922), except for the antirelativistic opposition, well described in Helge Kragh, "The Fine Structure of Hydrogen and the Gross Structure of the Physics Community, 1916-26," *HSPS*, 1985, 15: 67-125. Kemble's acknowledgment just referred to is to Sommerfeld's articles and appears at the end of Kemble's "The Bohr Theory." It may of course be argued that Kemble's isolation while writing his dissertation was lucky, because if he had seen Sommerfeld's sophisticated treatment of the spectroscopic problems, he might well have regarded his own approach as too naive and might have lost heart. (I thank A. Pais for this observation.)

86. Sopka, *Quantum Physics in America*, p. 1.54.

87. Millikan, "The Electron and the Light-Quant from the Experimental Point of View," in *Les Prix Nobel en 1923* (Stockholm, 1924), p. 1.
The sociological circumstances of scientific progress during periods of rapid advance in the first two decades of this century were somewhat different for experiment and for theory. That is, with respect to *experimental* physics, the mechanisms of acquisition of sound skills and certification of sound achievement worked well enough in a thinly spread community, i.e., even in the absence of many centers of experimental work in the United States as intense and populated as, say, the Jefferson Laboratory. A Michelson at Case or a Millikan at Chicago, with one or two collaborators, could evidently achieve superb results that were endorsed around the world. But major *theoretical* achievements appear to have depended more often on the local availability of a critical-size community within which new ideas could be acquired, tested, debated, modified, and eventually certified.

88. John S. Rigden, *Rabi: Scientist and Citizen* (New York: Basic Books, 1987), p. 36.

89. Even in the elaboration of his 1927 lecture on the complementarity principle, Niels Bohr could still write in 1928 that "the so-called quantum postulate, which attributes to any atomic process an essential discontinuity, or rather individuality, completely foreign to the classical theories" is characterized by "an inherent irrationality." N. Bohr, "The Quantum Postulate and the Recent Development of Atomic Theory," supplement to *Nature*, 14 Apr. 1928, *121*: 580. In addition, it is well documented that Bohr, longer than most major physicists, continued to be skeptical about the fundamentality of the light-quantum concept, for example saying at the 1921 Solvay Conference that it "presents apparently insurmountable difficulties from the point of view of optical interference." See John Hendry, *The Creation of Quantum Mechanics and the Bohr-Pauli Dialogue* (Dordrecht: D. Reidel, 1984), chap. 3. For an account of the history of the physics community's wrestling with the dualistic ideas from 1896 to 1925, see Bruce R. Wheaton, *The Tiger and the Shark: Empirical Roots of Wave-Particle Dualism* (New York: Cambridge Univ. Press, 1983). For Werner Heisenberg's report that until 1927 many in Bohr's circle were in "a state of almost complete despair," because "one would argue in favor of waves, and the other in favor of the quanta," see "Discussions with Heisenberg," in *The Nature of Scientific Discovery*, ed. Owen Gingerich (Washington, D.C.: Smithsonian Institution Press, 1973), pp. 567, 569.

90. Kemble regularly published research results in the 1920s and 1930s, beginning with the paper mentioned in note 62, and he also published two important monographs: *Molecular Spectra in Gases* (he had been appointed in 1923, at age thirty-four, to chair the commission preparing the report); and the influential survey -- the first in the United States -- of the new quantum mechanics, "General Principles of Quantum Mechanics," Part I, *Physical Review*, suppl. 1: 157-215, 1929, with E. L. Hill; Part II, *Review of Modern Physics*, 2: 1-59, 1930. This study became the basis for Kemble's widely used textbook, *The Fundamental Principles of Quantum Mechanics* (New York: McGraw-Hill, 1937). Indications of some of Kemble's contributions and their early visibility among European physicists are given in Sopka, *Quantum Physics in America*, and in Jagdish Mehra and Helmut Rechenberg, *The Historical Development of Quantum Theory*, vol. I (New York: Springer-Verlag, 1982).
Among other evidences that quantum physics was being "naturalized," see the pedagogically intended synopses starting to appear in 1920, including Edwin P. Adams, *The Quantum Theory*, National Research Council Bulletin no. 5 (Washington, D.C.: National Research Council, 1920), and L. Silberstein, *Report on the Quantum Theory of Spectra* (L. Hilger, 1920). (Silberstein immigrated to the U.S. in 1920.)

91. Among the persons attracted to work with Kemble were J. C. Slater, R. S. Mulliken, Gregory Breit, and Clarence Zener. See Sopka, *Quantum Physics in America*, p. 4.67. More details are given in the "Minute On the Life of Edwin C. Kemble," *Harvard Gazette*, 15 May 1987, pp. 8-9. For claims concerning Harvard's position among the early centers in quantum theory in the United States, see John C. Slater, "Quantum Physics in America between the Wars," *Phys. Today*, 1968, *21*: 45; Fredrich Hund, *The History of Quantum Theory* (London: George G. Harrap, 1974), p. 246; and Roger L. Geiger, *To Advance Knowledge* (Oxford Univ. Press, 1986), pp. 233-240 (where Kemble's role is noted). But the early handicaps for building up a major theoretical school at Harvard are hinted at in a reminiscence of Van Vleck: "The problem I worked on was trying to explain the binding energy of the helium atom by a model of crossed orbits which Kemble proposed independently of the great Danish physicist, Niels Bohr, who suggested it a little later. In those days the calculations of the orbits were made by means of classical mechanics, similar to what an astronomer uses in a three-body problem. The Physics Department at Harvard did not have any computing equipment of any sort, and to get the use of a small hand-cranked Monroe desk calculator, I had to go to the business school." (John H. Van Vleck, "American Physics Comes of Age," *Phys. Today*, June 1964, *17*: 22).

92. By the early 1930s they came to include the University of California, Chicago, Columbia, Cornell, Harvard, Johns Hopkins, Illinois, MIT, Michigan, Minnesota, New York University, Ohio State, Princeton, Wisconsin, and Yale, among others. See also Schweber, "Empiricist Temper Regnant," pp. 69-77; Spencer Weart, "The Physics Business in America, 1919-1940," in Nathan Reingold, ed., *The Sciences in the American Context* (Washington, D.C.: Smithsonian Institution Press, 1979); and Stanley Coben, "The Scientific Establishment and the Transmission of Quantum Mechanics to the United States," *American Historical Review*, 1971, *76*: 458.

For the development of an identifiably American approach to quantum mechanics, relatively free of metaphysical anguish and stressing the operational-instrumental interpretation of concepts, see Schweber, "Empiricist Temper Regnant"; Coben, "Scientific Establishment"; Holton, *Advancement of Science*, chap. 7; and Nancy Cartwright, "Philosophical Problems of Quantum Theory: The Response of American Physicists," in *The Probabilistic Revolution*, v.2, ed. Lorenz Krüger and Gerd Gigerenzer (Cambridge: MIT Press, 1987), chap. 16.

93. Van Vleck, "American Physics Comes of Age," p. 24.

ETHER/OR:
HYPERSPACE MODELS OF THE ETHER IN AMERICA

James E. Beichler

It is often stated that each generation of historians rewrites history. Insofar as such statements are true they reflect both new research in older areas of history and changing attitudes toward, or new interpretations of, older historical research. In this year, 1987, we are celebrating the one-hundredth anniversary of Albert Michelson's "ether-drift" experiment and reviewing both its historical significance with respect to the rise of American science and its impact on the international development of physics. In the case of Michelson, we have long known of and debated about the role that his ether detection experiments played in subsequent theoretical developments by Hendrik Lorentz, Henri Poincaré and Albert Einstein. These theories themselves bear a close historical relationship to the rising tide of relativism that followed Ernst Mach's exposition of the logical inconsistencies within Newton's concept of absolute space. Yet historians have been slow to recognize other developments in the concept of space that dealt primarily with the still earlier mathematical development of the non-Euclidean geometries.

Just one short year after the publication of Michelson's experimental results, Simon Newcomb, a world-renowned American astronomer, gave a short lecture before the Philosophical Society of Washington concerning the problems raised by such experiments and recent debates on "action-at-a-distance."[1] Newcomb was quite well acquainted with Michelson's and similar experiments. He had completed his own experiments to measure the speed of light and had corresponded directly with Michelson over the past several years,[2] offering advice to Michelson for improving his experimental apparatus. In his lecture, the staid American scientist Newcomb publicly suggested a model of the luminiferous ether based upon a hyperdimensional space. His intention was to overcome the philosophical and experimental obstacles presented by the recent inquiries.

A completely satisfactory explanation between matter and the ether is entirely wanting. The following mystery seems to surround it. Firstly, the ether transmits vibrations as if it were an elastic fluid yet it offers no resistance to the continuous motion of matter of any sort. It may be remarked in passing that this seeming contradiction may be accounted for by supposing that the ether is really an elastic three dimensional solid, which however does not fulfill the same medium space in which the material universe exists but, such a space removed from it by some infinitesimal amount which we conceive to be the radius of the repulsive force which envelopes the atoms of matter. We might even suppose two such solids enclosing the material universe between them. This hypothesis implies the possibility of motion in a fourth dimension of space, a motion of which no physical evidence has yet been obtained. In view of the old metaphysical superstition that the restriction of space to three dimensions is a priori necessary a brief digression upon this subject may be permitted.[3]

This model of our three-dimensional space, sandwiched between two four-dimensional solid ethers, was for Newcomb a way out of the dilemma of detecting the ether.

Although he may have been coerced to accept this radical hypothesis by the prevailing situations, Newcomb's name had long been associated with the development of non-Euclidean geometries in their earlier and simpler synthetic forms. His first publication on the matter, "Elementary Theorems relating to the Geometry of Three Dimensions and of Uniform Positive Curvature in the Fourth Dimension," appeared in *Crelle's* magazine in 1877.[4] In that article the subject was treated as a purely mathematical and non-analytical approach to single-elliptic geometry. The following year, Newcomb published a second article, entitled "Note on a Class of Transformations Which Surfaces May Undergo in Space of more than Three Dimensions," which appeared in the first volume of the *American Journal of Mathematics*.[5] Here, Newcomb developed the special case whereby a thin three-dimensional shell, or hollow enclosure, could be turned inside-out through a fourth dimension without tearing or rupturing the surface of the shell. Several other articles on the non-Euclidean geometries appeared in this newly established periodical, reflecting the philosophy and interests of its co-founder, James J. Sylvester.

Sylvester, who had come from England, was a recent addition to the faculty of Johns Hopkins University. He had come to Johns Hopkins at the request of Joseph Henry, specifically to build a strong mathematics program at an American university. Although he was not considered a geometer, Sylvester brought with him an interest in the new forms of geometry, a knowledge of recent research in England dealing with the new geometries and an openness to the possibilities of applying the new geometries to real physical situations.

It was Sylvester who, in 1869, first publicly announced the work of William Kingdon Clifford, a young English geometer,[6] regarding Clifford's application of a non-Euclidean geometry to explain the motion of matter in space. Sylvester clearly had privileged knowledge of Clifford's theories since they were never published. It would be inconceivable to think that Sylvester did not share those ideas with his new American colleagues, friends and students. If there had been any lack of interest in the non-Euclidean geometries in America before Sylvester's arrival, there can be no doubt that condition soon changed at Johns Hopkins. Sylvester's first pupil was George B. Halsted, who later became known throughout the world for his mathematical and historical researches as well as his popular expositions of the non-Euclidean geometries. Sylvester surrounded himself with a group of interested scholars including Halsted, Newcomb, Charles S. Peirce, William E. Story, Thomas Craig, and W. I. Stringham,[7] among others, all of whom later made contributions to the development of non-Euclidean geometries, and some of whom later argued for or defended the possibility that our physical space could be non-Euclidean.

Halsted was long associated with the popular exposition of these new geometries and the dissemination of knowledge pertaining to them. Within this context Halsted wrote a series of articles for the *American Mathematical Monthly* between 1894 and 1896.[8] The ideas put forth by Halsted in these articles were randomly and scathingly attacked by John N. Lyle, a professor of Natural Sciences in Missouri.[9] Lyle was unable to accept the fact that the non-Euclidean geometries could be separated from a real physical space. Despite the many attempts of pure mathematicians to abstract their geometrical concepts from real spaces, there always remained the possibility that the new geometries did in fact represent our real physical space. This situation left the new geometries open to attack by other scholars. Those who believed in the non-Euclidean geometries, whether they represented purely abstract notions or real physical spaces, could always claim that they only wished to represent abstract notions without referring their researches and hypotheses to the question of which geometry represented our experiential space. Halsted never developed his own theories or models of physical space, remaining loyal to the abstractions of pure mathematics. On many occasions he stated publicly that our space may indeed be non-Euclidean, but the establishment of that fact was well beyond any physical test and was therefore not a valid scientific question for the mathematician.[10]

Lyle's attacks on the pure mathematical side of the non-Euclidean geometries, based strictly upon his belief that they were being used to undermine the notion of our Euclidean physical space, were neither new nor unique. Johann B. Stallo, the American critic of nineteenth century science, in his book, *The Concepts and Theories of Modern Physics*, first published in 1881, levelled quite similar charges against the whole spectrum of non-Euclidean geometries, their founders and proponents. Stallo found the whole mathematical question of the new geometries to be profoundly tied to the concept of a real physical space:

> Here, at the outset, we find an assumption which obviously lies at the base of the whole theory; the assumption that space is a physically real thing -- not merely an object of experience, but an independent object of direct sensation whose properties may be ascertained by the aid of the ordinary instruments of physical and astronomical research -- whose degree of curvature, for instance, is to be determined by means of the telescope.[11]

Stallo was quite clearly reflecting his interpretation of the concepts which he had encountered within the early documents relating to the non-Euclidean geometries. His verdict on the new geometries, right or wrong, was that they were decidedly physical in content and not the pure mathematical abstractions that they were claimed to be. His chief worry seemed to be that science would endow space, which was merely a set of relations, with physical properties. If that were to be the case, then:

> We should be constrained to say that the only form or variety of objective existence is either space or matter (it being a mere question of nomenclature), and that all the properties we now attribute to matter are in truth and in fact properties of space.[12]

Stallo's charges were quickly and bluntly answered by Newcomb. Newcomb defended the mathematician's right to pure abstraction, claiming that no mathematician had ever entertained the notion of a real four-dimensional space. He found that Stallo's criticisms were "founded on an utter misapprehension of the scope and meaning of what he [was] criticizing," since the whole mathematical adventure of hyperspaces and non-Euclidean geometries was "avowedly hypothetical."[13] These criticisms of Stallo were made by the same man who several years later would propose a four-dimensional model of space to house the equally hypothetical luminiferous ether.

Newcomb was not alone in attempting to found a physical theory of space upon a non-Euclidean (hyperspace) hypothesis. Charles Peirce, another member of Sylvester's group at Johns Hopkins, took a far more positive and aggressive role in the application of a non-Euclidean geometry to the problems of physical space. After showing some initial interest in such geometries at Hopkins,[14] Peirce did not seem to develop any specific ideas toward their physical application until the 1890s. In fact, he later told a friend in the course of a letter he had put forth in 1870, "the suggestion -- without guaranteeing it," that space was Euclidean irrespective of the fact that it may be experimentally confirmed to be otherwise.[15] The reasons for his new-found interests in a non-Euclidean model of space at this particular time will remain forever unknown, but the possibility that the recent developments in experimental physics, i.e., Michelson's and similar experiments in America, helped to pave the path for his theories cannot be stressed too strongly, given Peirce's commitment to experimental observations in astronomy.

In an undated letter to Newcomb, Peirce sought moral and strategic support for his pursuit of the new theory. He wrote:

> I want to get into circumstances in which I can pursue certain researches. I want you to do certain things to aid me, and to that end, I want you first to remark how encouraging the figures look in regard to my attempt to make out a negative curvature of space.[16]

Peirce continued his letter with mathematical arguments and astronomical observations supporting his contentions before closing with a more general view of his theory:

> In my mind, this is part of a more general theory of the universe, of which I have traced many consequences,--some true and others undiscovered,--and of which many more may be deduced; and with one striking success, I trust there would be little difficulty in getting other deductions tested. It is certain that the theory if true is of great moment.[17]

Indeed it would be a discovery of great moment, and Peirce, working toward that discovery, publicized parts of his work throughout the decade.

In another letter to Newcomb, dated 21 December 1891, Peirce acknowledged his discovery that the astronomical figures did not support the conclusion that the curvature of space was positive, but were "rather against it."[18] Carolyn Eisele, who has done a remarkable job of tracing and publishing the diverse mathematical researches of Peirce, has placed this letter of December 1891 after the undated letter to Newcomb, announcing Peirce's theory, claiming, therefore, that Peirce dropped his theory soon after its inception. However, in the letter of 23 December, Peirce stated that his observations did not support a space with *positive* curvature.[19] A Lobachevskian space has a *negative* curvature. It would then seem that the 1891 letters to Newcomb were written before the letter in which Peirce declared his new theoretical leanings to Newcomb.

Seen in this context, the undated letter actually allows us to follow the evolution of Peirce's thoughts toward the adoption of his theory of a physical Lobachevskian space. Peirce must have attempted to measure the positive curvature of space, perhaps at the suggestion of Newcomb or with the foreknowledge of Newcomb's and other's views on the possibility that space was Riemannian, and discovered, to his satisfaction, that the opposite case was true.

Eisele states elsewhere that Peirce "entertained that idea [of the hyperbolic properties of space] for a time in 1891, only to give it up on further investigation."[20] This statement cannot be true. Peirce did not abandon his theory at the end of 1891. It is probable that he may not have completely formulated the theory until then, or during the early months of 1892. Since no complete records of Peirce's theory seem to have survived, the developmental sequence of the theory, as well as the nature and scope of his belief, can only be found in the fragmentary evidence that has survived.

The most noteworthy and complete documents to have survived concerning Peirce's theory of space are a paper that he read to the American Mathematical Society in 1894 and a lecture he gave as part of his Cambridge Lecture series in 1898. At a meeting of the American Mathematical Society, on 24 November 1894, Peirce read a paper entitled "Rough Notes on Geometry, Constitution of Real Space." From the title there can be no mistaking the exact intentions of his talk. The manuscript of this paper has never been found and the exact contents of the reading are not known. However, Eisele has located several pages of manuscript that she believes constitute Peirce's notes for that lecture.[21] After a short discourse on geometry, Peirce enunciated the properties of space. He then informed his audience that he had personally undertaken the task of determining the constant of space, which seemed to indicate "a hyperbolic space with a constant far from significant."[22] Then his discussion changed to include the concept of absolute motion as it applied to the Foucault pendulum experiment. This interpretation, regarding the reality of an absolute motion, with a statement tantamount to accepting the reality of an absolute space, flew directly in the face of recent physical thought and reflected Peirce's growing disenchantment with the rise of relativism in the physics of space. In a reference that was clearly meant to refute the Machian concept of relativity, Peirce stated that "All German attempts to escape this conclusion [of an absolute space] are meta-physical word-spinning."[23]

The conversion of Peirce to an anti-Machian, anti-relativist point of view has already been studied by Randall Dipert, who has found that Peirce adopted the concept of absolute space only after 1892.[24] This date corresponds quite well with the proposed adoption, by Peirce, of a Lobachevskian physical space.

The lecture at Cambridge was not meant as an exposition of Peirce's theory of physical space, but a detailed account of his logical system as regarded space. He first derived several principles that were related to his logical system. The sixth principle dealt specifically with the possibility of a fourth dimension. Peirce concluded that, given any number of dimensions from 1 to 4, "the numbers of dimensions ought to be just 4."[25] Although this conclusion was reached by rational methods, there is no doubt that Peirce thought of the fourth dimension as a legitimate hypothesis for physical space. At least as early as 1883, Peirce had stated his acceptance of a possible real fourth dimension:

> Although at no point of space where we have yet been have we found any possibility of motion in a fourth dimension, yet this does not tend to show (by simple induction, at least) that space has absolutely three dimensions.[26]

Ever the philosopher before the scientist, Peirce felt it necessary to found his physical theory on purely philosophical arguments and thus couched his acceptance of a hypothetical fourth dimension as well as his whole physical theory of space in philosophical terms. He concluded, in reference to his theory of space:

> I thus briefly stated one side of my theory of space. That is, without touching upon the question of the derivation of space and its properties, or how accurately it may be supposed to *fulfill its ideal conditions*, I have given a hypothesis from which those ideal properties may be deduced. Many of the properties so deduced are known to be true, at least approximately. Others, I am happy to say, are extremely doubtful. I say I am happy because this gives them the character of predictions and renders the hypothesis capable of experiential confirmation or refutation. One of the doubtful properties, the last mentioned, I have succeeded I think in proving to be true by calculations from the proper motions of stars. Another, that about atoms attracting differently in different directions, I have succeeded in making highly probable, from chemical facts. Still others have some evidence in their favor. The consequence most opposed to observation is the doubtful one of 4 dimensions.[27]

Peirce fully intended that physical space follow his philosophical deductions of what that space should be. This did not abrogate the necessity that his theory should be confirmed by experiment and observation, although Peirce clearly thought that observations would completely bear out his theory.

From the fragments and more complete documents a rough sketch of Peirce's physical theory of space has emerged. The elements of his theory included, at least, (a) four-dimensionality, (b) negative -- Lobachevskian -- curvature, (c) the reality of absolute space, (d) the possibility of experimental confirmation, and (e) a generality whereby the theory could explain "the characteristics of time, space, matter, force, gravitation, electricity, etc."[28]

During this same period, Newcomb was only moderately active with his views of hyperspace and non-Euclidean spaces. Although he defended the right of scientists to incorporate such hypotheses into their physical reasoning, he never again sought to elaborate upon his own physical model of space. He did, however, give major addresses and publish some papers concerning the concept of hyperspaces. On 28 December 1893 he delivered an address, "Modern Mathematical Thought,"[29] before the New York Mathematical Society and on 29 December 1897, he gave the presidential address, entitled "The Philosophy of Hyperspace,"[30] before the (by then renamed) American Mathematical Society. The first address was very general in its scope, but did include brief discussions of curved space and the fourth dimension. Newcomb noted that "it is a fundamental principle of pure science that the liberty of making hypotheses is unlimited"[31] before he concluded that,

> ...the question whether the actuality of a fourth dimension can be considered admissible is a very interesting one. All we can say is that, as far as observation goes, all legitimate conclusions seem to be against it.[32]

Here, we can see the same dismal prediction that Peirce made in 1898. For Peirce, the confirmation of a fourth dimension was becoming a "doubtful one" while for Newcomb, the "legitimate conclusions seemed to be against it."

However, in this 1894 speech Newcomb combined the essential elements of the hyperdimensional theories, or models, without overtly mentioning specifics. In so doing, he defended the rights of both the mathematician and the physical scientist to make hypotheses without committing themselves to abstract theories. A good knowledge of the background of the speech is necessary to properly understand its contents. If, as has been the considered opinion in the past,[33] there were no applications of the non-Euclidean hyperspaces to real physical situations until Einstein, then a large portion of Newcomb's address was at best meaningless and irrelevant to its audience. The thinly veiled comments made by Newcomb were well understood by those to whom he spoke since the concepts of hyperdimensional physical spaces were known, if not common, to the decade and era.

The 1896 address was directed more toward an exposition of the general characteristics of hyperspaces without reference to specific hyperspace models. Although Newcomb treated the four-dimensional and non-Euclidean spaces as separate concepts at this time, the same had not necessarily been true with the hyperspace models of the past. Regarding the possibility of a real four-dimensional space, Newcomb wrote that "There is no proof that the molecule may not vibrate in a fourth dimension,"[34] and later continued:

> The hypothesis of vibration in the fourth dimension merely suggests the possibility that this kind of motion may mark what is essentially

different from the motion of masses. Of course, such an hypothesis as this is not to be put forward as a theory. It must be worked out with mathematical rigor, and shown to actually explain phenomena before we can assign it such rank.[35]

Still later he added:

We have no experience to the motion of molecules; therefore, we have no right to say that those motions are necessarily confined to three dimensions. Perhaps the phenomena of radiation and electricity may yet be explained by vibration in a fourth dimension.[36]

It is quite obvious from the last line that Newcomb had not entirely given up all hope of explaining atomic phenomena with a hyperspace hypothesis. In this respect his ideas corresponded to Peirce's conception of the physical reality of the fourth dimension. Peirce had earlier written in "The Architecture of Theories," that "There is no room for serious doubt whether the fundamental laws of mechanics hold good for single atoms, and it seems quite likely that they are capable of motion in more than three dimensions."[37]

Newcomb's final word on the concept of hyperspaces appeared in the form of an article directed toward the popular audience entitled "The Fairyland of Geometry."[38] No new ideas were added in this article. The most striking aspect of this article was the prominent use of the term "fairyland" in both the title and the opening lines. Alfred Bork, who has made one of the few studies on the use of hyperspaces in the closing decades of the nineteenth century, has reached the conclusion that Newcomb did not take the concept of a physical hyperspace seriously. Newcomb's "interest must be classified as mathematical rather than physical, as suggested by the word 'fairyland.'"[39] Bork's interpretation relies too heavily on the word "fairyland" and its various connotations without emphasis on what Newcomb was saying in the articles in which he used the term. Newcomb actually took the notion quite seriously as a physical *hypothesis*.

The term "fairyland" was used by Newcomb in both of his papers to the Mathematical Society and can be traced still further back to at least the time of Newcomb's first article on the non-Euclidean geometry, which appeared in *Crelle's* magazine. In a letter to Newcomb dated 26 October 1877, Sylvester wrote:

Very many thanks for your presentation copy of your wonder moving fairytale. I shall hope now to gain a new insight into the subject and keep your memoir on my table at the University...[40]

It is possible that the use of the term "fairytale" in this context, or the equivalent metaphor of "fairyland" as penned by Newcomb, was a product of Sylvester's wit rather than Newcomb's mathematical imagination.

The connotation of "fairyland" also tends to focus the concept of hyperspace upon its more untoward aspects. Those very physical characteristics of the hyperspaces which made them ideal for the derivation of physical models also allowed their use by the popular spiritualist movement in search of a home for spirits and ghosts. This coincided, to a small degree, with a minority of people who attempted a mental "realization" of the fourth dimension. T. Proctor Hall published an account of this last type in the 13 May 1892 issue of *Science* magazine.[41] Whether Hall meant that a four-dimensional hyperspace could actually be realized or conceptualized with respect to real spatial objects is not important. He took far too much liberty with his hypothesis in the eyes of others, especially in stating that the realization of a hyperspace would be of use to both physicists for forces and theologians for the world of spirits.[42] T. Proctor Hall was not a spiritualist or theologian, but a legitimate mathematician. In succeeding issues of *Science*, Hall's article was criticized, sometimes rather scathingly, by W. P. Preble of New York, George B. Halsted, C. Staniland Wake of Chicago, and Edmund C. Sanford of Clark University.[43] There existed a specific hierarchy of what a scientist or mathematician could do and could not do. Hall overstepped the bounds of this hierarchy and incurred the wrath of his peers. In general, any statements concerning the use of physical hyperspaces were well guarded and it is difficult to gauge the temperament of the scientific community regarding these models and theories.

The attempt to realize a fourfold space was not an original idea of Hall's. His paper was based upon the earlier work of Charles H. Hinton. While a student at Oxford in England, Hinton wrote several essays dealing with the realization of four-dimensional spaces. These were later collected and published in the volume "Scientific Romances." Hinton's second book, "A New Era of Thought," published in 1888, served as the basis of Hall's paper. Hinton's essays were strictly philosophical, but he did use an analysis of how our three-dimensional world could be embedded in a four-dimensional space to derive a working model of the ether. He used his original model of 1880 and later expanded models to explain various physical phenomena. His early models were also used to explain ethics, art, mysticism and other nonscientific disciplines. After obtaining his M.A. from Oxford, Hinton came to America, teaching at Princeton for four years, then at the University of Minnesota, and finally obtaining a position at the Naval Observatory in Washington, D.C. Newcomb's model closely resembled that of Hinton, which had been developed several years earlier. However, Hinton later developed the mathematical analog for his model that had been missing from Newcomb's.

The first model was that of a three-dimensional ether, which formed a perfectly smooth sheet having some small thickness in the fourth dimension. On the hypothesis that our world has a four-dimensional existence that we are not conscious of, Hinton concluded that the thickness of the ether sheet in the fourth direction must be approximately that of "the ultimate particles of matter."[44] In another essay, published in the same book, Hinton elaborated upon this physical model, as he would for the next two decades. He likened electricity to minute threads in material bodies. When electricity was conducted by a metal wire the threads would twist and rotate in unison with their neighboring threads.[45] The threads would hold this twist in a conductor until they were discharged by grounding the conductor. The threads were actually ether particles and the twist occurred in the fourth dimension.

The ether sheet was then set "quivering and vibrating" in the direction of the fourth dimension, which became the cause of all motion in matter.[46] In this model, there was no hint of the sources upon which Hinton drew to develop his ideas. There are scattered similarities between Hinton's model and Clifford's, but Hinton attended Oxford while Clifford, Maxwell, Sylvester and others were associated with Cambridge. Hinton did show evidence of some minimal knowledge of non-Euclidean geometries by considering the term "curvature of space" in his treatment of the ether, but he rejected the concept of curvature as "perhaps the most mischievous ...expression."[47] His was a model of higher space, not of geometry. Still, he likened his smooth ether sheet, the "supporting body," to "a portion of a vast bubble."[48] This analogy alone is strikingly reminiscent of a Riemannian spherical surface with positive curvature.

After obtaining his M.A. from Oxford in 1886, Hinton finished the manuscript for his second book, *A New Era of Thought*. Once again, Hinton's approach to hyperspace was non-analytical and completely intuitive. He suggested that perhaps brain molecules had a power of motion in the fourth dimension that could not be comprehended by man as part of the external world, but as thought and imagination.[49] At this time, Hinton's ethereal sheet underwent a new mechanical change. "Instead of the aether being a smooth sheet serving simply as a support, it is definitely marked and grooved."[50] The grooves were permanent and fixed, while small enough to carry the atoms of matter. Thus, Hinton was to account for more mechanical motions, not just the natural manifes- tations of heat and light, which were carried by the vibrations of the ether in the sheet of his earlier model.

It was several years before Hinton came to America, and then by way of a teaching job in Japan. He published nothing more on his ether model until 1902 and then the character of his research changed drastically as his model became more analytical and less intuitive. He did publish a purely mathematical paper, "Hyperbolea and Solutions of Equations" in 1897,[51] and was later elected to the American Mathematical Society. The opportunity for contact with Newcomb and Peirce at meetings of the AMS cannot be discounted. It is quite certain that Hinton met Newcomb at some point, since he obtained a position at the Naval Observatory shortly after Newcomb's retirement from that establishment, and it is known that Newcomb still dealt with the observatory well after his retirement.[52] During Hinton's tenure at Princeton, he would have been afforded many opportunities to supplement his mathematical background, and in turn he was able to influence a young graduate student, Paul Renno Heyl, at the University of Pennsylvania.

Heyl's Ph.D. thesis, "The Theory of Light on the Hypothesis of a Fourth Dimension," was a direct application of Hinton's earlier ether model, and he acknowledged Hinton's "Scientific Romances" as the source for his inspiration. Heyl summarized his thesis in a paper presented before the AMS in October of 1987. He introduced his four-dimensional model using an analogy to a three-dimensional drumhead covered with dust. Light and heat would spread out as vibrations of the drumhead and would either be damped or absorbed by the dust particles, representing particulate matter and material bodies in the ether.[53]

The three-dimensional analogy presented by Heyl offered very real advantages over previous ether theories when applied to his four-dimensional model:

> The bearing of the hypothesis of a fourth dimension upon the theory of light should be obvious. It replaces the current idea of a tenuous jelly-like aether, *through* which we move by the idea of a rigid frictionless 4-dimensional aether, so thin in the fourth dimension as to be nearly a 3-dimensional solid, *upon* (or *in*) whose solid boundary we slip and slide around without feeling it. It replaces the somewhat vague idea of light being a "electro-magnetic stress" by the idea that it is a mechanical vibration. It replaces the conception of a ray of light as a complex of rays polarized in all possible planes by the idea of a vibration in but one direction, and that perpendicular to our solid space. The difficulty of the theory is all concentrated in the conception of our mutually perpendicular straight lines. Grant this, and the phenomena of light become capable of a mechanical explanation.[54]

The three-dimensional analogy and advantages of the hyperspace model out of the way, Heyl simply extended the differential equation for a vibrating solid in two and three dimensions, to a fourth dimension. Then, solving the resultant equations under different boundary conditions, Heyl predicted a slight departure from the law of the inverse square[55] and an apparent retroactive effect:

> We are led to the curious conclusion that in Hinton's aether the nature of the central disturbance *after* a given instant can influence

the form of the aether wave *before* that instant. In other words, the aether seems to be endowed with an uncanny faculty of foreknowledge.[56]

Such conclusions were reached by Heyl in proper mathematical fashion. There can be no doubting the legitimacy of his hypothesis and his ability to derive experimentally verifiable phenomena. Despite his highly mathematical analysis, Heyl made no mention of the non-Euclidean geometries that were unnecessary for his treatment. He was considering only the vibrations of particulate matter and not the overall macroscopic shape of the ether sheet or drumhead.

Nor did his early exposure to and support of a physical hyperspace theory hurt his career. Heyl became a respected physicist, eventually working for the Bureau of Standards. Heyl's thesis, as well as another paper on mathematical hyperspaces,[57] had been presented before the American Mathematical Society for the inspection of anyone who so wished. Therefore, it cannot be claimed that his work had been done in seclusion, safe from the wrath of a conservative scholarly establishment. Newcomb and Peirce would have nothing to fear in exposing their views on hyperspaces because they were both secure and well-respected scientists. Yet Heyl, the novice scientist, was not reprimanded by the scientific community in any way for using a hypothetical physical hyperspace because it was considered a legitimate hypothesis by many, and he had treated the hypothesis in a thoroughly professional manner as had been prescribed by Newcomb. In the meantime, Hinton must also have been honing and sharpening his mathematical skills. On 9 November 1901, he read a new paper, "The Recognition of the Fourth Dimension," before the Philosophical Society of Washington.[58] Not only did this paper reflect Hinton's new mathematical approach to his older hyperspace model, but several other new features were added: Both a total lack of philosophical speculation and the intuitive approach that had been Hinton's hallmark in earlier writings were missing; Hinton stated that "All attempts to visualize a fourth dimension are futile. It must be connected with a time experience in space,"[59] which was a direct refutation of his earlier philosophy. He now made clear references to other scientists and mathematicians, including James Clerk Maxwell, Lord Kelvin, and Simon Newcomb. The reference to Newcomb is especially revealing since Newcomb had been a long-standing member of the society, was probably in the audience for the reading, and may well have been a direct influence on the new role played by mathematics in Hinton's theory; and, at the very beginning of his paper, he acknowledged two ways of searching for four-dimensional phenomena, in the investigation of both the macroscopic and the microscopic worlds. This constituted an admission of the importance of non-Euclidean geometries and the curvature of space to his theory, which he had denied in an earlier essay.

The variations to the hyperspace model on which his new theory was built were both ingenious and subtle. Working directly with Maxwell's original model of ethereal vortices and a Kelvinesque representation of ether vortices, Hinton produced a model of double rotation about a sphere in the fourth dimension that could represent vortex atoms in the three-dimensional sheet. This double-rotation, "a kind of movement totally unlike any with which we are familiar in three-dimensional space"[60] allowed a direct mechanical explanation of the attractive and repulsive forces of static electricity. The vortex sheet also became equivalent to a collection of magnets where individual parts of the sheet corresponded to single magnets,[61] while an electric circuit became a rotation in the ether surrounding the current carrying wire.[62]

If we suppose the ether to be filled with vortices in the shape of four-dimensional spheres rotating with the A motion, the B motion would correspond to electricity in the one-fluid theory. There would thus be a possibility of electricity existing in two forms, statically, by itself, and, combined with the universal motion, in the form of a current.

. . .

To recapitulate:
The movements and mechanics of four-dimensional space are definite and intelligible.   A vortex with a surface as its axis affords a geometric image of a closed circuit, and there are rotations which by their polarity afford a possible definition of statical electricity.[63]

The motions A and B represented the mathematical portion of Hinton's presentation. Although the major portion of the presentation was later published in the *Bulletin of the Philosophical Society*, the mathematical portion was not included. Instead, it appeared, in part, in the *Proceedings of the Royal Irish Academy* under the title "Cayley's formulae of orthogonal transformations."[64] That part which appeared in the *Proceedings* was purely mathematical devoid of any physical interpretation, so a complete copy of Hinton's theory is not available. However, it is clear that rotations A and B were to be represented by co-Hamiltonians, a mathematical concept developed by Hinton via a change in notation for Hamiltonian quaterions when applied to the four-dimensional case.[65] When the mathematics is taken in conjunction with the hyperspace model that Hinton had finally settled upon, it becomes quite clear that Hinton had at last derived a complete theory of a physical hyperspace.

It is difficult to gauge the impact and influence of Hinton's theory. The decade when Hinton proposed his theory was one of great change and development in physics and the concept of space. At that time scientific knowledge was not disseminated so quickly as it is today, while new theories proposed by far more well-known scientists of the day would have eclipsed Hinton's contribution to science. The same was true of mathematics as it was of physics, and the quaterions that Hinton used to develop his theory also fell by the wayside and were replaced by vectors. An argument could be made that Hinton's theory would have become known in time, given that his earlier books had gained a wide and international audience. But such conjectures are not proper procedures for the historian of science, who should deal more with the reality of the era.

However, if all of the hyperspace models of the ether, four-dimensional and non-Euclidean, are taken together, it is the case that they were quite well known, both popularly and academically. This fact is mentioned in various articles and papers published throughout the era. Evidence of this can also be found in the popularity of foreign publications that were printed and sold in America. One of these, which is of special interest, was Walter W. Rouse Ball's *Mathematical Recreations and Problems (of past and present times)*. Between 1889 and 1914, this popular book went through six editions in the U.S. and Britain. In various editions, under various chapter numbers, Ball wrote on "Hyperspace" and "Matter and Ether Theories." In the chapters on "Hyperspace," Ball cited and gave a good deal of credit to Hinton's book *Scientific Romances*.[66] Ball had developed his own hyperspace model of the ether before finding that Hinton had proposed much the same model.[67]

A short review of the second edition of Ball's book was written by J. E. Oliver of Ithaca, New York, for the *Bulletin of the American Mathematical Society*. Oliver found the final two chapters of Ball's book, those on hyperspaces and on the constitution of matter, to be "of great philosophical interest."[68] The suggestions made by Ball concerning the hyperspace model, which were well documented in Oliver's review, appeared as

...the finest thing in the book. They are but little developed, and not always clear.... Nevertheless the possibilities of the proposed theory may be immense. Of course all depends upon the success with which it can be applied not only to philosophical generalities, but to the definite prediction of laws; still, *a priori*, the hypothesis would seem to be quite admissible.[69]

Both Newcomb and Peirce would have agreed with Oliver on the immense possibilities of such theories as well as the admissibility of the hypothesis.

Oliver also suggested that Ball might improve his hyperspace model by allowing the sheet of ether to "shade off" into the fourth dimension, "either through some kind of selection such as Mr. C. S. Peirce had suggested, or of quasi-attraction, a little as in Hinton's *Scientific Romances*."[70] This latest statement of Oliver's is of considerable importance in the study of hyperspace models because it throws new and unexpected light on an as yet unknown aspect of Peirce's theory, while testifying that Peirce's theory was known and taken seriously by others. And finally, Oliver acknowledged that some academicians during 1892 must have held a great deal of hope for the hyperspace models:

The rest of the account given of hyperspace and of the constitution of matter is well thought out and clear. It will help the general reader toward some truthful notions as to studies which may perhaps play an important part in the near future.[71]

CONCLUSION

This review of history has demonstrated the existence of a small but persistent scientific movement to develop hyperspace models of the ether. Such models were consistent with other scientific movements toward the end of the nineteenth century. These ether models were mechanical, as were other ether models, but unlike other models they utilized a four-dimensional absolute space. Insofar as the models were based on a fourth dimension of space, they dealt specifically with light and electromagnetic phenomena. Thus they complemented the work of other physical theoreticians of the 1890s who reacted to Michelson's negative results.

Hyperspace models of the ether and attempts to find a physical basis for non-Euclidean geometries have been largely ignored by historians. This fact seems curious when account is taken of the popularity of hyperspace speculations during the period and the successful use of a non-Euclidean geometry to explain gravitation a few decades later. Several simple reasons could be offered to explain this historical anomaly. First, and foremost, would be the prejudice against such theories arising from their association with the spiritualism of the period. It would be nearly as easy to find justification in an ignorance of such theories in the rush to accept relativity during the period as these theories were based on the older mechanistic concept of an absolute space. Yet other mechanistic theories of the ether, based on equally outmoded concepts, from the privileged point of view of hindsight, have found popular support as legitimate historical factors in the rise of modern science. Perhaps a misconception of the role of theoretical mathematics in physical science prior to the twentieth century has been propagated by historians. It is no coincidence that most of the information concerning these physical models of the ether has been found in traditional mathematical sources, even though the models were proposed by physical scientists.

In one of the few books dealing with the historical development of the concept of space, *Concepts of Space*, Max Jammer cites the "speculations" of Riemann and Clifford, crediting them as precursors of Einstein's general theory of relativity.[72] They were that and so much more. Riemann proposed a program in physics to be carried out in the future, while Clifford was the first to seriously consider the completion of that program. Einstein's general relativity was only a partial fulfillment of that program since it only addressed the macroscopic world and not the microscopic world. Riemann's program took root in America, with Newcomb, Peirce, Heyl, and Hinton all making contributions. Yet nowhere are they considered precursors of Einstein's relativity or heirs to the great nineteenth-century tradition of mechanical ether theories into which Albert A. Michelson precipitated a crisis in 1887.

NOTES

1. The typewritten manuscript of this talk, entitled "On the Fundamental Concepts of Physics," can be found in Box #94, Simon Newcomb Papers, Library of Congress. Newcomb begins the paper "What I have to say on this subject will best be introduced by some considerations respecting the question of *actio in distans* which has been discussed from many points of view in the Philosophical Magazine during the last few years."

2. The Michelson-Newcomb letters can be found in part in Box #32, Simon Newcomb Papers, Library of Congress. They have been reprinted, with comments, in Nathan Reingold, ed., *Science in Nineteenth-Century America: A Documentary History* (Chicago: Univ. Chicago Press, 1964), pp. 275-306.

3. Newcomb, "Fundamental Concepts," pp. 5-6.

4. Simon Newcomb, "Elementary Theorems Relating to the Geometry of a Space of Three Dimensions and of Uniform Positive Curvature in the Fourth Dimension," *Crelle's Journal für die reine und angewandte Mathematik*, 1877, *83*:293-299.

5. Simon Newcomb, "Note on a Class of Transformations Which Surfaces May Undergo in Space of More than Three Dimensions," *American Journal of Mathematics*, 1878, *1*:1-4.

6. Sylvester made the original announcement during his address to the mathematics and physics section of the British Association for the Advancement of Science meeting at Exeter in 1869. His address was subsequently published in the *Report of the British Association*, 1869, *39*:1-9, and reprinted, with explanatory notes, as "A Plea for the Mathematician," *Nature*, 30 December 1869, *1*:237-239.

7. This group of scholars and mathematicians was mentioned by Richard C. Archibald in the explanatory notes that accompanied his publication of the Newcomb-Sylvester correspondence in "Unpublished Letters of J. J. Sylvester and Other New Information Concerning His Life and Work," *Osiris*, 1936, *1*:85-154, on p. 140.

8. Halsted's series in the *American Mathematical Monthly* appeared under the continuing title "Non-Euclidean Geometry, Historical and Expository," in Volumes I and II. This series did not limit Halsted's contributions to that journal, as he was a regular contributor for many years.

9. John L. Lyle, Ph.D., was a professor at Westminster College, Fulton, Missouri, at the time that he contributed to the *American Mathematical Monthly* His contributions ended in 1897 upon his death. A biography of Lyle was published by F. P. Matz in April 1896, *3*:95-100. Matz seemed to excuse Lyle's anti-non-Euclidean stance, in a sense, as part of Lyle's rural upbringing; "John N. Lyle was born in Ralls County, Missouri, March 5, 1836. 'The space in which this county is located is trinally extended, and therefore objective. It has no curvature, either positive or negative. Here planes are flat, and perpendiculars to a transversal are equidistant.' If Lobachewsky had been born and raised in Ralls County, he would perhaps never have doubted *that two straight lines equidistant from each other may be drawn in the same plane*, nor written a theory of parallels in which this postulate of sound geometry is discredited." Matz, "Lyle," p. 95.

10. "Our replacement [for non-Euclidean geometry] is only confined in its free arbitrariness in that it should seem to snuggle to the seeming facts, and must introduce no logical contradictions. In this sense the ordinary triple-extended space of our experience is at present Euclidean or Bolyaian or Reimannian as you choose. Each is, up to the present day, in simple and perfect harmony with experience, with experiment, with the properties of the solid bodies and the motions about us." G. B. Halsted, "The Popularization of Non-Euclidean Geometry," *Am. Math. Monthly*, 1901, *13*:31-35, on p. 32. Halsted made similar statements in several other places including "Algebras, Spaces, and Logics," *The Popular Science Monthly*, 1880, *17*:516-522, on p. 521; "The Old and the New Geometry," *Educational Review*, 1893, 6:144-157, on p. 150; and "Newcomb's Philosophy of Hyperspace," *Science*, 1898, ns 7:212.

11. J. B. Stallo, *The Concepts and Theories of Modern Physics,* ed. Percy W. Bridgman (Cambridge, Mass.: The Belknap Press of Harvard Univ., 1960: This is a reprint of the first edition, New York: D. A. Appleton, 1881): pp. 228-229

12. Stallo, *Concepts*, p. 241.

13. Simon Newcomb, "Speculative Science," *The International Review*, April 1882, *12*:334-341, on p. 337.

14. Evidence of Peirce's initial interest in the non-Euclidean geometries while at Johns Hopkins can be found in the *Johns Hopkins University Circular*, 1879, nr.2:18. Peirce read a paper on the non-Euclidean concept of space, written by his student Miss Ladd, before the Metaphysical Club.

15. The letter in question was written to Francis Russell on 15 April 1909 and is reprinted in Charles S. Peirce, *The New Elements of Mathematics*, ed. Carolyn Eisele, Volume III: *Mathematical Miscellanea* (The Hague: Mouton Publishers, and Atlantic Highlands, New Jersey: Humanities Press, 1976), on p. 979.

16. The undated, handwritten letter from Peirce to Newcomb can be found in Box #34, Simon Newcomb Papers, Library of Congress. This letter, together with others that passed between Newcomb and Peirce, has been reprinted in "The Charles Peirce-Simon Newcomb Correspondence," ed. Carolyn Eisele, *Proceedings of the American Philosophical Society*, October 1957, *101*:409-433, on p. 421.

17. Eisele, "Correspondence," p. 422.

18. This letter also appears in the Newcomb Papers at the Library of Congress and is duplicated in Eisele's "Peirce-Newcomb Correspondence," p. 422.

19. Ibid., pp. 422-423.

20. Eisele in Peirce, *New Elements*, Volume II, p. ix.

21. According to Eisele: "Peirce not long thereafter read a paper on "Rough Notes on Geometry, Constitution of Real Space" at a meeting of the American Mathematical Society on Saturday, 24 September 1894. It was not printed later, but one of the incomplete papers in this collection (MS. 121) could well have been that talk, since Fiske had asked Peirce for a copy and Fiske's name is written in Peirce's hand on the back of it." Peirce, *New Elements*, Volume II, p. xiii. The paper that was found by Eisele is reprinted in Peirce, *New Elements*, Volume III, pp. 703-709.

22. Peirce, *New Elements*, Volume III, pp. 707-708.

23. Ibid., p. 708.

24. Randall R. Dipert, "Peirce and Mach on Absolute Space," *Transactions of the Charles S. Peirce Society*, 1973, *9*:79-94.

25. Excerpts from Peirce's Cambridge Lecture are reprinted in Charles S. Peirce, *Collected Papers of Charles Sanders Peirce*, Volume VI, eds. Charles Hartshorne, Paul Weiss and Arthur W. Burks (Cambridge: Harvard Univ. Press, 1935 and 1958), p. 63.

26. Peirce, *Collected Works*, Volume II, p. 458. This particular excerpt was taken from "A Theory of Probable Inference," written in 1883, and was repeated in an unpublished book, *Search for a Method*, in 1893, according to the editors of the *Collected Papers*.

27. Peirce, *Collected Works*, Volume VI, p. 64.

28. Charles S. Peirce, "The Architecture of Theories," *The Monist*, January 1891, *I*:161-176, on p. 176. Although this essay appeared before the date which I have established for Newcomb's conversion to a Lobachevskian physical space, I believe that the passage quoted, at the end of the essay, represents a general theory of philosophy on which Peirce was working before adopting the Lobachevskian hyperspace. In the article (pp. 173-174) Peirce discusses all three types of geometry, but does not commit himself to any one at the expense of others. The negative parallaxes that he discusses are "undoubtedly attributed to errors of observation." This would tend to support the hypothesis that Peirce did not adopt the Lobachevskian hyperspace until the very end of 1891, if not later. He probably then grafted his hyperspace model to the general philosophical theory of which he was lamenting at the end of the article. Perhaps he even found in the adoption of a Lobachevskian hyperspace the proper vehicle by which he could revive his general theory from the death to which he condemned it.

29. The address was subsequently published as Simon Newcomb, "Modern Mathematical Thought," *Nature*, 1 February 1894, *49*:325-329

30. The presidential address was published as Simon Newcomb, "The Philosophy of Hyperspace," *Bulletin of the American Mathematical Society*, 1898, *4*:187-195 and also in *Popular Astronomy*, September 1898, *6*:380-389.

31. Newcomb, "Modern Mathematical Thought," p. 328.

32. Ibid., pp. 328-329.

33. Morris Kline has argued that early in the nineteenth century, mathematics and science ended their age-old relationship of progressing hand-in-hand due to the advent of non-Euclidean geometries, negative and imaginary numbers. The non-Euclidean geometries were consistent within themselves but had no physical analog. Why would God create a system of mathematics without any physical analog? Mathematics, until the advent of these strange systems, was thought to be part of God's design. The mathematicians "forged ahead in the search for the mathematical laws of nature as though hypnotized by the belief that they, the mathematicians, were the annointed ones to discover God's design." Morris Kline, *Mathematics:The Loss of Certainty* (New York: Oxford Univ. Press, 1982), p. 71. This view of Kline's depends on the fact that there were no physical applications of the non-Euclidean geometries in the nineteenth century, which is in direct opposition to the ideas herein presented.

34. Newcomb, "Philosophy of Hyperspace," p. 192.

35. Ibid., pp. 192-193.

36. Ibid., p. 195.

37. Pierce, "Architecture," p. 164.

38. Simon Newcomb, "The Faryland of Geometry," *Harper's Monthly Magazine*, January 1902, *104*:249-252; and reprinted in Simon Newcomb, *Sidelights on Astronomy* (New York: Harper Brothers, 1906).

39. Alfred M. Bork, "The Fourth Dimension in Nineteenth-Century Physics," *Isis*, 1964, *55*:326-338, on p. 327.

40. This particular letter from Sylvester to Newcomb can be found in Box #41, Simon Newcomb Papers, Library of Congress.

41. T. Proctor Hall, "The Possibility of a Realization or Four-Fold Space," *Science*, 13 May 1892, *19*:272-274.

42. Hall, "Possibilities," p. 272.

43. W. P. Preble, "Four-Fold Space," *Science*, 27 May 1892, *19*:304-305; G. B. Halsted, "Four-Fold Space and Two-Fold Time," *Science*, 3 June 1892, *19*:319; C. Staniland Wake, "The Notion of a Four-Fold Space," *Science*, 10 June 1892, *19*:331-332; Edmund C. Sanford, "The Possibility of the Realization of Four-Fold Space," *Science*, 10 June 1892, *19*:332.

44. Charles Hinton, "What is the Fourth Dimension?" in *Scientific Romances* (London: Swann Sonnenschein, 1884-1885), reprinted in Charles H. Hinton, *Speculations on the Fourth Dimension*, ed. Rudolf v.B. Rucker (New York: Dover, 1980): 1-22, on p. 21.

45. Charles H. Hinton, "A Picture of Our Universe," in *Scientific Romances*, reprinted in Hinton, *Speculations*:41-53, on pp. 42-44.

46. Ibid., p. 52.

47. Ibid., p. 45.

48. Ibid., p. 52.

49. Charles H. Hinton, *A New Era of Thought* (London: Swann and Sonnenschein, 1888), p. 49, reprinted in part in Hinton, *Speculations*:106-119, on p. 110.

50. Hinton, *New Era*, p. 61; Hinton, Speculations, p. 111.

51. Charles H. Hinton, "Hyperbolea and Solutions of Equations," *Bull. Am. Math. Soc.*, 1897, *3*:309-320.

52. Rucker has so speculated in the introduction to Hinton, *Speculations*, p. xiv.

53. A summary of Heyl's presentation before the society was written by F. N. Cole, *Bull. Am. Math. Soc.*, December 1897, *3*:89. The original of Heyl's Ph.D. thesis is still in the library at the University of Pennsylvania, in Philadelphia. Paul Renno Heyl, *The Theory of Light on the Hypothesis of a Fourth Dimension* (Ph.D. thesis, Univ. Pennsylvania, 1897), p. 2.

54. Heyl, *Theory of Light*, pp. 2-3.

55. Ibid., p. 30.

56. Ibid., p. 37.

57. The other paper was presented at the February 1898 meeting of the society. A summary of the paper, "The measure of the bluntness of the regular figures in four dimensional space," was written by F. N. Cole, *Bull. Am. Math. Soc.*, April 1898, *4*:293-294. A third paper by Heyl was published as "Properties of the Locus r=Constant, in Space of n Dimensions," *Publications of the University of Pennsylvania*, Mathematics, 1897, *nr. 1*:33-39.

58. Charles H. Hinton, "The Recognition of the Fourth Dimension," *Bulletin of the Philosophical Society, Washington, D.C.*, 1902, *14*:179-203, reprinted in Hinton, *Speculations*, pp. 142-162. The paper was republished as the final chapter, "A Recapitulation and Extension of the Physical Argument," in Hinton's last book, *The Fourth Dimension* (London: Swann Sonnenschein, 1904), pp. 203-230.

59. Hinton, "Recognition," p. 183; Hinton, *Speculations*, p. 145.

60. Hinton, "Recognition," p. 196; Hinton, *Speculations*, p. 156.

61. Hinton, "Recognition," p. 202; Hinton, *Speculations*, p. 161.

62. Hinton, "Recognition," pp. 201-202; Hinton, *Speculations*, pp. 160-161.

63. Hinton, *The Fourth Dimension*, p. 230.

64. Charles H. Hinton, "The Geometrical Meaning of Cayley's Formulae of Orthogonal Transformation," *Proceedings of the Royal Irish Academy*, 1902-1904, *24*:59-65.

65. Hinton, *The Fourth Dimension*, p. 230.

66. For example, Walter William Rouse Ball, *Mathematical Recreations and Problems of Past and Present Times* (London: MacMillan, second edition, 1892), p. 190.

67. W. W. R. Ball, "A Hypothesis Relating to the Nature of the Ether and Gravity," *Messenger of Mathematics*, 1891, *21*:20-24.

68. J. E. Oliver, "Review of Mathematical Recreations," *Bulletin of the New York Mathematical Society*, December 1892, *2*:42-46, on p. 45.

69. Ibid., pp. 45-46.

70. Ibid., p. 46.

71. Ibid.

72. Jammer wrote of "Riemann's anticipation" of Einstein, stating that "Riemann's allusions were ignored by the majority of physicists. His investigations were deemed too speculative and theoretical to bear any relevance to physical space, the space of experience" in *Concepts of Space* (Cambridge:Harvard Univ. Press, 1954), p. 160. He further stated that Clifford's "speculations aroused great opposition among academic philosophers who still adhered to the Kantian doctrine..." Jammer, *Concepts*, p. 161.

AN ECLECTIC OUTSIDER:
J. WILLARD GIBBS ON THE ELECTROMAGNETIC THEORY OF LIGHT

Ole Knudsen*

The theoretical development of electromagnetism and optics in the period after Maxwell is an intricate subject, for several reasons. First, the physicists of the period were faced with a great number of phenomena and quantitative data that even in modern physics can only be accounted for by long and complicated calculations. Second, most of these phenomena involve, in a detailed and essential way, the interaction of electromagnetic fields with matter, or -- to use the language of the period -- the interaction between ether and matter, which everyone regarded as the great unknown of physics. Finally, and this is the main concern of this paper, there existed no commonly accepted foundation for electromagnetic theory, nor, a fortiori, for optics.

The various electromagnetic theories -- and quite a number were proposed between 1870 and 1900 -- may be classified as belonging to either of two traditions, the Maxwellian or the Continental. These have been amply described by Buchwald, who argues that the difference between them was so profound that "no one not educated in Britain or directly from the *Treatise* ever did grasp the basic structure of Maxwellian theory."[1] The only exception to this rule, according to Buchwald, is J. Willard Gibbs. The purpose of this paper is to describe Gibbs's unique position with respect to the foundational problems of electromagnetic theory. This requires a recapitulation of the principal features of the two traditions.

The Continental tradition retained the basic notion of the two electric fluids consisting of imponderable particles exerting distance forces on each other. For two particles at rest these forces were given by Coulomb's law; this led to a satisfactory mathematical theory of electrostatics. In the 1840s, W. Weber showed that by adding certain terms, depending on relative velocity and acceleration, to Coulomb's law one could account for electromagnetic phenomena as well, thus obtaining a unified theory of electrostatics and electrodynamics. In the mathematical theory the concept of electrodynamic potential played a central role. For two closed circuits, l, and l', F. E. Neumann had defined this as the function

$$V_N = -\frac{A}{2} I\, I' \oint_l \oint_{l'} \frac{d\mathbf{l} \cdot d\mathbf{l}'}{r} \tag{1}$$

where A is a constant, I and I' are the currents in l and l', respectively, and $d\mathbf{l}$ and $d\mathbf{l}'$ are infinitesimal circuit elements, separated by the distance r. Neumann showed that the spatial derivatives of this function gave the components of the electromagnetic forces between the two circuits, while its time derivative gave the electromotive force induced in one circuit by the other (modern readers will recognize the integrand in eq. [1] as the so-called Neumann coefficient of mutual induction). It was soon found that Weber's force

* I wish to thank Mr. Vincent Giroud, Curator of Modern Books and Manuscripts, The Beinecke Rare Book and Manuscript Library, Yale University, for permission to publish extracts from the Gibbs manuscripts in the Library's possession.

law also led to an electrodynamic potential, which however had a different mathematical form:

$$V_W = -\frac{A}{2} I\, I' \oint_l \oint_{l'} \frac{(\,d1 \cdot e_r\,)(\,d1' \cdot e_r\,)}{r} \tag{2}$$

where e_r is a unit vector in the direction from dl' to dl. The difference between the two integrands in equations (1) and (2) is a total differential that vanishes when integrated along the closed curves l and l'. Since all electromagnetic experiments dealt with closed currents, the two expressions accounted equally well for all known electromagnetic phenomena, and in this respect the two expressions for the electrodynamic potential might be regarded as equivalent. There was nevertheless a tendency to regard these integral expressions as arising from more fundamental differential formulae expressing the interaction between infinitesimal current elements. The two obvious candidates for such "elementary potentials," as they were called, were of course those obtained by simply removing the integral signs in equations (1) and (2), respectively, i.e. either

$$V_N = -\frac{A}{2} I\, I' \frac{d1 \cdot d1'}{r} \tag{3}$$

or

$$V_W = -\frac{A}{2} I\, I' \frac{(\,d1 \cdot e_r\,)(\,d1' \cdot e_r\,)}{r} \tag{4}$$

A great deal of discussion, involving lengthy and complicated mathematical calculations, was devoted to the question of whether these two expressions would lead to different predictions in situations in which one part of an electric circuit is moving in sliding contact with the rest of the circuit, and, if so, whether such differences would be experimentally observable.

This mathematical electrodynamics reflected a physical picture in which an electric charge consisted in an accumulation of one of the electric fluids, and a conduction current consisted in a flow of the two fluids in opposite directions. All electrostatic and electromagnetic phenomena were in principle to be accounted for by forces acting directly at a distance between the particles of the electric fluids. The electrostatic and electrodynamic potentials were mathematically convenient, intermediate steps between macroscopic phenomena and the fundamental microscopic forces.

The Maxwellian theory meant a total rejection of this physical picture. To Maxwell and his British disciples, an electric charge was not to be thought of as an accumulation of something, nor an electric current as a flow of something. The primary concept of the theory was that of the electromagnetic field, a particular physical state existing in every point in space, outside as well as inside material bodies. Space was filled with an absolutely continuous substance, the ether, whose primary qualities were inertia and elasticity. The electromagnetic field consisted in some kind of mechanical state, a state of motion and/or strain, in the ether. The energy of an electromagnetic system was localized in all those parts of the ether where the field was present, not in the material bodies carrying charge and current distributions. Hence the electrodynamic potential was not a true representation of electromagnetic energy, and discussions about the fundamental correctness of one or the other differential formula for the potential of two current elements were totally irrelevant. Even the very notion of a current element as a basic physical entity was meaningless.

The Maxwellians did have a concept called "electricity" or, sometimes, "electric quantity." This did not, however, denote anything physical; one might say that its only function was to serve as object for the verb "to displace."

The concept of electric displacement was perhaps Maxwell's most important invention. It served as the basis for his mathematical description of the electrostatic behavior of dielectrics, and by postulating that displacement currents had magnetic effects like ordinary conduction currents, Maxwell obtained a set of field equations that were consistent with the continuity equation for electric charge and that led to the wave equations that became the foundation for the electromagnetic theory of light.

The Maxwellians regarded displacement current as a physical process taking place in dielectric materials and in the free ether, a process of essentially the same nature as a conduction current. This could not be reconciled with the notion of electric fluids confined to material bodies. The old conservation law of electric charge, which was the chief justification for the concept of the electric fluids, therefore had to be given a new interpretation. Maxwell did this by defining the *true* or *total* current as the sum of the conduction and displacement currents and showing that the continuity equation expressing charge conservation could be transformed to the equation

$$\nabla \bullet \mathbf{j}_T = 0 \tag{5}$$

where $\mathbf{j}_T$ denotes the density of the total current. To the Maxwellians, "electricity" denoted that mysterious entity of which the true motion was given by $\mathbf{j}_T$. Equation (5) showed clearly that this true motion was like that of an incompressible fluid, hence there could be no accumulation of "electricity" anywhere.

In 1870, three years before Maxwell published his monumental *Treatise on Electricity and Magnetism* but several years after he had published the essence of his theory in article form, Hermann Helmholtz wrote a lengthy article on electrodynamics.[2] Helmholtz had sensed the importance of Maxwell's result, that light may be regarded as electromagnetic waves and that the velocity of light may be identified with the ratio between the electromagnetic and electrostatic units of charge; but it is clear that he did not accept Maxwell's field concept, that he found Maxwell's concept of displacement obscure and his notion of electricity unintelligible. Instead he created a generalized action-at-a-distance theory of which the following features are essential for this paper.

He began by combining Neumann's and Weber's differential expressions for the electrodynamic potential in the following way

$$V = \frac{1+k}{2} V_N + \frac{1-k}{2} V_W \tag{6}$$

The integrated potential V he showed, would give a valid description of closed circuit electrodynamics for any value of the constant k. Hence k must be regarded as an undetermined quantity, the value of which could only be fixed by sufficiently precise experiments on open circuits. (The expressions given by Neumann and Weber correspond to choosing k equal to plus or minus one, respectively).

An important step in Helmholtz's theory was introduced by a discussion of the influence of dielectric materials. He described the basic process as a microscopic separation of the two electricities, resulting in an electric polarization, in every volume element, proportional to the electric force acting on that element. This was just a recapitulation of a well-known theory of material dielectrics. Following Maxwell, Helmholtz argued that since a change of polarization would involve a motion of the two electricities, a time-varying polarization would be equivalent to an electric current. He then interpreted Maxwell's theory as stating that the ether was a polarizable medium.

Such an assumption was not unnatural to a Continental physicist, because it was quite common to speculate that the ether might consist of a neutral combination of the two electric fluids. Helmholtz was now able to derive a set of differential equations describing the electric and magnetic polarization of the ether. These equations contained the unknown constant k and another unknown constant, ϵ_0, characterizing the electric polarizability of the ether. Helmholtz showed that if he chose k equal to zero and made ϵ_0 tend to infinity, his equations described transverse waves of polarization, propagating with the velocity of light; hence Helmholtz denoted this particular choice of values ($k = 0$, $\epsilon_0 = \infty$) as the "Maxwell limit" and claimed to have developed a general electrodynamics that encompassed the theories of Neumann, Weber and Maxwell as special cases.

It was almost certainly during his years of study in Europe, 1866-69, that Gibbs got his first introduction to electrodynamics and optics at an advanced theoretical level. As I have shown elsewhere,[3] Gibbs became thoroughly familiar with electrodynamic potential theory and the wave theory of light, and he became aware of the Danish physicist L. V. Lorenz's attempt at creating an electrodynamic theory of light. It would, of course, have been possible for him to learn about Maxwell's theory as well, since Maxwell had published his two first versions of it before Gibbs went to Europe, but there is no mention of Maxwell in his notebooks from this period.

Sometime after his return, and after the publication of Helmholtz's theory,[4] Gibbs began an extensive study of Maxwell's theory. Exactly when this happened is impossible to pin down. Gibbs left more than 1,000 manuscript pages on electromagnetism and optics, which are now in the possession of the Beinecke Rare Book and Manuscript Library at Yale University. Most of these are undated, and are grouped together as Gibbs left them, not chronologically, but according to subject.

What is probably an early group of notes is found in two folders labelled (by Gibbs himself) "Critique of Maxwell's Method" and "Electrical Theories. Notes & Queries. Fundamental Equations. Critique on Sundry Authors."[5] These notes reveal that Gibbs made a detailed study of Helmholtz's paper and compared it critically with Maxwell's 1865 article,[6] which contains the version of Maxwell's theory that Helmholtz referred to in his paper. There are also many notes on Maxwell's *Treatise*.[7]

To show the nature of these notes, let me quote some of those on Maxwell's 1865 article (see Fig. 1). The first group show Gibbs's understanding of Maxwell's basic concepts (the numbers in brackets are in the original and refer to paragraphs in Maxwell's paper):

> Maxwell starts with the desire to get rid of forces at distance, esp. because these are functions of velocity.
> Thinks probable that there is rotation in magnetic field.
> Assumes [22ff] apparently that momentum and energy lie in the electromagnetic field, in particular in the distribution of the magnetic effects, as distinguished from supposed linear conductors.
> [33] Energy resides not only in circuits but in surrounding space. Does he not really make the electrical motions have no energy in themselves?
> [55] Total motion of electricity (p'q'r') = current (pqr)+d/dt displacement (fgh).
> Transmission is the same as current.

While these notes simply recapitulate without criticism some of the most essential features of Maxwell's theory, other notes show that on the technical level Gibbs was not completely satisfied:

[68] sign in G is wrong, as seen from H A & C.

[70] Maxwell's remarks on the number of equations are very crude. One equation is superfluous.

[91] It is observable that Maxwell does not eliminate the extra variables from his 20 Equations, & thus find the general differential equation of Optics.

It is interesting to see that Gibbs caught the inconsistency in sign between Maxwell's equations G, H, A, and C, which has received some attention in recent historical literature.[8] And Gibbs's dissatisfaction with the mathematical structure of Maxwell's theory of light was to be a dominant theme in his own work, as we shall see.

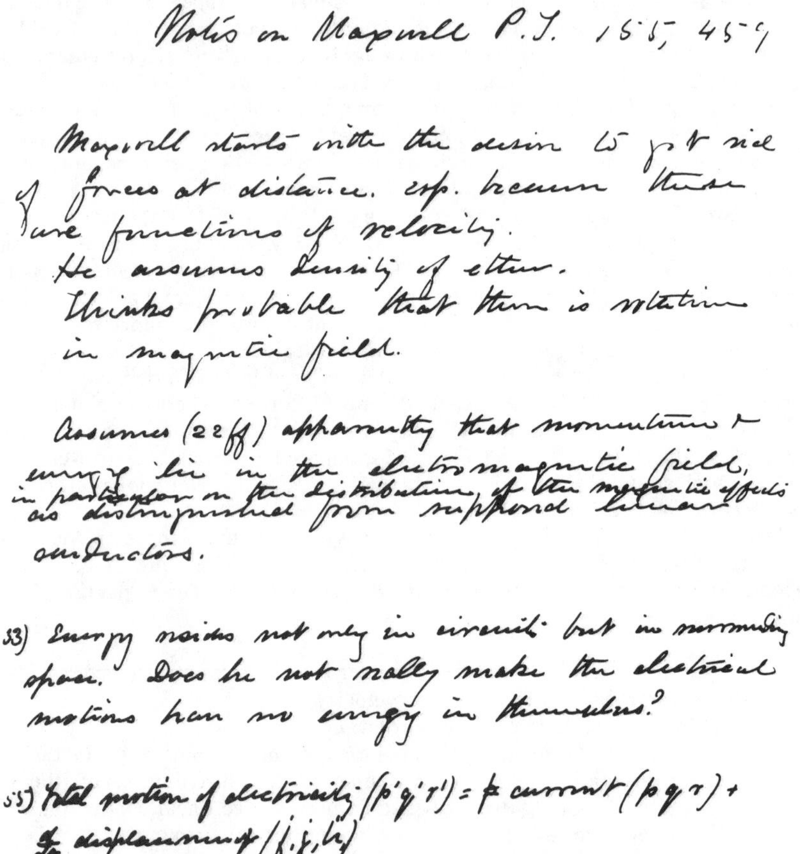

Figure 1
Gibbs's Notes on Maxwell

The notes on Maxwell's *Treatise* confirm the impression that despite his familiarity with Continental theories Gibbs had no difficulty in assimilating Maxwell's field-theoretical view, but that he found the mathematical structure of Maxwell's theory of light unnecessarily complicated.

In comparison, Gibbs's notes on Helmholtz's theory show a much more critical attitude towards the substance of some of Helmholtz's statements. He was obviously not convinced of the correctness of Helmholtz's assertion that the theory would reduce to Maxwell's for the particular choice of k equal to zero. He apparently tried to follow the role played by k through Helmholtz's paper, and found one instance where it seemed to him that it would rather be by setting k equal to unity that one would make Helmholtz's expression equivalent to Maxwell's. "Helmholtz is entirely wrong," he wrote on the bottom of one page. Gibbs's judgement was entirely sound, for although Helmholtz's end result about the velocity of transverse waves may be brought to coincide with Maxwell's, the mathematical and conceptual discrepancies between Helmholtz's modified action-at-a-distance theory and Maxwell's field theory are such as to make them virtually incompatible.[9]

Gibbs's interest in electromagnetism led on the one hand to his purely mathematical work on vector analysis, which will not be dealt with in this paper, but it also led him to publish in 1882-83 three short papers on the electromagnetic theory of light, in which he tried to develop an electromagnetic foundation for optics that was both more simple and more general than those of Maxwell and Helmholtz. He did this by introducing a truly Maxwellian concept, a vector field U called the total electric displacement, defined so that its time derivative was equal to Maxwell's total current:

$$\partial U / \partial t \; = \; j_T \tag{7}$$

The simplification of the theory came about because Gibbs completely eliminated the magnetic field from his considerations and expressed everything in terms of the one vector U. Gibbs apparently regarded the magnetic field vectors as just unnecessary complications. In fact, he never used the word "electromagnetic," but spoke consistently of "electrodynamics" and the "electrical theory of light." In his two first papers[10] he set up expressions for the electromagnetic energy in terms of U, while in the third paper[11] he established a general equation of motion for U. This equation, or modified versions of it, is found in many of Gibbs's notes, where it is often called "the fundamental equation of optics."

Gibbs also made his theory more generally applicable by assuming that matter is a "fine-grained" molecular structure that modifies the electric motions described by U. He showed how this assumption might be described by splitting U into a regular part, which might be conceived as the spatial average of U over volumes small in comparison with the wavelengths of light, and an irregular part. The unknown relation between the irregular and regular parts of U would then introduce certain unknown linear operators into the fundamental equation, and Gibbs demonstrated how, by plausible choices of the form of such operators, the theory might account for various optical effects. He regarded it as a particular advantage that his account did not depend on particular assumptions about the detailed structure of matter and the ether or about their interaction. "The consideration of the processes which we may suppose to take place in the smallest parts of a body through which light is transmitted, farther than is necessary to establish the general equation given above, is foreign to the design of this paper," Gibbs wrote in a footnote to his third paper.

As I have shown elsewhere,[12] this way of treating the optical properties of matter is rather similar to the procedure that L. V. Lorenz had used in his optical papers and may well reflect an inspiration Gibbs had received during his years in Europe. It is also remarkable that Gibbs expressed his doubt about the correctness of one of Maxwell's most fundamental assertions expressed in equation (5) above. Gibbs regarded this as an unproved assumption, not verified by experiment. On this point he agreed with Helmholtz, in whose theory the right hand side of equation (5) became a function of the undetermined constant k and, as Gibbs noted in one of his unpublished notes, did not even reduce to zero in the "Maxwell limit."

In 1888 and 1889 Gibbs published two papers[13] in which he utilized the energy methods of his first two papers to argue that the "electrical" theory of light was superior to the elastic theory, which was still being actively pursued by the British physicists R. T. Glazebrook and William Thomson. Gibbs's argument consisted in showing that while the electrical theory could account in a natural way for optical phenomena, the elastic theory could only do so by making assumptions that were evidently artificial and unphysical.

The five short papers mentioned here were all that Gibbs ever published on the electromagnetic theory of light. After 1887, when Hertz began to publish his famous experiments on electromagnetic waves, a number of physicists, including H. A. Lorentz, Drude, and Hertz himself, felt that the time was ripe for a critical review of the different electromagnetic theories. Gibbs was no exception, for among his manuscripts there is a folder entitled "On the possible contributions to Electrical Theory which are afforded by the phenomena of Optics" and containing the beginning of a projected paper with this title. The manuscript begins with an introduction that is worth quoting in full, because it gives a clear impression of Gibbs's views on the fundamental issues:[14]

> Nothwithstanding the great development of the mathematical theory of electicity, & the precision wh may be attained in electrical measurements, there are many questions of a fundamental character in this science, for wh it has appeared very difficult, if not quite impossible, to obtain any answer from such experiments as are ordinarily termed electrical.
>
> Among such questions are the following:
>
> I. Is there only one kind of electrical 'fluid,' or are there two, with opposite qualities?
>
> II. Has electricity inertia, like ordinary matter?
>
> III. The properties of insulating bodies indicate that electrical fluxes may take place within them subject to an elastic resistance. Whether the same is true of the so-called vacuum, might be regarded as an open question before the recent experiments of Professor Hertz. If we now regard this question as decided, there still remains that of the quantitative determination of the electrical elasticity of a vacuum, & of the proportion of a charge given to an insulated conductor wh actually remains upon it.
>
> IV. The laws of electrodynamic induction have been established by experiments on solenoidal currents. What is the law for electrical motions wh are not solenoidal?

V. Are there any electrical motions wh are not solenoidal?

Some of these questions are now actively discussed. Others have been the subjects of famous controversies, & if less discussed at present, it is not that they are regarded as settled, but on account of the conviction that the data for deciding them are not furnished by electrical experiments, & the growing preference for such expressions of the laws of electricity as contain no more than can be verified by experiment.

The evidence available for the determination of these questions will be vastly increased, if we regard the motions of light as electrical, & admit the evidence of optical phenomena to supplement that of those ordinarily denoted by the term electrical.

It is the object of this paper to consider these questions in the light of the evidence afforded by optical phenomena. It is therefore assumed that the motions of light are electrical, but any farther assumption of Maxwell's theory of light will be avoided, especially the solenoidal hypothesis.[*] In the domain of electricity also, only such principles will be assumed as are generally admitted, either as ultimate verities, or at least as convenient expressions of the known facts. We may thus arrive at the principle features of the modifications of the theory of electricity wh are necessary in order to include the phenomena of optics.

[Footnote:]

> [*]It is not here intended to express any opinion whether the electrical theory of light is or is not proved beyond a reasonable doubt, but only to assume the electrical nature of light, for the sake of argument, in order to see what consequences will necessarily follow. Neither is it intended to imply that any results will be obtained wh are essentially different from Maxwell's. We shall in fact find the fundamental principles of his theory a necessary consequence of admitting the electrical nature of light.

Gibbs may have felt, as indicated in the footnote on Maxwell's theory, that Maxwell's fundamental principles were vindicated by optics, but he was nevertheless very far from being an orthodox Maxwellian. In fact, no contemporary Maxwellian would have seriously posed any one of his five questions. The electrical fluids had been expressly banished by Maxwell himself, so the first question would make no sense to a Maxwellian. As for the second question, Weber's electrical particles had possessed inertia, and H. A. Lorentz had ascribed inertia to electrical motions in material conductors and dielectrics and had been rebuked for his efforts by none other than Gibbs himself with a typically Maxwellian argument.[15] And the three last questions cast doubt on one of Maxwell's most cherished notions, that "electricity" moves like an incompressible fluid and does not accumulate on charged conductors. While these questions would again have made no sense to a Maxwellian, they are highly relevant from the standpoint of Helmholtz's theory. Gibbs's "electrical elasticity of a vacuum" is just another phrase for Helmholtz's polarizability of the ether, characterized by the undetermined constant ϵ_0, and the two last questions would, in Helmholtz's theory, be answered by the determination of the unknown constant k.

The manuscript also shows that Gibbs believed that optical data would be more successful than purely electrical ones in solving the fundamental problems. In Berlin, Gibbs had followed the lectures of G. H. Quincke, who specialized in optical precision measurements, and among his manuscripts there are tables comparing numerical values of phase shifts in metallic reflection, calculated by Gibbs from his theory, with the corresponding experimental values measured by Quincke. Gibbs was also familiar with Michelson's determination of the velocity of light[16] and had obtained direct information from Michelson on one specific point.[17] He was thus well aware that optical measurements had reached a higher degree of precision than electromagnetic ones, and this is probably one of the grounds for his faith in the importance of optics.

It is true that Gibbs, as he stated in this manuscript, to some degree accepted Maxwell's fundamental principles, but the questions he raised certainly owed more to the Continental tradition than to Maxwell. Just as in thermodynamics he combined elements from different approaches in order to derive his own "fundamental thermodynamic equation"[18] and in order to "find the point of view from which the subject appears in its greatest simplicity,"[19] so also here he combined Maxwellian and Continental elements in order to create a simple and general mathematical theory, based on the "fundamental equation of optics."

If Gibbs's "electrical theory of light" cannot be classified as either typically Maxwellian or typically Continental, his attitude must be characterized as uniquely eclectic for his time, which incidentally may be one reason why his papers seem to have had no influence whatsoever on his contemporaries. It is tempting to relate this eclecticism to the external circumstance that Gibbs, like all American physicists of the period, was an outsider, attempting to follow, digest, and possibly take part in, the development of a scientific discipline whose main centers were located outside his own national culture. The outsider has obvious disadvantages; but it is perhaps easier for him to avoid becoming so deeply immersed in one particular paradigm as to be completely unable to understand a different one. The contrast between the ease with which Gibbs could profit from the work of both traditions, and Hertz's self-acknowledged failure to grasp the essentials of Maxwell's theory, may be seen as an illustration of this point.

In a recent paper,[20] Klein has given a persuasive interpretation of Gibbs's style in science as an integral element of his "retiring disposition," which made him spend his whole career in isolation with nothing but the most superficial contact with colleagues anywhere. I do not think there is any conflict between this view and the point I have tried to make. A person may be an outsider for both external and internal reasons, and I think that in Gibbs's case the qualities of his personality were enhanced by his situation as an American outsider. These speculations do suggest, however, that it might be of interest to look for similar traits of eclecticism among other American physicists of the earlier part of the Michelson era.

NOTES

1. J. Z. Buchwald, *From Maxwell to Microphysics: Aspects of Electromagnetic Theory in the Last Quarter of the Nineteenth Century* (Chicago and London: Univ. Chicago Press, 1985), p. 188.

2. H. Helmholtz: "Ueber die Bewegungsgleichungen der Elektricität für ruhende leitende Körper," *Journal für die reine und angewandte Mathematik*, 1870, 72:57-129.

3. O. Knudsen, "The Influence of Gibbs's European Studies on His Later Work," in *From Ancient Omens to Statistical Mechanics*, ed. J. L. Berggren and B. R. Goldstein, *Acta Historica Scientiarum Naturalium et Medicinalium*, vol. 39, Copenhagen University Library 1987, pp. 271-280.

4. Helmholtz, "Ueber die Bewegungsgleichungen der Elektricität für ruhende leitende Körper."

5. The Beinecke Rare Book and Manuscript Library, Yale University, MS Vault Gibbs 9.

6. J. C. Maxwell, "A Dynamical Theory of the Electromagnetic Field," *Philosophical Transactions of the Royal Society*, 1865, 155:459-512.

7. J. C. Maxwell, *Treatise on Electricity and Magnetism*, 2 vols. (Oxford: Clarendon Press, 1873).

8. J. L. Bromberg, "Maxwell's Electrostatics," *American Journal of Physics, 1968, 36*:142-151.

9. Buchwald, *From Maxwell to Microphysics*, pp. 177-186.

10. J. W. Gibbs, "On Double Refraction and the Dispersion of Colors in Perfectly Transparent Media," *American Journal of Science*, 1882, ser. 3, *23*:262-275; and "On Double Refraction in Perfectly Transparent Media Which Exhibit the Phenomena of Circular Polarization," ibid., 460-476.

11. J. W. Gibbs, "On the General Equations of Monochromatic Light in Media of Every Degree of Transparency," *Am. J. Sci.*, 1883, ser. 3, *25*:107-118.

12. Knudsen, "The Influence of Gibb's European Studies on His Later Work."

13. J. W. Gibbs, "A Comparison of the Elastic and the Electrical Theories of Light with Respect to the Law of Double Refraction and the Dispersion of Colors," *Am. J. Sci.*, 1888, ser. 3, *35*:467-475; and "A Comparison of the Electric Theory of Light and Sir William Thomson's Theory of a Quasi-labile Ether," *Am. J. Sci.*, 1889, ser. 3, *37*:139-144.

14. The Beinecke Rare Book and Manuscript Library, Yale University, MS Vault Gibbs 7.

15. Cf. O. Knudsen, "Electric Displacement and the Development of Optics after Maxwell," *Centaurus*, 1978, *22*:53-60.

16. J. W. Gibbs, "Reviews of Newcomb and Michelson's 'Velocity of Light in Air and Refracting Media' and of Ketteler's 'Theoretische Optik,'" *Am. J. Sci*, 1886, ser. 3, *31*:62–67.

17. J. W. Gibbs, "On the Velocity of Light as Determined by Foucault's Revolving Mirror," *Nature*, 1886, *33*:582.

18. M. J. Klein, "The Scientific Style of Josiah Willard Gibbs," in *Springs of Scientific Creativity: Essays on Founders of Modern Science*, ed. R. Aris et al. (Minneapolis: Univ. Minnesota Press, 1983), pp. 142–162.

19. J. W. Gibbs, letter to the American Academy of Arts and Sciences, 10 Jan. 1881, quoted in L. P. Wheeler, *Josiah Willard Gibbs: The History of a Great Mind*, rev. ed. (New Haven: Yale Univ. Press, 1952), pp. 88–89.

20. Klein, "The Scientific Style of Josiah Willard Gibbs."

MICHELSON-MORLEY, EINSTEIN, and INTERFEROMETRY

Loyd S. Swenson, Jr.

In a 1983 article on "The Origin of the Universe," Victor F. Weisskopf, one of the doyens of quantum electrodynamics, nuclear structure and elementary particle physics, made a remark that may symbolize the significance of the Michelson-Morley ether drift experiment more profoundly than he intended. In showing various linkages between particle physics and astrophysical cosmology today, Weisskopf wrote:

> It is remarkable that we are now justified in talking about absolute motion, and that we can measure it. *The great dream of Michelson and Morley is realized.* They wanted to measure the absolute motion of the earth by measuring the velocity of light in different directions. According to Einstein, however, this velocity is always the same. But the 3°K radiation represents a fixed system of coordinates. It makes sense to say that an observer is at rest in an absolute sense when the 3°K radiation appears to have the same frequencies in all directions. Nature *has* provided an absolute frame of reference. The deeper significance of this concept is not yet clear.[1]

Weisskopf may be alone in asserting such a bold statement these days. "Big Bang" cosmology has many other roots and many other branches. Most professional astronomers now seem less interested in the absolute motion problem than in the isotropic background radiations (microwave 30K, X ray, and gamma ray). However, Weisskopf assured me in a personal note in late 1985 that he had no reason to regret this paragraph. His remarks quoted here point up the fact that ongoing developments of interferometry in both macrophysics and microphysics ensure a perennial significance of the highest order to the researches that A. A. Michelson, E. W. Morley and their successors embarked on a century ago.

This paper is a memoir to set in order some lessons I've learned from 27 years of intermittent work and worry over the relationships between so-called 'crucial' experiments and theories in science generally, and specifically about the relationship between the 'crucial' Michelson-Morley ether drift experiment of 1887 and the advent of Albert Einstein's first theory of relativity in 1905. My two books in the social history of physics already seem dated in some ways yet seem corroborated in most respects by recent scholarship. In the first, *The Ethereal Aether*, I took a biographical approach to the whole life of the classic optical tests for a relative ether-wind against the motions of the earth. That book surveyed the many failures of Michelson's dream from its conception in 1880 until its practical death in 1930 (Michelson died the following year while still trying to perfect his techniques for determining the velocity of light in a mile-long vacuum tube).[2]

If the biography of the whole life of the Michelson-Morley experiment was the story of a celebrated failure, then my second, related book, *Genesis of Relativity*, was a history of a fabulous set of successes. It narrated Einstein's growth by tracing the five decades from the 1870s to the 1920s as a progression of periods alternately characterized by syntheses and analyses. Thus, the context for Einstein's productive scientific life was reduced to a manageable scale, and the Michelson-Morley experiment's role was also reduced to a more proper context. For example, the contemporary and equally crucial experiments of Heinrich Hertz with "ether waves" around 1887 were detailed and seen as surrounding young Einstein's consciousness, too. From the Maxwellian synthesis of the 1870s, through analyses of electric radiation in the 1880s, through stillborn attempts at syntheses of the ether, electrons, atoms and energy in the 1890s, thence into plural analyses of relativity in the 1900s, and finally to the Einsteinian synthesis of the 1910s -- these were the themes of my more mature judgment. [3]

Recent scholarship has resurrected an old issue about the exact relationship between Michelson-Morley and Einstein's earliest work on relativity.[4] Almost all who have studied this issue seriously -- including Bernard Jaffe, Robert S. Shankland, Gerald Holton, Tetu Hiroshige, Horst Melcher, Jagdish Mehra, Arthur I. Miller, Stanley Goldberg, Lewis Pyenson, Hans J. Haubold, D. Ted McAllister, Martin Klein, Nathan Reingold, Albert Moyer, and myself -- have had to deal with the problem of character versus reputation in history: what influential people believe to be true is often more important in the course of human events than what was (or is) the actual case. It is certain that Einstein's chief mentors believed that the Michelson-Morley experiments were definitive, that the null results from the ether-drift tests were anomalous and therefore demanding theoretical revisions. Still debatable, for many historiographical reasons, is the issue of precisely what Einstein knew or imagined and when.[5]

The question is less interesting to those of us with a long-term historical perspective, because we see the breadth and depth of influences on scientific change as mediated through contexts as well as, and often more than, through texts. Historians of science who have short-term and internalist perspectives might like to find in the forthcoming second volume of *The Collected Papers of Albert Einstein*, or wherever else, unequivocal evidence that the young Einstein was directly affected in his scientific thinking by immediate contact with the null results of the Michelson-Morley experiment. Scholars with broader and externalist perspectives may be satisfied to know, in Stephen Kern's work and words, that *The Culture of Time and Space* pervaded the period from the 1880s to the 1920s. The leading physicists of the older generation to whom Einstein looked for guidance and inspiration -- especially Mach, Kirchhoff, Helmholtz, Hertz, August Föppl, Paul Drude, Wilhelm Wien, and H. A. Lorentz -- were virtually obsessed with the failures to find evidence for the relative motion of the earth against the luminiferous ether.[6]

Many scholars have dealt with the "before and after 1905" problem in relation to the Michelson-Morley experiments and in reference to Einstein's miraculous year. But so far as I know, few have analyzed yet the intermixed problems of generation gaps, communication gaps, and the gaps between disciplines so closely related as physics and astronomy. Why and how do certain experiments catch hold of public and private imaginations? This seems to occur despite obvious empirical design limitations, and it leads to widening gaps between generations, specialties and elites among scientists. The Michelson-Morley experiment is a good example of these phenomena. The fact that it continues to be intriguing despite its supposedly having been settled once and for all many times since 1887 testifies to that.[7]

At the 1981 centennial of the original Michelson ether-drift experiment in Berlin and Potsdam, I reviewed the histories of Michelsoniana and Einsteiniana and suggested in conclusion five primary problems for mid-range historical research that might illuminate the relationship between Michelson-Morley and Einstein. So far as I know these are still unanswered questions. But my primary thesis at Potsdam, taken from Holton, who was not present, was that we should be celebrating a centennial for the birth of optical interferometry. Should not Michelson's interferometer, given its perennial usefulness and seminal adaptability, be ranked with the telescope, microscope, barometer and thermometer in the pantheon of scientific instruments?[8]

In re-reviewing some primary documents and secondary evidence, particularly Michelson's own public records, for this 1987 centennial, I am again impressed by

Michelson's modesty, but even more by Michelson's evolution -- overlooked before -- toward understanding that interferometric techniques can be applied to *all* microscopy and telescopy, therefore by implication to *all* transverse wave motion and to *all* electromagnetic radiation. It would seem that Michelson's talent for meticulous measurements led from his ether-drift instrument to his wave-theory insights. His and Morley's interferometers created interferometry.[9]

If that assumes too much, because too many other people were involved, because Michelson was more versatile than that implies, or because it is too nearly a tautology, then perhaps the Nobel Prize citation for physics in 1907 was properly worded. It was awarded to Michelson simply "for his precision optical instruments and the spectroscopic and metrological investigations conducted therewith." Yet one may wonder which is ultimately more important, relativity or interferometry, especially now that laser interferometry is being applied to test for gravitational waves predicted by general relativity theory.[10]

The lineage of optical instruments to make the first explicit use of interference phenomena for practical purposes dates back at least to the French school of optical experts in the 1850s. Michelson usually credited Jules Jamin and his "interferential refractometer" of 1856 for being his prototypical inspiration to develop his "differential refractometer" in the 1880s. This "Michelson-type interferometer," as it came to be known, was useful at first simply for separating the split-beams from a single source at right angles to each other, thus permitting many kinds of measurements to become astonishingly precise. As a current-meter for the proper motion of the earth through space or for the relative motion of the earth through the local or larger plenum, it was a dismal failure (except in Dayton C. Miller's hands). As an air-thermometer independent of atmospheric pressure, however, it first began to show its versatility as early as 1882.

Michelson never patented it, and so by the 1890s the split-beam interferometer was commercially produced by several scientific instrument makers. It soon became standard equipment in almost all high-tech labs, whether academic or industrial, because it was extremely valuable as a tool for determining coefficients of expansion and elasticity, for the testing of precision screws, and especially for obtaining extremely precise spectroscopic analyses. Michelson went on to design, and often to develop, many other types of interferometers, such as comparators, and planetary, stellar, vertical, and earth-tide measuring devices. Countless competitors and emulators appeared to make interferometry a globally known tool for scientific and technological advancement. Other big names in the field--notably Fabry-Perot, Lummer-Gehrcke, Mach, Sagnac, Majorana, Twyman-Green, Meggers-Peters, Righi, Zehnder, Ives, and Joos, for examples -- helped to make optical interferometry even more important while Michelson was still alive.[11]

There can be little doubt that the "great dream of Michelson and Morley" to which Weisskopf alluded regarding absolute motion was more the dream of Michelson and Miller than of Morley. Furthermore, the dream had less to do with measuring the so-called absolute motion of the earth than it did with learning more about the proper motion of the earth and the effects of the motions of media on the behavior of light. "Absolute" meant relatively absolute already by 1887, and it should not be forgotten that the classic Michelson-Morley paper of that year begins with a discussion of astronomical aberration and a devastating critique of the emission, or particle, theory of light. So, the quest in 1887 was for the relative motion of the earth *in orbit*, or maybe for the motion of the solar system as a whole, but certainly for an answer to the question whether the "luminiferous ether" is at rest relative to the earth's surface, or not. We must force ourselves to remember that the problem of the solar apex was in its infancy and not clarified until the advent of galactic astronomy in the 1930s.[12]

It is true that Michelson began with the hope of measuring the velocity of the earth's motion through the ether as a whole and that Miller maintained the same hope after 1925. But between 1880 and 1925 the cosmos changed. Michelson and Morley collaborated for only about five years; Morley and Miller collaborated for about fifteen. Michelson and Morley gave up on ether-drift testing together after only one season; Morley and Miller experimented to de-bug the classic design for at least five years. Michelson had learned celestial navigation, among other trades, early in life; Morley had served as a chaplain while becoming a chemist; Miller had earned a Princeton Ph.D. in astronomy before becoming expert in optics and acoustics. Only Miller followed through to completion for all seasons of the year under comparable conditions the classical 1887 experimental design. Since that was not done until 1926 and since Miller claimed to have found the long-sought absolute motion of the earth, the results led to a prize and then into a scientific scandal of long standing.[13]

That scandal prompted about five years of feverish experimental activity, including several elaborate attempts by Michelson to respond to Miller's challenging findings. Michelson was basically sympathetic to Miller's results, for he recognized this as unfinished business that he and Morley and Miller had started. But the fact that Michelson and his new crews of experts around Mount Wilson were unable to corroborate Miller's observations gave even greater weight to the accumulating evidence from other crews of experimentalists who were also still finding null results optically in their often novel and intricate retests of the classic Michelson-Morley ether-drift experiment.[14]

By the late 1920s and early 1930s physics and astronomy had changed so much and grown so much more sophisticated that it is not surprising that Michelson and Miller in their later years should have seemed to be reactionaries to many of the younger generation of specialists. What is surprising, to me at least, is how flimsy was the actual character of the 1887 Michelson-Morley experiment compared with its widespread and formidable reputation already by 1905. Scientists are not supposed to be good historians, but they are supposed to be honest. Michelson, Morley, Miller, and Einstein were honest in checking their sources and in speaking and writing cautiously about their precursors, but many of those who wrote and spoke about the significance of the experiment after them were not. For they too often assumed that what happened to the ether-drift tests between 1925 and 1930 had already happened, say, between 1900 and 1905.[15]

Between 1880 and 1905 there were only about ten reports of ether-drift experiments or of experiments closely related to Michelson's quest for evidence of relative motion of earth and ether. Between 1905 and 1925 there were about ten more such primary documents of published experimental tests. More than a dozen such basic papers appeared in the late 1920s and early 1930s. This public production record seems to show that the Michelson-Morley experiment grew into an *experimentum crucis* only after Miller made it so. And in the midst of Miller's challenge Einstein is supposed on good authority to have uttered the famous German words to the effect that "Subtle is the Lord, but malicious He is not."[16]

Many physicists know, or believe at least, that the final word on the interferometric embarrassment to relativity theory posed by D. C. Miller's experiments in the 1920s was given at last by Robert S. Shankland and his colleagues in Cleveland during the 1950s. In 1955, the year that Einstein died, Shankland et al. published an elaborate analysis of Miller's work, judging his anomolous, small but positive results to have been caused by inadequate temperature control.[17]

I first made the acquaintance of Professor Shankland in the early 1960s as I was finishing my dissertation on Michelson-Morley. I was eager to meet the man who had so recently settled the Miller issue and who had gained Einstein's confidence in so doing. But more than a decade would pass before Shankland and I could talk in person about the characters and reputations of Michelson, Morley, Miller, and Einstein and about interferometry. The only aspect of my doctoral research that Shankland recognized seriously at first was the discovery of letters by Michelson from 1881 that reported to his patron, Alexander Graham Bell, that the Berlin professor Hermann Helmholtz had warned Michelson about the pitfalls of temperature control. To Shankland this seemed to be historical confirmation that Michelson's mentor expected -- or at least had a premonition of -- a temperature effect in this delicate second-order experiment, an effect that could, would and, in the 1920s, indeed did happen. Miller's spurious results, Shankland assured me by phone and postcards in the 1960s, were attributable in part to wishful thinking and to mistakes in experimental design as well as a few erroneous calculations. Miller's own body-heat or infrared radiation probably interfered with his observations of interference fringes. Temperature gradients across the interferometer Shankland judged to be the ultimate culprit.

At last in August 1974 I was privileged to spend several days with Shankland in Cleveland while doing an oral history interview on his life and distinguished career for the Neils Bohr Library of the American Institute of Physics. Much of our talk naturally concerned Miller's relationships to Michelson and Morley, and interferometry, and Shankland's talks with Einstein about the same subjects. On the verge of retirement Shankland was intellectually vigorous, quite candid and forthright, eager to talk about these momentous topics, and a most gracious host. All quotations that follow are from the transcript of the Shankland interview on deposit in taped and typescript form at the Bohr Library in New York City.[18]

Shankland came to Case in 1925 as a freshman to study physics with Professor Miller, then at the height of his fame not only as president of the American Physical Society but also as that year's prizewinner from the American Association for the Advancement of Science. Although Shankland was too young to work directly with Miller on the seasonal observations with the interferometer atop Mount Wilson overlooking Los Angeles, during those years he matured rapidly as an undergraduate. But he was maturing in a relativistic world, had already studied Einstein's theories, and said that he could never understand why Professor Miller persisted in pursuit of an absolute motion for the earth. Soon, however, Shankland learned to love optics and acoustics through an ever closer association with Miller. Earning his B.S. in 1929 and M.S. in 1933 at Case, Shankland went to the University of Chicago to study with Arthur H. Compton for his Ph.D., which was conferred in 1935. Except for government service and industrial consultancies, he served continuously on the Case faculty from his first appointment in 1930. Miller died in 1941, and Shankland inherited the legacy of Michelson, Morley, and Miller in direct succession after World War II.

Shankland had an institutional memory and a pride in the tradition of physics in Cleveland at the two institutions that became Case Western Reserve University. He did not show me a strong chronological sense but his historical articles do show a meticulous sense for accuracy and nuance. Most of my two-day interview with him covered his own career, primarily in acoustics, but there were about six occasions when he waxed eloquent about the three M's. I shall summarize those passages in what follows, but first, to finish this introduction, it is important to note Shankland's conservatism, politeness, and deferential attitudes. Whenever we touched upon the subjects of this paper, Shankland spoke with pride and grace. He talked about Einstein's encouragements at length and about how those conversations were published. He also revealed much about the attitudes and personalities concerned with Miller in the 1920s and Einstein in the 1950s. Shankland admired and respected both men so much that his labor of love in re-analyzing Miller's "great piles of data sheets" by assembling a team to track down the cause of the problem was virtually an act of reconciliation. Privately there had never been any need for reconciliation between Miller and Einstein, but publicly Cleveland and Case Western Reserve University may have needed reassurance.

One of Shankland's most pertinent comments relating to this symposium came in response to my probing about the first Michelson-Morley collaborative effort in 1886, the so-called Fizeau ether-drag experiment, retesting for the motion of the medium (in this case, Morley's distilled water) on the velocity of light. Shankland said:

> I think one of the reasons that Einstein was so taken with the Fizeau experiment was that it gave a number. You see, these null experiments, important as they are, are always subject to the question: Well, was there something missing in the experiment that didn't reveal it? Michelson to the end of his days was worried about this point. But when you have a number, and the Fizeau experiment had a number -- and another number that Einstein was so interested in was the aberration constant -- those not only would be stimuli to a theory, but they would check against a theory in a way that a null experiment could not.[19]

Thus Shankland argued that Einstein was as much interested in the Michelson-Morley repetition of the Fizeau experiment as he was in "what we all had considered the great Michelson-Morley experiment."

Shankland emphasized that he was already a true believer in Einstein and relativity before he came to college, that he worked with Miller more and more without arguments or honest discussions, almost as if Miller were a grandfather, and that he learned to admire and respect Miller as a teacher, professor, experimenter, and interpreter of his own results. Yet he never could understand how this experiment could be in conflict with relativity. And so, as Miller passed away, having bequeathed his data sheets to Shankland's locked closet, and as the years passed by and the pressure from colleagues built up, Shankland decided to tackle the problem and try to solve the mystery. Is God subtle? Was Nature malicious to Miller?

Shankland admitted that, soon after his World War II work with sonar and hydrophones, he discovered that the task of making a new analysis of Miller's data was too big to handle alone. Sidney W. McCuskey, a fellow physicist at Case who had known Miller well, was first to join the job. The calculations of many sorts were formidable, but the advent of computers big and little at Case in the early 1950s gave hope and led Fred C. Leone, a programmer, to be enlisted. Shankland sought and gained a series of interviews with Einstein regarding tactics and strategy for the new analyses of Miller's data. These encouraging conversations were indispensible for continuation of the project, said Shankland, but even more expertise was needed. So finally Gustav Kuerti, another colleague who was expert in rational mechanics, vibration analysis, and magnetostriction, joined the team.

The four men worked several years clarifying the issues before research began to show, from meticulous thermometer readings that Miller himself had carefully recorded, that minute temperature gradients noticeable only through extensive data processing might be at fault. Shankland told Einstein, when asked why Miller had not noticed the malicious temperature gradients, that Miller of course had no electronic computer.

The grand significance of Michelson's ether-drift idea and of his instrumental design that began right-angled interferometry Shankland never doubted, but the very real importances of Morley's contributions to Michelson's developments he felt had always been undervalued:

> The great advance made by the Cleveland experiment in 1887 was that their observed fringe-shift was only 1/40th what the theory predicted, whereas in Potsdam it was 1/2 what the theory predicted. The difference between being 40 points off and 2 points off is a big difference. Nobody could deny the validity of the Michelson-Morley

experiment, although they could deny the validity of the Potsdam experiment.[20]

Throughout the interview Shankland emphasized his experimentalist view of the importance of the Michelson-Morley experiment in setting the "climate of opinion" for a half-century of radiation studies. Provoked in the first-place by the interests of Maxwell, Kelvin and Rayleigh, the performance of the interferometer in the summer of 1887 was assured attention from the highest level of physical thinkers such as Lorentz, Einstein and Eddington. Not even such theorists could afford to ignore the null results, partial though they were, of 1887:

> Now, if you forget for a moment that we are in Cleveland, and near the site of the Michelson-Morley experiment, and [if] you look back at this long progression of experiments, you will certainly have to say that the first experiment that measured to the quantity of V over c squared, V2 over c2 [Velocity of earth in orbit squared over velocity of light squared] had a new importance in the whole stream of things.[21]

In this penultimate analysis of Shankland's role in the whole stream of things, since he died in July 1982, perhaps he should be added to the list of characters who have evoked this symposium. Shankland's role toward Miller is analogous to Miller's role toward Michelson. Beyond them all hovered Einstein, and between them all was the linkage of interferometry.

Because Michelson became a Nobel laureate almost a decade and a half before Einstein and was 27 years older, Michelson, a paragon of experimental optics, probably evoked as much respect in the 1920s as did Einstein, a (and perhaps already the) paragon of theoretical physics. Michelson's life had been full and varied, dominated by a disciplined search to learn ever more about the nature of light by interference methods. His studies of the velocity of light, diffraction, refraction, interference, and effects of the motion of the medium on the velocity of light, which is what he finally called the ether-drag and ether-drift experiments, were admired as definitive exemplars of the highest sort of experimental physics. Yet toward the end of his life, Michelson had to admit the inadequacies of interferometry to deal with certain questions in the face of relativity theory.[22]

Since the 1930s, so much more has happened over two more generations at work in interferometry and relativity research that no adequate summary is possible here. Interferometry is ubiquitous in both ground-based and space research and in both micro-microphysics and macro-macrophysics. Suffice it to say that the instrumental ideas evoked in the quest for the relative motion of the earth and the ether have evolved into some of the most advanced tools and techniques for tomorrow's world observatories and laboratories.[23]

The Einstein centennial celebrations of 1979 stimulated much new literature on Einsteiniana, and there appeared many more references to the relationship between the Michelson-Morley experiment and restricted relativity theory. Such references were for the most part not much more enlightened historically than those of half a century ago. Symposia like this one on Michelson-Morley and the best of the Einstein celebrations, plus the editing of Einstein's *Collected Papers* in perhaps as many as 40 volumes, promise to change all that.[24]

Meanwhile, three quotes from leading astrophysical scientists of the 1980s may indicate the current mood toward Michelson-Morley and Einstein if not toward the quest for absolute motion. Martin Harwit in his 1981 book, *Cosmic Discovery*, quoted a sentence from Michelson's 1903 book, *Light Waves and Their Uses*, to illustrate late Victorian overconfidence in the virtual completion of the edifice of physical science, but later Harwit writes about logarithmic phase space as follows:

> Our universe appears to be isotropic, homogeneous, and slowly evolving. Statistically this means that we observe pretty much the same kind of behavior whether we look north or south in the sky, and we would observe phenomena rather similar to those seen locally if we went to a galaxy five hundred million light years away. We also see rather similar behavior this week as last. The absolute time, and the absolute direction in which our telescope is pointed, therefore, is not terribly important. We see the same phenomena, statistically speaking, wherever and whenever we look. In the wavelength domain, however, this is no longer true. We see quite different phenomena when we go from one wavelength to another that is several thousand times longer or shorter. Different physical processes are observed in these widely spaced wavelength ranges, and the wavelength therefore appears as a dimension in our phase space of observations, whereas absolute time does not.[25]

Stephen W. Hawking in his recent popularization of current cosmology cites the work of Michelson and Morley in 1887, saying

> They compared the speed of light in the direction of the earth's motion with that at right angles to the earth's motion. To their great surprise, they found they were exactly the same! Between 1887 and 1905 there were several attempts, most notably by the Dutch physicist Hendrick Lorentz, to explain the result of the Michelson-Morley experiment in terms of objects contracting and clocks slowing down when they moved through the ether. However, in a famous paper in 1905, a hitherto unknown clerk in the Swiss patent office, Albert Einstein, pointed out that the whole idea of an ether was unnecessary, providing one was willing to abandon the idea of absolute time.[26]

Finally, Edward Harrison in his beautiful book on the riddle of the dark night sky describes the transition in thinking between 19th century ether voids and 20th century curved space as follows:

> Albert Einstein...became the most outstanding scientist of our time and perhaps of all time. Early in this century he brought together the various lines of investigation by many physicists and formulated the theory of special relativity. According to this new theory, everything exists in a universal spacetime, and this common spacetime decomposes into the different spaces and times of things in relative motion. Furthermore, the structure of the four dimensional continuum of spacetime accounts for the constancy of the speed of light for all observers, demonstrated in the experiments by Albert Michelson and Edward Morley.[27]

I end this brief memoir by reminding the reader of Weisskopf's bold paragraph at the beginning. If indeed the "great dream" of Michelson and Miller (whether or not

Morley shared it seriously) has been realized by the discovery of 3°K background radiation, and if Nature *has* provided an absolute frame of reference, then it should be helpful to know more exactly what actually happened to change our minds for a time. Michelson-Morley, Einstein's relativity of simultaneity, and interferometry as an art that became a science were each a complicated part of an even more complex whole. Hindu cosmographers as well as a few Westerners, such as Johann Heinrich Lambert (1728-1777), have long envisioned the problems of universal motion, absolute and relative, as might be seen from the center of centers. But only in the past century was an instrument developed that promised to give a grip on the vacuum of space. The grips it gave were different from the one expected. The elusive ethereal ether faded away into superfluity as Einstein's insights into space-time, mass-energy, and gravity-inertia matured. If now absolute motion is being reincarnated, then surely the deeper significance of this piece of the history of science is not yet clear either.[28]

NOTES

1. Victor F. Weisskopf, "The Origin of the Universe," *American Scientist*, Sept.-Oct. 1983: 477 (emphasis added). Letter, Weisskopf to Swenson, 8 Oct. 1985. Of related interest, see the discussion of Michelson-Morley by Larry Abbott, "The Mystery of the Cosmological Constant," *Scientific American*, 1988, *258*: 106-113, on p. 113. See also Martin Harwit, *Cosmic Discovery: The Search, Scope, and Heritage of Astronomy* (New York: Basic Books, Inc., 1981), pp. 147-152.

2. See my *The Ethereal Aether: A History of the Michelson-Morley-Miller Aether-Drift Experiments, 1880-1930* (Austin: Univ. Texas Press, 1972). Cf. Rom Harré, "The Impossibility of Detecting the Motion of the Earth," in his *Great Scientific Experiments: 20 Experiments that Changed Our View of the World* (Oxford: Phaidon, 1981), pp. 124-134.

3. See my *Genesis of Relativity: Einstein in Context* (New York: Burt Franklin & Co., Inc., 1979), esp. pp. 59-69, 75-92.

4. Albert Einstein, "How I Created the Theory of Relativity," a translation of a lecture given in Kyoto on 14 December 1922 by Yoshimasa A. Ono, in *Physics Today*, Aug. 1982. Cf. John Stachel, "Einstein and Ether Drift Experiments," *Phys. Today*, May 1987: 45-47.

5. For a better translation of Einstein's Kyoto address, see Hans Joachim Haubold and Eiichi Yasui, "Jun Ishiwaras Text über Albert Einsteins Gastvortrag an der Universität zu Kyoto am 14 Dezember 1922," *Archive for History of Exact Sciences*, 1986, *36*: 271-279. See also R. S. Shankland, "Michelson's Role in the Development of Relativity," *Applied Optics*, 1973, *12*: 2280-2287, as well as John Earmon, Clark Glymour, Robert Rynasiewicz, "On Writing the History of Special Relativity," *PSA* 1982, *2*: 403-416. Cf. my *Ethereal Aether*, *passim*, and *Genesis of Relativity*, chaps. 2 and 4.

6. See John Stachel, ed., *The Collected Papers of Albert Einstein, (Vol. 1) The Early Years, 1879-1902* (Princeton Univ. Press, 1987), for one example, and for another, Stephen Kern, *The Culture of Time and Space, 1880-1918* (Harvard Univ. Press, 1986). See my *Genesis of Relativity*, pp. 99 ff., 145 ff.

7. See, e.g., Gerald Holton, *Thematic Origins of Scientific Thought--Kepler to Einstein* (Harvard Univ. Press, 1973); and my article "The Michelson-Morley-Miller Experiments Before and After 1905," *Journal for the History of Astronomy*, 1970, *1*. See also Lewis S. Feuer, *Einstein and the Generations of Science* (New York: Basic Books, 1974) for a study in the sociology of physics; or M. A. Handschy, "Re-examination of the 1887 Michelson-Morley experiment," *American Journal of Physics* 1982, *50*: 987-990.

8. See my article for the first Michelson Centennial held in East Berlin and Potsdam in April 1981: "The Michelson-Morley-Miller Experiments and the Einsteinian Synthesis" in *Astronomische Nachrichten*, 1982, *303*: 39-45. See also Gerald Holton's thema in praise of the interferometer in his "Foreword" to my *Ethereal Aether*.

9. Not merely wave theory for acoustics and optics but for the enlarging electromagnetic spectrum seems to have been Michelson's intuition. See Michelson's Lowell Lectures at Harvard in 1899, later published as his first book, *Light Waves and Their Uses* (Chicago: Univ. Chicago Press, 1902); cf., his second and last book, *Studies in Optics* (Chicago: Univ. Chicago Press, 1927), and note S. Chandrasekhar's cautious preface, p. v.

10. See my article "Michelson and Measurement" for the special issue "Michelson-Morley Centennial" of *Physics Today*, 1987, *40*: 23-30. See also my article on Michelson's interferometer entitled "Measuring the Immeasurable" in *American Heritage of Invention and Technology*, 1987, *3*: 43-49. Cf. Andrew D. Jeffries, Peter R. Saulson, Robert E. Spero and Michael E. Zucker, "Gravitational Wave Observatories," *Scientific American*, 1987, *256*: 50-58 B.

11. See *Ethereal Aether*, pp. 228-245, and Harvey B. Lemon's "Bibliography of Publications of Albert Abraham Michelson," *A.J.P.*, 1936, *4*: 1-11.

12. For facsimile reprints of the classic Michelson and Michelson-Morley papers of 1881, 1886 and 1887, see *Ethereal Aether*, Appendices A, B, and C. That some sort of "ether" was necessary to make possible the production of interference phenomena Michelson never doubted. Re some of this confusion over the kinematics of the experiment, see Stanley Goldberg's review of *Ethereal Aether* in *J. Hist. Astron.*, 1973, *4*: 196-199, on p. 199.

13. For the cosmological changes, see Richard Berendzen, Richard Hart, Daniel Seeley, *Man Discovers the Galaxies* (New York: Science History Publications, 1976). For the personal changes, see my *Ethereal Aether*, esp. pp. 32 ff., and chaps. 10 and 11. For Miller's summation of this controversy, see his "The Ether-Drift Experiment and the Determination of the Absolute Motion of the Earth," *Reviews of Modern Physics*, 1933, *5*: 203-234.

14. See my summation in *Ethereal Aether*, chaps. 10 and 12.

15. The controversy over the rise of relativity theory before the fall of ether theories explains much of this; see *Ethereal Aether*, chaps. 8 and 9.

16. See my chronological list of primary reports of ether-drift and related experiments at the beginning of my bibliography for *Ethereal Aether*, pp. 295-298. Abraham Pais chose the first half of this epigram by Einstein in response to Miller for the title of his biography *'Subtle is the Lord...': The Science and the Life of Albert Einstein* (Oxford: Oxford Univ. Press, 1982), p. 113.

17. See R. S. Shankland, S .W. McCuskey, F. C. Leone, and G. Kuerti, "A New Analysis of the Interferometer Observations of Dayton C. Miller," *Revs. Mod. Phys.*, 1955, *27*: 167-178.

18. Transcript, Interview with Robert S. Shankland by Loyd S. Swenson, Jr., for AIP Bohr Library Archives, taken at Case Western Reserve University, Cleveland, Ohio, on 20-21 August 1974, pp. 1-201.

19. Transcript, Swenson interview of Shankland, p. 147.

20. Transcript, Swenson interview of Shankland, p. 145.

21. Transcript, Swenson interview of Shankland, pp. 195-196.

22. See his *Studies in Optics*, pp. 154, 161. See also Jean M. Bennett, D. Theodore McAllister, and Georgia M. Cabe, "Albert A. Michelson, Dean of American Optics--Life, Contributions to Science, and Influence on Modern-Day Physics," *Applied Optics*, 1973, *12*: 2253-2279. Cf. Dorothy Michelson Livingston's biography of her father, *The Master of Light: A Biography of Albert A. Michelson* (New York: Charles Scribner's Sons, 1973).

23. See e.g., Bruce Schechter, "Searching for Gravity Waves with Interferometers," *Phys. Today*, Feb., 1986: 17-18. See also Wallace and Karen Tucker, *The Cosmic Inquirers: Modern Telescopes and Their Makers* (Harvard Univ. Press, 1986) or the anonymous article "Interferometry in Space," *Sky and Telescope*, 1983, *66*: 500.

24. Cf. e.g., the four essays by myself, C. P. Snow, Howard Stein, and Ilya Prigogene, respectively, in *Albert Einstein: Four Commemorative Lectures* (Austin: Univ. Texas Humanities Research Center, 1979). See also Harry Woolf, ed., *Some Strangeness in the Proportion* (Princeton's Institute for Advanced Study Symposium) (Reading, Mass: Addison-Wesley, 1980).

25. Harwit, *Cosmic Discovery*, pp. 7-8, 181.

26. Stephen W. Hawking, *A Brief History of Time: From the Big Bang to Black Holes* (Toronto: Bantam Books, 1988), p. 20.

27. Edward Harrison, *Darkness at Night: A Riddle of the Universe* (Cambridge: Harvard Univ. Press, 1987), pp. 170-171.

28. For Lambert and Einstein, see my *Genesis of Relativity*, pp. 16-17, 69, and 188-222. For some details on ether experiments and physical theory before 1970, see *Ethereal Aether*, pp. 228-245. For more historical perspective, see Edward Grant, *Much Ado About Nothing: Theories of Space and Vacuum from the Middle Ages to the Scientific Revolution* (Cambridge Univ. Press, 1981), and G. N. Cantor and M. J. S. Lodge, eds., *Conceptions of Ether: Studies in the History of Ether Theories, 1740-1900* (Cambridge Univ. Press, 1981).

IV

American Science in the Age of Michelson and Beyond

PHYSICS AND NATIONAL POWER, 1870-1930

Daniel J. Kevles

When Albert Michelson began his career in research, in the last quarter of the nineteenth century, American colleges and universities were altogether capable of producing physicists. The elective system had been introduced into the undergraduate curriculum, which permitted students to major in technical subjects, and so had laboratory instruction. Graduate programs had been established in both scientific and nonscientific subjects; at least twenty-five institutions awarded the Ph.D., notably the new Johns Hopkins University, which graduated a sizable fraction of the thirty-three physics Ph.D.'s earned at American institutions by 1890. Still, in 1883, in an address to the American Association for the Advancement of Science, Henry A. Rowland, who was professor of physics at Hopkins, felt compelled to ask "what must be done to create a science of physics in this country, rather than to call telegraphs, electric lights, and such conveniences by the name of science. . . ?" He added, "The cook who invents a new and palatable dish for the table benefits the world to a certain degree; yet we do not dignify him by the name of chemist."[1]

The America of Rowland's day -- an America absorbed by agriculture and industrialization -- was a nation that made mythic heroes of the farmer on the land and the inventor in the factory. Physicists advanced knowledge and remained obscure; inventors appropriated their findings and became rich and famous. In the burgeoning electrical industry, the type-case technological innovator was, of course, Thomas Edison, the self-taught genius whose spectacular success was generally taken as proof that in business, college training was not only unnecessary but a liability. In 1884, the trade journal *Electrical World* said, "Edison's mathematics would hardly qualify him for admission to a single college or university . . ., but we would rather have his opinion on electrical questions than [that] of most physicists." [2]

Science in the federal government had grown considerably since the Civil War, but emphasis went to disciplines such as geology that were pertinent to the settlement and development of the land. In the mid-1880s, physicists and astrophysicists urged that the government should create a bureau of electrical standards and a "physical laboratory" for spectral research. The recommendation came to nothing, not least because most politicos agreed with the opinion of *The Nation*: "The Government should not keep a school for original research, and it is not advisable to see established at Washington laboratories of chemistry, physics, biology, etc., etc., intended only for such work as can be done elsewhere." Moreover, the watchword of federal science was practicality. In the 1880s, one government scientist typically insisted to a Congressional investigating committee that he was not engaged in "vain theorizing," and another -- the director of the Coast and Geodetic Survey -- declared: "We are not fomenting science. We are doing practical work for practical purposes." [3]

However, despite Rowland's protestations, there was a science of physics in America. It was just not one that, in its several branches, Rowland altogether admired. Rowland entitled his address "A Plea for Pure Science," by which he meant both scientific subjects without immediate practical utility -- for example, the branches of heat, light, electricity, and magnetism that were then the centerpieces of high physics -- and scientific purpose uninfected by the profit motive. Pure research was done -- or so its advocates said -- solely for the love of truth. (About a quarter of a century later, Albert Michelson told one of his students at the University of Chicago, at least as the student recollected, that he would be "prostituting [his] training and ideals" by going off to industry.)[4] Yet a good deal of research went on in late-nineteenth century American (as well as European) physics that, though it may not have met Rowland's criteria, was something more than mere invention. It may have taken its cues from practical challenges, but it was physics nonetheless.

Such physics was pursued in the bureaus of federal science, which were home to one out of six of the productive American physicists of the period. These physicists worked in branches of their discipline related to the earth sciences, notably the emerging fields of geophysics and meteorology. The small Geophysical Laboratory in the U.S. Geological Survey was headed by Carl Barus, who sought to determine how the melting point of rocks varied with pressure and became an international authority on pyrometry. It also included Robert S. Woodward, whose work bore on the debate that Lord Kelvin had stimulated on the age of the earth.[5]

Several meteorological physicists were employed by the Weather Service in the Army Signal Corps, housed on G Street in the capital. This small civilian staff was headed by Cleveland Abbe, who, in 1880, had persuaded the Chief Signal Officer to spend money for the study of meteorology. The enlistment of college graduates in the Weather Service was encouraged by exempting them from regular military duties; civilian scientists were added to the staff and given the titles of professor and assistant professor. In the building on G Street, Abbe created a Study Room for his professors -- there were seven or eight of them, plus ten assistants -- and opened a physical laboratory in the basement. Weather Service scientists explored the nature of storms and tornadoes, investigated the mysteries of the atmosphere, and, in all, nudged ahead the science of meteorology.[6]

All the while, highly trained representatives of some of the most central fields of physics were finding a place in the electrical industry. The telegraph operators and inventor-entrepreneurs who had pioneered the electric light, power, and communications industries were being joined by men with advanced training in pertinent branches of physics, notably electricity and magnetism. As early as the mid-1870s, Edison's laboratory had included a Ph.D. in mathematical physics. In 1880, one of Rowland's own students, William Jacques, went to work for the Bell Telephone Company, becoming head the next year of the firm's first formal laboratory, the Electrical and Patent Department. Other physics Ph.D.'s followed Jacques into the electrical business, whose managers felt an increasing need for them.[7] They brought their expertise to bear on the region between the pure physics of electromagnetism and the practical demands of developing power generators, lighting grids, and electrical instruments, and joined in helping to form a new profession -- electrical engineering.

At a number of colleges and universities, physics departments trained students for electrical enterprise, until such efforts were spun off, later, into independent electrical engineering departments.[8] Typical of these activities were those of the young physics instructor at the Case School of Applied Science, in Cleveland, Albert A. Michelson. The details are provided by A. Michal McMahon, in his informative study of the electrical engineering profession:

> Michelson developed a program of studies including, besides basic courses in classical physics, a number of classes in the engineering fields. Drawing came during the first two years, with mechanical and civil engineeering taught during the third and fourth years. These were replaced in the physics curriculum, in 1885, with specialized electrical topics. A course entitled 'Electricity and Magnetism' was offered in the second year, along with related laboratory work. More electrical courses came the following year and attention was given to 'theory and practice' and such 'practical problems' as could be illustrated through the study of batteries; the measurement of currents, resistances, and electromotive forces; the

'location of faults in telegraph circuits; laws of electromagnetics; intensity of magnetic fields; efficiency of electric lamps and dynamo-machines.' . . . Finally, in 1887, the year before he departed for Clark University, Michelson offered a full 'Course in Electrical Engineering.' Electricity and Magnetism was introduced in the second and third years, and in the fourth: thermodynamics, engineering construction, details of practice and design, electrotechnics, and laboratory work in electrical testing.

According to the school catalogue, the electrical engineering course was offered because of the "important advances in the application of Electricity and Magnetism to electric lighting, electrometallurgy, and electric transmission of power."[9]

In most colleges and universities of the day, emphasis went to teaching rather than to research. Still, during this period, in the United States, more than 200 people did physics research, publishing some 800 articles between 1870 and 1890, in the *American Journal of Science* as well as in European journals and the proceedings of local learned societies. The output was sufficiently large to warrant the establishment, in 1893, of *The Physical Review*, the nation's first journal in the discipline. Five out of six of these articles were produced by some 40 American physicists -- the leaders in the profession -- and their work covered the principal fields of interest on both sides of the Atlantic, including heat, light, electricity, and magnetism.[10] Several -- among them Rowland, Michelson, and Josiah Willard Gibbs -- were outstanding, gaining attention and well-deserved respect in the European physics community.

In 1899, almost forty American physicists assembled at Columbia University to form the American Physical Society and elected Henry Rowland as their first president. Several months later, in his presidential address, Rowland, who was mortally ill with diabetes, blessed the new society with a final testament, declaring, "Much of the intellect of our country is still wasted in the pursuit of so-called practical science. . . . But your presence here gives evidence that such a condition is not to last forever."[11]

Rowland was, of course, right that pure physics was turning a corner in the United States, but he was mistaken in suggesting that the change was coming at the cost of practical science. At the turn of the century, manufacturing was yearly supplying some 30 percent more of the national income than mining and agriculture combined. Exports passed the billion-dollar mark and exceeded imports for the first time in American history.[12] A good many of the reasons for the transformation in American physics were to be found in such mundane matters -- in the statistics of trade and commerce, concerns about the competitive position of American manufactures, and ruminations upon the importance of physics to national economic power.

The force of such considerations was manifest in the turn-of-the-century agitation to establish a federal agency for physical and chemical standards. The governments of Germany and England had recently established handsomely equipped laboratories for the determination of standard physical and chemical units. The lack of such a laboratory in the United States was said to be humiliating -- and commercially costly. The increasing technical complexity of industry made more pressing the need for uniform standards and instruments for measuring chemical and electrical quantities. At Congressional hearings on the subject in 1900, Secretary of the Treasury Lyman G. Gage averred that in applied science and trade, "in all the great things of life," the country was in competition with the older and more thoroughly established nations of the world. On 3 March 1901, legislation was passed that created a National Bureau of Standards, with a $250,000 laboratory and authority to engage in whatever research might be necessary to establish standards for the entire realm of physics and chemistry. Soon, some fifty physicists were at work in the Bureau's well-equipped facilities.[13]

Awareness of mounting competition was clear-cut in the burgeoning electrical and communications industries, where the value of manufactured goods rose from $19 million in 1889 to $335 million in 1914. As a result of consolidations and mergers, the industry now included several large firms -- among them, General Electric and AT&T -- ambitious to dominate their national markets and made aware by their Ph.D. physicists and physical chemists that scientific research would help equip them to achieve that goal. Research promised the improvement of existing products, the creation of new ones, and the establishment of patent rights to fend off competitors -- both foreign and domestic -- in primary markets and confer monopoly rights in new outlets.[14]

During the first decade of the twentieth century, the laboratories of General Electric and AT&T, which had been given over for the most part to routine testing, were transformed into research laboratories. Physicists were hired and set to work -- at General Electric, on improving features of the tungsten light bulb; at AT&T, on developing the new vacuum tube. Their efforts were hugely successful. For General Electric, they yielded the ductile-tungsten lamp, which was more rugged, more efficient, and less costly than any other incandescent lamp and which turned the company's share of the American lighting market from 25 percent in 1911 to 71 percent in 1914. For AT&T, they accomplished the development of highly efficient vacuum-tube amplifiers, which were essential to the extension of long-distance telephone service and which gave AT&T powerful entry to the emerging radio industry.[15]

The research laboratory at AT&T grew from a staff of 23 in 1913 to 106 in 1916; its budget, from $71,000 to $249,000. By 1916, when the General Electric laboratory moved to new quarters, it had one of the best physics research facilities in the country, including 60,000 square feet of space, generators and transformers that could deliver electric power at currents as high as 12,000 amperes and potentials up to 200,000 volts. In 1915, industrial research laboratories were the source of one in seven papers published in *The Physical Review*; in 1920, of about one in five.[16] Between 1913 and 1920, when the number of people in the American Physical Society doubled, representation of industrial physicists went from one-tenth to one-quarter of the membership.[17]

Still, industrial physics had its drawbacks to physicists who considered themselves heirs to Henry Rowland's pure-science outlook. It may have been more lucrative than an academic career, but it offered less professional freedom. At the General Electric and AT&T laboratories, the work norm tended to be group rather than individual research. Research problems had of necessity to be chosen from among those of interest to the company, which were not necessarily those of central importance to the world physics community. As Frank Jewett, the head of the AT&T laboratory, once said, "Our research and development organization is a truly scientific body, but one in which the results of science are designed to be of utility to fit into the orderly progress of electrical communications."[18]

The operations of industrial laboratories were also marked by a good deal of proprietary secrecy towards the rest of the world and a degree of compartmentalized confidentiality even within the facilities themselves. To be sure, Willis R. Whitney, the director of the General Electric laboratory, encouraged open communication among his staff, and the scientists and engineers at AT&T as well as General Electric were permitted to attend scientific meetings. However, they went to the meetings more to listen than to contribute. The GE work on ductile tungsten was kept under close wraps for several years and tight secrecy was imposed at AT&T on wireless work there. An AT&T staff member recalled:

> When the radio tests began, we were ordered to complete secrecy,
> not even our wives were to be told of our jobs. The penthouse on

> the Hotel Dupont [the receiving end of tests conducted between
> Montauk and Wilmington] was not occupied until heavy window
> shutters had been installed and woe to us if we didn't keep them
> closed. Even at [the laboratory] things were under lock and key.

The papers published from industrial laboratories inadequately revealed the technical progress made in them. AT&T scientists tended to keep to themselves their considerable knowledge of vacuum-tube physics and for several years their theory of radio sidebands was confined entirely within the company. Industrial physics thus disadvantaged physicists eager for status and reputation in the larger profession.[19]

Only a small fraction of new Ph.D.'s in physics went into industry, and the proportion reached no more than about a fifth of the group by 1920. The large majority entered the academic world, where the research environment was open and latitudinous and where employment opportunities started mushrooming at the turn of the century. With the accelerating electrification of homes and factories, and then the creation of the radio industry, electrical engineering enrollments multiplied yearly, spilling industry-bound students into physics courses and confronting department chairman with over-crowded laboratories and overtaxed staffs.[20]

The chairmen, a breed of builders eager to promote research, realized that the need for more staff and more facilities for undergraduates promised dividends for investigations in physics. At Wisconsin, California, Michigan, and Illinois, at Pennsylvania, Harvard, Columbia, and Cornell, they won appropriate dispensations for new laboratories as well as for more professors. The quality of high-school physics teaching had improved considerably. And with the increasing engineering enrollments, a growing number of students not only took physics courses but became aware of the widening career possibilities in the discipline. Through the 1890s, American universities had graduated a total of fifty-four physics Ph.D.'s; they turned out twenty-five in 1909 alone. That year membership in the American Physical Society reached 500, a fivefold increase since 1901, and in 1914 it climbed to some 700.[21]

At an increasing number of academic institutions, students could take courses in, as an MIT catalogue read, the "Constitution of Matter in the Light of Recent Discovery." At the centers of physics graduate training, which now comprised Hopkins, Cornell, Yale, Harvard, and Chicago, doctoral candidates pursued work in key fields of this new physics -- the physics of atoms and electrons, radioactivity and radiation. At Chicago, students learned about electrons and quanta from Robert A. Millikan, who would, of course, become, after his senior colleague Albert Michelson, the second American to win the Nobel Prize in physics.[22]

In the university of the early twentieth century, research was beginning to receive as much emphasis as teaching. Academic administrators, with an eye to practical-minded industrial corporations and state legislatures, celebrated scientific investigation for the economic benefits it could bring. American physicists took advantage of the new dispensation. In the quarter century after 1890, some 650 physicists published research, almost a hundred of them -- about twice as many as in the late nineteenth century -- doing so at a reasonably productive rate. Many of their papers appeared in *The Physical Review*.[23] Unlike Henry Rowland, physicists of the day recognized that their arrangements with practical culture could pay handsome dividends indeed.

In 1914, a Congressman expostulated in a meeting of an appropriations subcommittee: "What is a physicist? I was asked on the floor of the House what in the name of common sense a physicist is, and I could not answer." The outbreak of World War I may not have taught him what a physicist was, but it certainly drove home the significance of science to national power. The British blockade cut the United States off from its accustomed imports of German dyes, scientific instruments, or optical glass, and American corporations possessed little of the know-how required to produce or synthesize these materials. *Scientific American* issued timely advice: For a "comparatively small outlay," research scientists could develop new processes and make many American businesses "absolutely independent of Europe." Elaborating on that theme, trade journal editors devoted numerous pages to the advantages of industrial research, including the profitable opportunities in the present and the challenge of the peacetime future. After the conflict of arms, the predictions went, the United States would face a trade war with a Europe whose industry was honed to superb efficiency. Research, proclaimed Willis R. Whitney, the director of the research laboratory at General Electric, was now a "necessity to any people who are ever to become a leading nation or world power."[24]

After the United States entered the war, science was mobilized on an unprecedented scale. The conflict was rightly known as a chemist's war, but physicists performed key services in industrial and governmental laboratories, developing means for the detection of submarines, for example -- this was only a "problem of physics pure and simple," one of them said -- contributing to the solution of aeronautical problems, and managing technical troops in the field. Army and navy projects even found their way into the laboratories of some forty of the nation's colleges, and the campuses operated under the constraint of tight security regulations for the first time in American history. For thoughtful military observers the technical meaning of the wartime experience was clear: The advance of defense technology required the organized efforts of scientists and engineers whose first steps often had to be, as a naval official said of submarine detection, "in a sense, backwards into the unexplored regions where fundamental physical truths and engineering data were concealed."[25]

Despite the dramatic identification of science with military power, the postwar retreat into isolationism and economizing prevented a significant peacetime effort in defense research and development. During the 1920s, the military's technical bureaus were cut back from their wartime levels, and the National Bureau of Standards was compelled to pursue a prudently practical research program. William F. Meggers, one of the leading atomic physicists at the Bureau, complained that his colleagues could get plenty of money for research on "economies in automobile manufacture," but little if any for spectroscopy. Still, the Bureau remained home to a sizable number of physicists. Moreover, officers in the technical bureaus of the army and navy, which had hitherto been concerned primarily with testing, had come to appreciate the advantages of research -- and of research scientists. During the 1920s, a small but growing cadre of physicists was to be found in agencies such as the Army Signal Corps and the new Naval Research Laboratory.[26]

In the industrial sector, where postwar competition from abroad was expected to be ferocious, the pressures were against economizing in research. An industrial physicist remarked, "Bankers have become science-minded," recognizing that "in commercial warfare . . . research supplies the ammunition." [27] During the 1920s, an increasing number of industrial corporations opened research laboratories; by 1931, there were some 1,600 of them, employing nearly 33,000 people, in the United States. To be sure, the large majority of these installations were devoted to routine testing or development rather than research. The principal industrial installations for physics remained General Electric and AT&T, whose research effort became the Bell Telephone Laboratory, in 1924. Leonard Reich, a leading historian of industrial research, has described the size and scope of the new Bell Labs:

> In its first year, Bell Laboratories employed approximately 3,600 people, with 2,000 on the technical staff, and had a budget of over $12 million. Researchers worked in the areas of radio, electronics, chemistry, magnetics, optics, applied mathematics, speech and

hearing, conversion of energy between electrical and acoustical systems, generation and modification of electric currents, instruments of all kinds, paints and varnishes, and problems relating to the aging and preservation of wood (for telephone poles, of course.)[28]

The award of two Nobel Prizes -- to Irving Langmuir and Clinton J. Davisson, for work done, respectively, at General Electric and Bell Labs -- would soon symbolize the degree to which industrial research laboratories might become institutions of high and serious science. Then, too, in the 1920s, at Bell Labs, management eased restrictions on publication, holding, in the words of a 1922 report from the Engineeering Department, that scientific publication was "of very great importance in creating in the minds of a very influential group of the public a favorable attitude toward the Bell System." Nevertheless, Bell scientists were discouraged from publishing results that might jeopardize the company's competitive position. In the 1920s, as in the prewar era, industrial research did not attract the large majority of new physics Ph.D.'s. They continued to opt for the academic world, where there was not only comparatively more freedom in research but where industrial capitalism was enriching the research environment.

The industrial bounty did not come to academia directly from industrial corporations; in the 1920s, both law and preference demanded that they make their investments in pure research in their own laboratories, where technologically promising discoveries could be patented before they were published.[29] The source of funds from the accumulation of industrial wealth made its way into academic science through philanthropic organizations, especially the Rockefeller philanthropies, whose programs were shaped to a significant extent by the postwar identification of science with national economic -- and cultural -- power.

Beginning in 1919, the Rockfeller Foundation provided funds to the National Research Council for postdoctoral fellowships in several sciences, including physics. Under the guidance of Wickliffe Rose, a thoroughgoing enthusiast of science, the Rockefeller General Education Board shifted to a policy of raising the quality of advanced training and research, especially in the physical and biological sciences. By 1932 the Board had single-handedly enlarged the funding of academic science by some nineteen million dollars -- about three times what it had awarded the humanities, and some six times what the total endowment for science in the United States had been at the turn of the century.[30]

In the dozen years after 1920 twice as many Americans, some thousand in all, took Ph.D.'s in physics as in the half century between Appomattox and Sarajevo. About 130 of them won National Research Council fellowships and the majority of that group spent their postdoctoral year or two at a university in the United States. Between 1919 and 1932 at least fifty, a significant fraction of the most promising young American physicists, made their way to Europe for postdoctoral study on fellowships from the National Research Council, the International Education Board, or the new Guggenheim Foundation. Still, American physics was becoming decidedly attractive to Europeans. By 1931 American universities had hired at least fifteen Continental physicists, which led a British scientist to remark that the United States was "constantly" attracting "some of the ablest men in Europe." By 1932, most American physicists had returned home to become active members of a profession which -- almost three times as large as in 1919 -- included some twenty-five hundred members.[31]

By that time, physicists in America were publishing prolifically in the major areas of their discipline, especially the new quantum mechanics. And the world physics community was paying more attention to the work appearing in *The Physical Review* than it was to that in the *Annalen der Physik*, the leading German journal.[32] The lamentations of Rowland's day were gone from American physics, and so was the sense of provincial inferiority. When, in 1932, it was announced that Albert Einstein would join the new Institute of Advanced Study in Princeton, New Jersey, the French physicist Paul Langevin was stimulated to note that the Pope of physics had moved and to predict that the United States would now become "the center of the natural sciences." [33]

In all, the ascendance of American physics depended less upon the nation's embrace of Henry Rowland's values than upon the development of a mutually advantageous accommodation between the desires of the pure physicist and the requirements of practical culture. From the side of academic physicists, the accommodation involved the pioneering enrichment of new fields -- meteorology, geophysics, electrical engineering, aeronautics -- with, so to speak, the human capital of trained physicists. [34] It also hinged on the supply to industry of engineering students equipped with physics instruction. From the side of the practical culture, it involved the emergence of an appreciation of the economic value of pure science. It also depended upon an attendant willingness to foster the advancement of knowledge, not only within the walls of industrial corporations but outside them, through gifts and endowments to academic science. Still, Rowland's plea for pure science sounded a theme that continues to echo in the discourse of American physicists -- a discourse marked by a search for identity and support in a culture that for the most part values physicists less for their understanding of nature than for what they can contribute to American national power.

NOTES

1. Daniel J. Kevles and Carolyn Harding, "The Physics, Mathematics, and Chemistry Communities in America, 1870-1915: A Statistical Survey," *Social Science Working Paper No. 136, California Institute of Technology, October, 1976*, Table 7; Henry Rowland, "A Plea for Pure Science," *The Physical Papers of Henry Augustus Rowland* (Baltimore: Johns Hopkins Univ. Press, 1901[?]), p. 594.

2. *Electrical World*, IV (20 Sept. 1884), 96.

3. Daniel J. Kevles, *The Physicists: The History of a Scientific Community in Modern America* (Cambridge: Harvard Univ. Press, 1987), pp. 47,55,59.

4. Spencer Weart, "The Physics Business in America, 1919-1940: A Statistical Reconnaissance," in Nathan Reingold, ed., *The Sciences in the American Context: New Perspectives* (Washington, D.C.: Smithsonian Institution Press, 1979), p. 303.

5. John W. Servos, "To Explore the Borderland: The Foundation of the Geophysical Laboratory of the Carnegie Institution of Washington," *Historical Studies in the Physical Sciences*, 1983, *14*, Part I:152-154, 166, 179.

6. Kevles, *The Physicists*, pp. 9, 48-49.

7. Leonard S. Reich, *The Making of American Industrial Research: Science and Business at GE and Bell, 1876-1926* (Cambridge, England: Cambridge Univ. Press, 1985), pp. 143-4, 147; A. Michal McMahon, *The Making of a Profession: A Century of Electrical Engineering in America* (New York: The Institute of Electrical and Electronics Engineers, 1984), pp. 3, 17.

8. McMahon, *The Making of a Profession*, pp. 43-45, 51, 67.

9. Ibid. pp. 44-45.

10. Kevles, *The Physicists*, p. 26

11. Ibid., p. 77

12. Ibid., p. 66.

13. Ibid., pp. 66-67

14. Reich, *The Making of American Industrial Research*, pp. 35, 37, 39-40.

15. Ibid., pp. 80-81, 161-62, 201-202.

16. Ibid., pp. 176, 92, 251.

17. Weart, "The Physics Business in America," p. 302.

18. Reich, *The Making of American Industrial Research*, pp. 7-8, 102, 217.

19. Ibid., pp. 110, 186-191, 195.

20. Kevles, *The Physicists*, p. 77; Weart, "The Physics Business in America," pp. 303-304.

21. Kevles, *The Physicists*, pp. 77-78.

22. Ibid., pp. 79-81, 90.

23. Kevles and Harding, "The Physics, Mathematics, and Chemistry Communities," Tables 2 and 4.

24. Kevles, *The Physicists*, pp. 96, 102-103.

25. Ibid., pp. 121, 138.

26. Ibid., pp. 145-48, 190; David Kite Allison, *New Eye for the Navy: The Origins of Radar at the Naval Research Laboratory* (NRL Report 8466; Washington, D.C.: Naval Research Laboratory, 1981), pp. 14, 37, 42, 85, 92, 138.

27. Weart, "The Physics Business in America," p. 302.

28. Reich, *The Making of American Industrial Research*, pp. 2-3, 184.

29. Kevles, *The Physicists*, pp. 186-88.

30. Ibid., pp. 149-50, 191-93.

31. Ibid., pp. 200-202, 220.

32. Weart, "The Physics Business in America," p. 298.

33. Quoted in Robert Jungk, *Brighter than a Thousand Suns*, trans. James Cleugh (London: Penguin, 1960), p. 51.

34. See, on geophysics, Judith R. Goodstein, "Waves in the Earth: Seismology Comes to Southern California," *HSPS*, 1984, *14*, Part 2: 221; John W. Servos, "To Explore the Borderland: The Foundation of the Geophysical Laboratory of the Carnegie Institution of Washington," p. 151.

SCIENTIFIC ELITE REVISITED:
AMERICAN CANDIDATES FOR THE NOBEL PRIZES IN PHYSICS AND CHEMISTRY 1901-1938

Elisabeth Crawford

The creation of a stratified scientific community in which standards of excellence are set by a small elite holding positions of power and influence has been seen by many as important for the development of American science in the 20th century. In analyzing how the physical sciences began their ascendancy to first rank internationally by the turn of the century, Kevles stresses the need to accommodate the elitism inherent in high-quality science to the democratic assumptions and geographic pluralism that heretofore had dominated the American scene.[1] Here he reflects the viewpoints of contemporaries such as Henry Rowland and Simon Newcomb, who advocated a best-science elitism in which training would be concentrated to a handful of first-class institutions, and standards of excellence would be set by those whose ability was nationally recognized, for instance, through election to a reinvigorated National Academy of Sciences.[2]

No doubt reflecting the extent to which these ideals had been realized, in the 1960s the twin notions of social stratification and elitism became the cornerstones of the Columbia Program in the Sociology of Science directed by Robert K. Merton and funded by the National Science Foundation. In the works produced by Merton, J. and S. Cole, and Harriet Zuckerman, among others, the social determinants of the positions of individuals in the stratification system, and the consequences for the quantity and quality of their work as well as for its communication and reception within the scientific community were charted using a panoply of measurements: prestige rankings of institutions and awards, citation counts, and productivity data. The most obvious consequence of stratification, elitism, was celebrated in Zuckerman's *Scientific Elite: Nobel Laureates in the United States.*[3]

In her book, Zuckerman takes as the crowning-point of elitism, the "ultra-elite" of the ninety-two Nobel laureates in physics, chemistry and physiology or medicine selected for work done in the United States between 1907, when Michelson was awarded the first prize, and 1972, the cutoff of her study. The laureates are the top of the pyramid that she sees as characteristic of American science, the base being made up of the half-million individuals who described themselves as "scientists" in the 1974 national census. How the laureates arrived at their elevated positions and what the consequences were for their scientific and professional lives are the main subjects of her inquiry.

The data she presents on laureates' training, careers, scientific work, and recognition through honors other than the Nobel Prizes bear out her presupposition of an ultra-elite whose members distinguish themselves early on and who continue, throughout their careers, to accumulate advantages that place them above their peers. To take some examples: laureates are more likely than other productive scientists, who have not entered the ultra-elite, to have received their doctorates and their first appointments at one of thirteen elite universities[4]; they earn their doctorates young and therefore carry out their most significant scientific research at an earlier age than their peers; they are rewarded at an accelerated rate through appointments to chairs, positions of responsibility in scientific societies, and honorific awards, of which the Nobel Prize is the crowning glory. Since more than half of the Nobel laureates studied under other laureates, for Zuckerman the ultra-elite is self-perpetuating, with new recruits resembling their elders in terms of fields of research, scientific styles and career patterns. Despite frequent references to what she calls the *noblesse oblige* that reigns in the ultra-elite, its members are not simply coopted into the aristocracy of science but have earned this status through their contributions to the advancement of knowledge.[5]

Although the existence of such contributions and the fact that they were significant for science are repeatedly stressed in the book, hardly anything is said about what they actually entailed for the cognitive, social and institutional maps of American physics, chemistry, and medicine in the sixty-some years covered by Zuckerman. Instead, she goes to considerable length to prove her point about the accumulated advantages of the scientific ultra-elite of Nobel Prize winners. As it emerges in her study, this elite is university-based, engages almost exclusively in basic research -- primarily in theoretical microphysics and biochemistry -- and controls access to resources and rewards through peer-group review and pressure. Since Zuckerman disregards historical development, the ultra-elite emerges full-blown as the dominant group in 20th-century American science. She overlooks that it was created through successive decisions by the Swedish Nobel awarders, decisions that reflect what the members of the award committees considered important for the advancement of their disciplines at different times.[6]

One way to find out if the Nobel Prize winners represented not just an elite-- which hardly needs proving -- but the *ultra-elite*, according to Zuckerman's definitions, would have been to compare them with the nonwinning candidates. This route was not possible, for at the time she collected her data the names of the candidates were kept secret.[7] The secrecy rule was relaxed in 1974, when the statutes of the Nobel Foundation were changed to permit access to materials in the Nobel archives at least fifty years old for purposes of historical research. This made it possible to conduct a census of nominees and nominators for the prizes for physics and chemistry; the first edition, covering the years 1901 and 1937, was published in 1987.[8]

How did the candidates who won the Nobel Prize differ from their nonwinning colleagues in terms of advantages, those possessed from the outset as well as those accumulated in the course of a career? The three areas -- education, employment, and rewards in the form of honors -- that marked out Zuckerman's ultra-elite are also the ones I will use to compare the prize winners and the nonwinning candidates. My population of forty candidates for the prizes in physics and chemistry 1901-1938, thirteen of whom were prize winners, is more restricted in time and scope than Zuckerman's ninety-two laureates in physics, chemistry and physiology or medicine 1907 to 1972. (The Appendix lists my candidate population and gives basic information about each individual.) Since Zuckerman does not present separately the statistical data on the laureates of the three different prizes she studies, my comparisons with her prize winners during the longer period will have to include those in physiology or medicine as well. Before exposing in detail the elite attributes held by the prize winners and the nonwinning candidates, I will describe the involvement of American scientists with the Nobel institution. These scientists were not only candidates but also nominators for the prizes, and their support of their own countrymen -- own-country nominations made up about 65 percent of those received by Americans before World War II -- determined the makeup of the population of the prize winners and nonwinning candidates.

AMERICAN SCIENTISTS AND THE NOBEL INSTITUTION

The beginnings of the American involvement with the Nobel institution were inauspicious. Among the nominations for the prizes of 1901, the two that concerned Americans, both in physics, had been handed in by nominators who proposed themselves and were hence disallowed.[9] The one was by and for H. A. Rowland, who may have put into practice his own version of the elitism he had preached for American science; the other was by and for R. H. Thurston, who directed the Sibley College of Mechanical Engineering at Cornell University.

In the two following years, there were no American candidates for the physics prize. In chemistry, Gibbs -- not Josiah Willard but Woolcott, the grand old man of American chemistry, then in his eighties -- was put forth in 1902 by C. E. Munroe of the Columbian University of Washington, D.C., who called him "by far the most distinguished of American chemists." In addition, Munroe nominated two younger chemists, E. W. Morley and T. W. Richards, for their determinations of atomic weights.

The physics prize awarded A. A. Michelson in 1907 for his interferometer was the first for an American citizen but it was also an American prize in that the work rewarded, that is, the construction and use of precision instruments, dominated physics research at the turn of the century.[10] That Michelson received the prize with very few nominations was largely due to the strong position of measuring physics in Sweden, particularly in the Uppsala physics department, whose members made up the majority of the Nobel Committee for Physics.

The period up to World War I has been called the "Michelson era" in American physics; in Nobel selections, once Michelson had received the prize, it became the "Hale era." G. E. Hale was nominated for the first time in 1909, received strong support for each of the five years 1913 through 1917, and was proposed intermittently until 1934. The great advances made in astrophysics through Hale's invention of the spectroheliograph put him in a strong position for a prize since they found favor with both the specialized precision physicists on the committee and those more broad-minded -- such as Svante Arrhenius, who had an interest in merging micro- and macrophysics into one large entity, cosmical physics. The practicalities of Hale's award -- should he receive the prize alone?, together with Henri Deslandres who had contributed significantly to the development of the spectroheliograph?, or perhaps with still other astrophysicists?-- presented the committee with so many problems, however, that although members were favorably disposed, the matter was put in abeyance until after World War I. When the question of an award for Hale was taken up again in the early 1920s, committee priorities had changed and astrophysics was pronounced no longer part of physics.[11]

The involvement of American physicists changed character in the interwar period as the quantity and quality of work, and the institutional support it received, was put on a par with and eventually came to surpass that of major scientific powers, Germany in particular. Atomic and theoretical physics were the beneficiaries of much of this support; they were also the areas of most interest to the Nobel Committee for Physics. It is not surprising, then, that the string of awards made to Americans after World War I concerned these areas.

The first postwar prize went to Robert Millikan in 1923 for his oil-drop experiments establishing the elementary charge of the electron. It was followed by those to A. H. Compton (1927), C. D. Anderson (1936), C. J. Davisson (1937), and E. O. Lawrence (1939). In contrast to the prewar years, alongside the prize winner and the lone nonwinning candidate, there was now a field of contenders. Some of them, L. H. Germer and I. S. Bowen, for instance, were cited for collaborations with the eventual prize winner, an indication of the new importance of teamwork in American science.

That America was indeed becoming the new center of physics[12] was shown most strikingly in the international support given not just the winners but also such a perennial nonwinner as R. W. Wood. In the case of Davisson, Anderson, and Wood, such support accounted for close to eight-tenths of their "vote," whereas for Hale and Millikan, it amounted only to one-third and one-fifth, respectively. The inclusion among the foreign nominators of most of the great names in atomic and theoretical physics is a telling illustration of Americans' coming to the fore of the international discipline of physics.

The chemists' involvement with the Nobel institution reflects the more eclectic nature of work in this discipline both in America and elsewhere. The candidates also show that the discipline had more breadth than is indicated by the names of the five prize winners whose candidacies were put forth before World War II (T. W. Richards, I. Langmuir, H. C. Urey, W. F. Giauque, and W. M. Stanley). In physical chemistry, for instance, the names of the award winners, I. Langmuir and W. F. Giauque, should be supplemented by those of W. D. Harkins, who made important contributions to surface chemistry, and G. N. Lewis, "the only chemist in America who ranked with Irving Langmuir."[13] Lewis garnered a record thirty-five nominations during the thirteen years he was a candidate for the prize.

In inorganic chemistry, T. W. Richards was preceded in his lifework -- determining atomic weights -- by E. W. Morley. Another candidate who did important work on atomic weights was a government scientist, W. F. Hillebrand, who served both in the U.S. Geological Survey (1880-1908) and in the Bureau of Standards, where he was chief chemist (1908-1925).

Although no American organic chemist won a Nobel Prize before the Second World War, this was not for a lack of candidates. Among these were: Arthur Michael, "the best organic chemist in the United States," who worked not at one of the research universities that were emerging in the late 19th century, but at Tufts College[14]; Edward Curtis Franklin, who built up the chemistry department at Stanford University; and Moses Gomberg at Michigan, who specialized in the chemistry of free radicals. Finally, the population of candidates includes the biochemist, W. M. Stanley, whose 1946 prize for his work isolating the tobacco mosaic virus in pure crystalline state inaugurated the string of awards to Americans for work in this specialty after World War II.[15]

ELITE ATTRIBUTES OF AMERICAN CANDIDATES

In the following, I will examine prize winners and nonwinning candidates with respect to the following: (1) education and training; (2) first jobs and subsequent employment; and (3) rewards in the form of honors. These were the areas for which Zuckerman found that her laureates constituted a group apart from other scientists, an ultra-elite, as she called them.

Education and Training

Zuckerman's demonstration of her thesis that laureates are set apart by the advantages they accumulate throughout their careers naturally starts with their education. She finds that "the clumping of future members of the scientific ultra-elite in elite institutions begins early in the selective educational process." While all the laureates she studies "went to college, of course,"[16] this was also the case for eight-tenths of the nonwinning candidates. The ones without college education generally belonged to the generation of self-taught scientists born in the middle of the 19th century. More important, the nonwinners and the winners (in both Zuckerman's extended population and our more restricted one) are alike in that more than half attended Ivy League or other elite colleges.

As the future members of the elite moved from undergraduate to graduate education, the laureates and the nonwinning candidates converged on the thirteen elite universities that, as Zuckerman's tabulations show, granted degrees to 85 percent of the laureates and 80 percent of the members of the National Academy of Sciences elected from 1900 through 1967.[17] Still, there are important differences, for these data refer only to those who, *one*, held doctorates, and, *two*, had been granted these by *American* educational institutions. Among the laureates studied by Zuckerman, all with the exception of A. A. Michelson, held doctorates, and 80 percent had earned them in the United States. Almost half of the eighteen laureates who held foreign doctorates were emigré scientists who had fled the rise of fascism in Europe. Of the candidates, both winners and nonwinners, half held doctorates from American institutions, one-fourth from foreign ones, and another one-fourth had no doctorate at all. The higher portion of those holding foreign doctorates or none at all is explained by the population of non-winning candidates reaching back to the late 19th century. Graduate education was very different then from what it had become early in the 20th century, when the future laureates of the interwar period, the real starting point of Zuckerman's study, received their doctorates.

This is shown most clearly when the candidates -- winners and nonwinners -- are divided into three groups according to when they received their doctorates and/or entered the profession: (1) before 1890; (2) between 1890 and 1914; and (3) after 1914. Of the fifteen individuals in the first group, one (T. W. Richards) held an American doctorate, seven held foreign ones, all of them from German universities, and another seven none at all. The fifty-fifty split between Ph.D.'s and non-Ph.D.'s is also extant in Kevles's much larger group of productive physicists, chemists, and mathematicians, 1870-90. The same congruence is found to apply to the second group of candidates, those graduating between 1890 and 1914.[18] By this time, American universities were much better equipped for graduate education; ten candidates thus held domestic Ph.D.'s; three, foreign ones, and three, none at all. After 1914, all the candidates had earned their Ph.D.'s from American universities.

Irrespective of when the doctorates were awarded, the overwhelming majority were earned at elite universities; ten out of thirteen of these being represented among the candidates. There was no institutional concentration of the elite -- except possibly at Harvard University, which accounted for four of the eighteen doctorates, all earned in the period 1890 to 1914. Others on the list -- Chicago, Princeton, Michigan, and Berkeley, for instance -- are among those that Kevles found "acquired depth" in the period 1890-1914 in that they had three or more members from the groups of productive physicists and chemists on their staff.[19]

The winners among the candidates also have in common with Zuckerman's larger population of laureates that they received their doctorates at an early age and thus were off to a head start in their careers. The median age of 26 years compares well with that of 25 for Zuckerman's laureates. By contrast, the nonwinners were a median 28 years old when they received their degrees, which is the same as Zuckerman's larger population of members of the National Academy of Sciences, 1900-1967. According to Zuckerman "the run of doctorates in science in 1957" had a median age of 29.5 years.[20] These differences cannot be explained solely by a selective educational process or by career strategies, as Zuckerman seems to think, but must also contain an element of native ability, which she does not discuss.

First Jobs and Subsequent Employment

The concentration in elite institutions carried over to the candidates' first jobs. Slightly over half of the candidates received their first appointment at an elite institution; for another half again, this was also the institution that had granted them their doctoral degree. In the case of the laureates studied by Zuckerman, the corresponding figures are 65 and 40 percent. By the time the future laureates had been appointed to full professor, however, 78 percent had their appointments in elite institutions. The stronger concentration of Zuckerman's laureates at elite institutions reflects their having come of age professionally in the 1930s and 1940s, when American research universities had matured into institutions that could not only educate and train scientists but also launch them on their careers.

This was not so during the earlier period, when many of the candidates embarked on their careers. Instead, they show a variety of first-job experiences illustrated, for example, by G. E. Hale, who upon having received his B.A. from MIT went back to his hometown Chicago to found the Yerkes Observatory and the *Astrophysical Journal*; by R. W. Wood, who started out teaching at the University of Wisconsin before he was called to succeed H. A. Rowland at Johns Hopkins University; or by W. D. Harkins, who taught chemistry at the University of Montana while preparing for his doctorate at Stanford University (1907). Alongside these, there were of course those who, like Zuckerman's laureates, started out as instructors at the universities where they had received their Ph.D.'s and then stayed to become full professors: I. S. Bowen at the California Institute of Technology, Moses Gomberg at Michigan, Theodore Lyman at Harvard, and M. I. Pupin at Columbia University.

On the whole, the physicists and chemists studied here do not reflect the practices of faculty inbreeding at American elite institutions that Geiger found had "increased to its maximum extent at the end of the 1920s."[21] Geiger's data, however, applies to all disciplines, a majority of which had well-established practices of recruitment and promotion. In physics and chemistry, new departments and laboratories were still being opened in the 1920s. This naturally created opportunities for scientists away from their "home-universities," particularly for those who had made a sufficient mark on their disciplines to be proposed for the Nobel Prize.

A telling example is provided by the triumvirate G. E. Hale, R. Millikan and A. A. Noyes, all candidates for the prizes in the early 1920s (only Millikan was to win one), when they joined forces to turn the Throop College of Technology in Pasadena into a major research university, the California Institute of Technology. Here Hale, who had had his base at the Mount Wilson Observatory since early in the century, enrolled Millikan and Noyes -- from Chicago and MIT, respectively -- in an ambitious program that drew on private and public sources of financing to advance basic science.[22]

The fruits of their labor were apparent in the 1930s, when Caltech had become one of five centers -- the others being Berkeley, Princeton, Chicago, and Harvard -- where promising young physicists and chemists got started on their careers by holding National Research Council fellowships.[23] This route into the ultra-elite became the pattern for the postwar laureates in Zuckerman's study. Among the prewar candidates for the prizes, only those getting their doctorates starting around 1930 followed this route: for instance, C. D. Anderson, E. O. Lawrence, W. M. Stanley and H. C. Urey, all future prize winners.

Whereas only six out of ninety-two, or 7 percent, of the laureates studied by Zuckerman had a nonacademic affiliation, this was the case for ten out of forty, or 25 percent, of the candidates. If the latter figures were to include the three most prominent American inventors of the early 20th century -- Thomas Alva Edison, and Orville and Wilbur Wright, who were discounted as self-employed inventors -- they would of course be even higher. The nonacademics were not all nonwinners, since they included the prize winners C. J. Davisson of Bell Telephone Laboratories and Irving Langmuir of the research department of the General Electric Co. The ten nonacademics among the candidates were evenly split between those working in government research offices, principally the U.S. Geological Survey and the Bureau of Standards, and in industrial research laboratories, such as Bell Telephone or General Electric. The former tended to be those who had received their doctorates or entered the profession before 1914, the latter belong to the interwar period.

The work in basic science of the two industrial scientists rewarded with Nobel Prizes was not typical of that carried out in their laboratories, which tended to be more applied in character.[24] Inventions, whether authored by industrial or university scientists, were also put forth much more frequently for the prizes than indicated by the actual awards. The nonwinning candidates thus include: W. D. Coolidge, who invented the high-vacuum, high-voltage, heated X-ray tube; A. E. Kennelly, who together with Oliver Heaviside postulated the reflecting atmospheric layer of ionized gases (later confirmed experimentally by E. V. Appleton), that proved to be of prime significance for radiotransmission; and M. I. Pupin, whose work on electrical resonators led to the finding that the insertion of inductance coils in telephone lines ("pupinized" lines) improved their performance. They did not win because the Nobel Prize juries showed only limited interest in awards in the area of technology.[25]

Honorific Awards

According to Zuckerman's model, the superior performance of laureates shown throughout their careers -- they publish early and copiously, they carry out their Nobel Prize-winning research when in their late thirties or early forties, they continue to be more productive than other scientists of their age after they have won the prize -- is confirmed by the rewards they receive in the form of promotions, positions of authority, and honorific awards. The latter have been singled out here since they are less influenced by the vagaries of time, institutional practice, and academic politics than promotions or elections to scientific societies. How then do the laureates and the nonwinning candidates compare with respect to the three major forms of honors studied by Cole and Zuckerman -- prizes and medals, honorary doctorates, and election to domestic and foreign academies of sciences?

Scientific achievement was, and probably still is, recognized first by election to one's national academy of science, in the present case the National Academy of Science (NAS). Foreign membership in one of the great academies of science, for instance, the Royal Society of London, the Académie des Sciences de Paris, and the Prussian Academy of Sciences in Berlin usually came later. With one exception -- F. Sanford of Stanford University -- the laureates and the nonwinning candidates were all members of the NAS. Furthermore, as Zuckerman has pointed out, in the case of the laureates, election to the NAS almost always preceded the award of the Nobel Prize.

The candidates' elections to foreign academies confirm the supposition that national honors usually precede international ones. Here again both winners and nonwinners are honored; of the nine candidates who were members of the Royal Society of London-- the highest ranking foreign honor, in the opinion of many American scientists -- three were laureates and six were not.[26] Among both groups, membership in foreign academies was much more frequent in the era before World War I caused irremediable damage to international exchanges of honors.[27] Membership in foreign academies are thus concentrated in a handful of individuals who represent the first generation of candidates: G. E. Hale, who held eight such memberships, having been elected to the Paris Académie des Sciences in 1908 and the following year to the Royal Society; H. A. Rowland with the same number, also including Paris and London; T. W. Richards and A. A. Michelson, with seven memberships each.[28]

In the areas of prizes and medals, and to a lesser degree, honorary doctorates, the winners distinguish themselves from the nonwinners by collecting awards earlier in their careers and particularly by the international character of these. The only awards that count for the laureates are of course those received before the Nobel Prize, since after the prize they were generally showered with honors.[29] Whereas the large majority of the chemists held the prestigious J. Willard Gibbs Medal of the American Chemical Society, for instance, the fact that the future laureates received this award before the Nobel Prize, that is, generally before age fifty, is another indication of their being more precocious than their colleagues. The same applies, although to a lesser degree, to both the Cresson Medal of the Franklin Institute and the Comstock Prize or the Henry Draper Medal of the NAS. At the time they were awarded the Nobel Prize, many of the future laureates had also received the more prestigious honorary doctorates -- those from Harvard, Princeton, Columbia and Yale, for instance -- which was not the case for the nonwinning candidates at the same age.

The winners and the nonwinners differed most significantly when it came to international awards, mainly the medals that the Royal Society of London bestowed on prominent foreign scientists: the Copley, the Rumford, the Davy and the Hughes medals. The majority of the winners and the two runners-up, G. E. Hale and G. N. Lewis, all received one or more of these honors in the years of active candidacy for the Nobel Prize. A. A. Michelson, for instance, received the Copley in 1907, only a month before the announcement of his Nobel award; Hale also received the Copley but much later in life; C. J. Davisson, I. Langmuir, E. O. Lawrence, R. A. Millikan, all were honored with the Hughes Medal before their Nobel Prizes; and G. N. Lewis and T. W. Richards with the Davy one, both while they were being considered for the Nobel. It would be interesting to know if the prize winners' higher profile internationally only applies to honorific awards or if it is rooted in their more international publishing and citation patterns. This could be established, at least for physics, by comparing the entries for the winners and the nonwinners in the *Physics Citation Index 1920-1929*.[30]

THE SELF-PERPETUATING ELITE: MASTERS AND APPRENTICES

The mechanism that makes Zuckerman's ultra-elite self-perpetuating is the master-apprentice pattern she identifies in which laureates or laureates-to-be train their heirs, who in turn go on to win prizes and train future laureates. Of the ninety-two American laureates, she finds that forty-eight, or slightly more than half, had been trained by actual or future laureates. This figure is not far off the mark for laureates of every nationality, 41 percent of whom "have had at least one laureate master or senior collaborator." The percentage for the Americans would have been even higher, she claims, had it not been for some solitary masters such as P. W. Bridgman, who restricted the number of his thesis students and thus "only" had one future laureate, John Bardeen (Physics, 1972) as an apprentice. Serving one's apprenticeship to a Nobel master was most frequent among the physics laureates (61 percent), somewhat less in chemistry (58 percent), and the least so in physiology or medicine (43 percent).

This pairing off between Nobel masters and apprentices was not only characteristic of the postwar world of "big science" but could be extended back, Zuckerman claims, to cover several "generations" of scientists. One such chain in her book involves five generations of physicists and chemists, starting with Wilhelm Ostwald (Nobel Prize for chemistry in 1909) and running through Walther Nernst (Chemistry, 1920), R. A. Millikan (Physics, 1923), C. D. Anderson (Physics, 1936), to end (as of 1977) with Donald Glaser (Physics, 1950).[31]

My comparison between the winning and the nonwinning candidates reveals that the winners were only slightly more likely than the nonwinners to have had a Nobel laureate "master," defined as the thesis advisor. True, some physics winners -- C. J. Davisson, A. H. Compton, and C. D. Anderson -- did their theses under the guidance of Nobel masters (O. W. Richardson in the case of the two former, R. A. Millikan for the latter) but so did the nonwinner I. S. Sprague (R. A. Millikan). Among the chemists, I. Langmuir was the only winner with a future laureate (W. Nernst) as a thesis advisor, the others were the nonwinners G. N. Lewis (T. W. Richards) and A. A. Noyes (W. Ostwald). One is also struck by the number of laureates whose masters were not very prominent: W. F. Giauque, for instance, did his thesis at Berkeley under G. E. Gibson rather than

G. N. Lewis; T. W. Richards at Harvard under J. P. Cooke rather than W. Gibbs, and E. O. Lawrence followed his master, W. F. G. Swann, a specialist in cosmic ray studies, from the University of Minnesota to Chicago and then to Yale, where he received his doctorate in 1925.

Zuckerman's definition of a "master" covers not only thesis advisors but also those sought out for postdoctoral study. This provides a few more Nobel masters for the laureates among the candidates in the period up to 1938: T. W. Richards, who spent a postdoctoral semester with Ostwald in Leipzig and Nernst in Gottingen (1895); and H. C. Urey, who did postgraduate work 1924-25 in Bohr's Institute in Copenhagen. This extended definition also has the effect, however, of increasing the number of nonwinning candidates apprenticed to Nobel masters. G. N. Lewis, for instance, excelled by having three Nobel masters: T. W. Richards, his thesis advisor, and W. Ostwald and W. Nernst, with whom he spent a postdoctoral year (1901).

To fit so many of the laureates onto the Procrustean bed of master-apprenticeship relations, Zuckerman has to go beyond her stated definition of an apprentice as a graduate student or postdoctorate and assume that a mere coincidence in space and time constituted a master-apprentice relation. One example suffices here: A. A. Michelson, who is shown as having been apprenticed to G. Lippmann (Physics, 1908). During the winter (1881-82) that Michelson spent in Paris, however, he probably saw more of other French physicists specializing in optics -- M. A. Cornu and E. E. N. Mascart, in particular -- but of course they did not qualify as Nobel masters. Furthermore, Michelson was hardly an apprentice at this stage, his reputation being in fact so well-established that when he appeared in Paris the French physicists thought he was the son of the "famous Michelson."[32]

In Zuckerman's view, socialization into the ultra-elite is the primary function of apprenticeship, or to use her own words, it "helped shape [the future laureates'] style of scientific work, their conception of the role of the working scientist, and their self-image in that role."[33] This view is largely based on her interviews with the laureates or their autobiographies. Both usually showered praise on Nobel masters: "a great scientist" (an anonymous physics laureate on R. A. Millikan), "a champion" (another physics laureate on E. Fermi), "an outstanding teacher at the critical stage in my scientific career" (H. Krebs on O. Warburg) and of the benefits that they had reaped from associating with them.[34] This kind of *noblesse oblige*, to use Zuckerman's term, extends to professors being given more prominence in the biographies that the prize winners contribute to *Les Prix Nobel* when the former were also Nobel laureates. It is thus possible that the materials on which Zuckerman builds her model of the self-perpetuating ultra-elite contains a systematic bias exaggerating the apprenticeships served under Nobel masters. These need to be reexamined using other materials than scientists' own accounts for what they really involved of working relationships and influences on choices of problems.

CONCLUSIONS

In Zuckerman's opinion, the ultra-elite reflects immanent features -- elitism, stratification, and the accumulation of advantages through the reward system of science -- that have determined the organizational structure of American science in the 20th century and that, furthermore, are functional to progress in science. I hold the alternative viewpoint that the ultra-elite can be seen as reflecting a segment of the American scientific enterprise -- academic, basic research in theoretical microphysics and biochemistry-- during a limited historical period. Even viewed in this restricted sense, the notion of an ultra-elite composed of Nobel laureates contains some basic flaws.

The comparison between winning and nonwinning candidates in physics and chemistry during the period 1901 to 1938 provides support for the alternative viewpoint presented above and this on three different counts:

First, the data on the training and employment of the population of nonwinning candidates, which reaches back into the 19th century, show different patterns from those that came to predominate after World War I, and particularly after World War II. Even among the winners, it is not possible to lump together, as Zuckerman did, A. A. Michelson and such products of post-World War II "big science" as C. N. Yang and T. D. Lee (Physics, 1957).

Second, the nonwinning candidates include industrial scientists whose work was put forth more frequently for the prizes than indicated by the actual awards. The data on these scientists suggest that they differed from academic scientists, for instance, by garnering fewer honorific awards. After 1972, the cutoff for Zuckerman's study, a string of awards, particularly in physics, has gone to scientists working in industrial research laboratories. It would be interesting to know if the Nobel Prizes are unique in this respect or if other honors making up the reward system of science now extend beyond academe.

Third, the comparison between winners and nonwinners points up the stronger international bent of the former, which was observed here only with respect to honorific awards. It would not be surprising, however, if the winners were found to evince greater international visibility in publication and citation patterns as well. This would be a logical consequence of the prize juries' concern to remain in the mainstream of international specialty orientations in physics and chemistry. There is a fundamental ambiguity, however, about Zuckerman's use of the prize adjudications by scientific corporations in Sweden, even when they had a strong international bent, to delineate a national ultra-elite in the United States.

The similarities between the winners and the nonwinners are more striking, however, than the differences. This confirms Zuckerman's observation that both the laureates and the ones she identifies as the runners-up are members of the scientific elite. The same can be said of the members of the NAS. Zuckerman would have contributed more to elite studies, if rather than investigating an ultra-elite defined by the prize juries in Stockholm, she had taken up the intricate matter of how to define American scientific elites in relation to the mass of scientists that surrounds them. Even when extended to the nonwinning candidates, the study of this particular elite does not tell us much either about scientific elites in general, or about how their work advanced scientific knowledge. Nor does it suggest that there were historical developments in the disciplines of physics and chemistry that involved both profound conceptual changes and scientists' participation in the activities of the government, the military and industry during and after two world wars. One has to conclude then that for the history and sociology of science, the notion of elites has limited explanatory value.

Appendix: American Candidates for the Nobel
Prizes in Physics and Chemistry, 1901-1938[*]

Nobel Prize

ABEL, John Jacob (1857-1938)		
ANDERSON, Carl David (1905-)	Physics	1936
BARUS, Carl (1856-1935)		
BOWEN, Ira Sprague (1898-1973)		
BRIDGMAN, Percy Williams (1882-1961)	Physics	1946
CAMPBELL, William Wallace (1892-1962)		
COMPTON, Arthur Holly (1892-1962)	Physics	1927
COOLIDGE, William David (1873-1975)		
DAVISSON, Clinton Joseph (1881-1958)	Physics	1937
FRANKLIN, Edward Curtis (1862-1937)		
GERMER, Lester Halbert (1896-1971)		
GIAUQUE, William Francis (1895-1982)	Chemistry	1949
GIBBS, Woolcott (1822-1908)		
GOMBERG, Moses (1866-1947)		
HALE, George Ellery (1868-1938)		
HARKINS, William Draper (1873-1951)		
HILLEBRAND, William Francis (1853-1925)		
KENNELLY, Arthur Edwin (1861-1939)		
LANGMUIR, Irving (1881-1957)	Chemistry	1932
LAWRENCE, Ernest Orlando (1901-1958)	Physics	1939
LEWIS, Gilbert Newton (1875-1946)		
LOOMIS, Alfred Lee (1887-1975)		
LYMAN, Theodore (1874-1954)		
MICHAEL, Arthur (1853-1942)		
MICHELSON, Albert Abraham (1852-1931)	Physics	1907
MILLIKAN, Robert Andrews (1868-1953)	Physics	1923
MORLEY, Edward Williams (1838-1923)		
MORSE, Harmon Northrop (1848-1920)		
NOYES, Arthur Amos (1866-1936)		
POWER, Fredrick Belding (1853-1927)		
PUPIN, Michael Idvorsky (1858-1935)		
RICHARDS, Theodore William (1868-1928)	Chemistry	1914
ROWLAND, Henry Augustus (1848-1901)		
SANFORD, Fernando (1854-1948)		
SCHMIDT, Carl Frederick (1893- ?)		
STANLEY, Wendell Meredith (1904-1971)	Chemistry	1946
STERN, Otto (1888-1969)	Physics	1943
UREY, Harold Clayton (1893-1981)	Chemistry	1934
WASHBURN, Edward Wight (1881-1934)		
WOOD, Robert Williams (1868-1955)		

---

   [*] Included here are the candidates receiving two or more nominations. A few
inventors and technologists (T. A. Edison, the Wright Brothers) who received more than
one "vote" are excluded. Three individuals -- C. Barus, H. A. Rowland, and W. Gibbs--
from Kevles's list of productive physicists and chemists (see Note 18) are included,
although they only received one nomination.

NOTES

1. Daniel J. Kevles, *The Physicists: The History of a Scientific Community in Modern America* (New York: Alfred A. Knopf, 1978), pp. 43-44, 79-80, 219-220.

2. Henry Augustus Rowland, "A Plea for Pure Science," *The Physical Papers of Henry Augustus Rowland* (Baltimore: Johns Hopkins Univ. Press, 1902), pp. 593-613; Simon Newcomb, "Abstract Science in America, 1776-1876," *North American Review*, 1876, *121*: 104, 108, 111.

3. Harriet Zuckerman, *Scientific Elite: Nobel Laureates in the United States* (New York: The Free Press, 1977); Jonathan R. Cole and Stephen Cole, *Social Stratification in Science* (Chicago: Univ. Chicago Press, 1973); Robert K. Merton, *The Sociology of Science: Theoretical and Empirical Investigations* (Chicago: Univ. Chicago Press, 1973).

4. The following made up Zuckerman's list of elite universities during the 1920s to 1940s, the period under examination in her study: California at Berkeley, California Institute of Technology, Chicago, Columbia, Cornell, Harvard, Illinois, Johns Hopkins, MIT, Michigan, Princeton, Wisconsin and Yale.

5. Zuckerman, *Scientific Elite*, pp. 37-50, 163-207.

6. Elisabeth Crawford, *The Beginnings of the Nobel Institution: The Science Prizes, 1901-1915* (Cambridge: Cambridge Univ. Press; Paris: Editions de la Maison des Sciences de l'Homme, 1984); Robert Marc Friedman, "Americans as Candidates for the Nobel Prize: The Swedish Perspective," in this volume.

7. The indiscretions committed in the essays on the prizes in physiology or medicine, and chemistry in the 1962 edition of the official history of the institution, *Alfred Nobel: The Man and His Prizes*, ed. H. Schück et. al. (Amsterdam: Elsevier, 1962), enabled Zuckerman to draw up a roster of mainly biomedical scientists whose work was referred to as prizeworthy but who had not all been proposed for the prize. There were too few Americans among those receiving "honorable mentions" to permit their being systematically compared with the laureates with respect to elite attributes. A selective comparison was carried out, instead, for members of the National Academy of Sciences elected 1900-1967, who were found to constitute an "extended elite," and a sample of "ordinary" scientists, who turned out to be just that.

8. Elisabeth Crawford, John L. Heilbron, and Rebecca Ullrich, *The Nobel Population: A Census of Nominees and Nominators for the Prizes in Physics and Chemistry, 1901-1937* (Berkeley, Cal.: Office for History of Science and Technology, Univ. California, Berkeley, 1987; Uppsala: Office for History of Science, Uppsala Univ., 1987).

9. Para. 7 of the Statutes of the Nobel Foundation (1900) states: "A direct application for a prize will not be taken into consideration."

10. Kevles, *The Physicists*, pp. 26-32.

11. Crawford, *The Beginnings of the Nobel Institution*, pp. 56-59, 173-175; Friedman, "Americans as Candidates for the Nobel Prize."

12. Kevles, *The Physicists*, chap. 14, pp. 200–221.

13. Ibid, p. 225.

14. Daniel J. Kevles, "The Physics, Mathematics, and Chemistry Communities: A Comparative Analysis," in *The Organization of Knowledge in Modern America*, ed. Alexandra Oleson and John Voss (Baltimore: Johns Hopkins Univ. Press, 1979), pp. 139–172.

15. Lily E. Kay, "W. M. Stanley's Crystallization of the Tobacco Mosaic Virus, 1930–1940," *Isis*, *77*: 450–472.

16. Zuckerman, *Scientific Elite*, p. 83.

17. For a list of elite universities, see Note 4.

18. Daniel J. Kevles and Carolyn Harding, "The Physics, Mathematics, and Chemical Communities in America, 1870–1915: A Statistical Survey," *California Institute of Technology, Social Science Working Paper No. 136* (1977), Tables 5–12.

19. Kevles, "The Physics, Mathematics and Chemistry Communities: A Comparative Analysis," p. 154.

20. Zuckerman, *Scientific Elite*, p. 89.

21. Roger L. Geiger, *To Advance Knowledge: The Growth of American Research Universities, 1900–1940* (Oxford: Oxford Univ. Press, 1986), pp. 223–226.

22. Ibid, pp. 183–189.

23. Kevles, *The Physicists*, pp. 219–220.

24. Ibid, pp. 188–189.

25. Crawford, *The Beginnings of the Nobel Institution* pp. 160–161, 164–166.

26. The members of the RLS were: J. J. Abel, W. W. Campbell, W. Gibbs, G. E. Hale, I. Langmuir, A. A. Michelson, T. W. Richards, H. A. Rowland, and R. W. Wood. The Royal Society ranked eighth in terms of visibility and tenth in terms of prestige (the Nobel Prize being of course first) of the ninety-eight honorific awards that Cole submitted for judging to a sample of some 1,300 academic physicists in the late 1960s. (Cole, *Social Stratification*, pp. 270–275).

27. Elisabeth Crawford, "Internationalism in Science as a Casualty of World War I: The Relations between German and Allied Scientists as Reflected in Nominations for the Nobel Prizes in Physics and Chemistry," *Social Science Information*, 1988, *27*(2):163–202.

28. I am indebted to John May, University of California, Berkeley, for providing me with data concerning the Academy membership of the American candidates.

29. Zuckerman, *Scientific Elite*, pp. 236–238.

30. *Physics Citation Index, 1920-1929*, 2 vols. (Philadelphia: Institute for Scientific Information, 1981).

31. Zuckerman, *Scientific Elite*, pp. 99-106.

32. Dorothy Livingston Michelson, *The Master of Light: A Biography of A. A. Michelson* (New York: Scribner's, 1973), pp. 86-88.

33. Zuckerman, *Scientific Elite*, p. 124.

34. Ibid., pp. 124-132.

AMERICANS AS CANDIDATES FOR THE NOBEL PRIZE: THE SWEDISH PERSPECTIVE[*]

Robert Marc Friedman

Scientists, historians, and the general public tend to regard the Nobel Prize as an indicator of ultimate excellence in science. Only recently, after the opening of the relevant archives related to proceedings of the Nobel committees and the Royal Swedish Academy of Sciences (*KVA*), for materials over fifty years old, has documentation to begin making sense of the prize decisions become accessible. Initial studies have shown clearly that in a number of ways the Swedish awarders of the prizes must be the focus of any meaningful attempt to comprehend the prize decisions.[1] In addition, these studies show that the prize should be understood as an institution that developed and evolved. The Nobel Prize cannot be regarded as a constant, fixed entity: interpretation of the statutes, committee procedures and priorities, scientific criteria for selecting winners, the degree of reliance on nominators, and relations between the respective committees and the academy all changed over time. Individual committee members and their relations with one another could influence the outcome of the process by which the respective five-member committee arrived at a consensus for awarding the annual prizes. Generalizations on the criteria for success in receiving a prize, therefore, cannot readily be claimed. Examination of the candidacies of two American physicists, A. A. Michelson and G. E. Hale, reveals the significance of changing Swedish concerns for the outcome.

DISCIPLINARY CONCERNS RELATED TO THE NOBEL PRIZE

To understand the Nobel Prize -- including the decision making -- as an institution, a disciplinary perspective must be taken. Prior to World War II -- and especially during the interwar period -- deliberations related to awarding Nobel Prizes, including interpretation of the relevant statutes, became enmeshed in the processes by which a number of factions and personalities within the Swedish physical science community attempted to define and to legitimize their particular notions of physics and chemistry. Some committee members sought at times to shape the growth of their own national disciplines and to respond to and influence developments abroad. That is, the Nobel Prize should be comprehended historically in the pre-World War II period as part of the general problem of disciplinary development in Sweden. Although the latter shared some features and followed in part international developments, specific local characteristics proved significant. Limited materiel and personnel resources heightened the ecological relationship among research specialties and institutions: restricted availability of funds and qualified researchers implied that the growth or introduction of new specialties often had to be

[*] Institute for Studies in Research and Higher Education, Norwegian Research Council for Science and the Humanities, Munthesgt. 29, N-0260 Oslo, Norway. I am indebted for support for my present research on the Nobel Prizes to the National Science Foundation (Program in History and Philosophy of Science, SES-8512431), American Institute of Physics Grant-in-Aid Program, and the History of Science Society Unaffiliated Scholars Program. I thank the Department of History of Science, Johns Hopkins University, and especially Tore Frängsmyr and the Office for History of Science, Uppsala University, for institutional support. Elisabeth Crawford provided useful comments on an earlier draft.

accomplished at a cost to others. Especially after World War I, those scientists who sought to establish vital research schools, especially in new specialties such as atomic physics or biochemistry, devised strategies to compete for resources. Some Nobel committee members gradually recognized that the prize institution could be recruited to assist in these efforts: to endow authority, to hinder or provide legitimation of research programs and methods, and to offer a source of funding.

When it became known following the death of Alfred Nobel that the Royal Swedish Academy of Sciences was named to make awards related to the fields of physics and chemistry, the reaction among the Academy's members was far from universally positive. Oscar Widman, who was to become a member of the chemistry committee, noted that he and others with whom he had spoken were inclined to prefer that the Academy not accept such a difficult task, a task that surely will lead to much "humbug."[2] Others, such as Svante Arrhenius, while recognizing the many difficulties entailed in implementing Nobel's testament, also understood the possibilities available to the local scientific communities. Not only would leading international researchers come to Stockholm and perhaps remain for a few weeks, and even months, but the Swedish prize-givers would be received abroad with greater respect.[3] To assess the relative merits of the different candidates, large, model laboratories were to be erected -- the so-called Nobel Institutes. Prize winners and other international researchers, it was hoped, might spend up to several months at such laboratories. During the first years following Nobel's death, members of the small Stockholm Högskola hoped that these laboratories could be appended to the Högskola, thereby enriching their fledgling institution. From the start, the Nobel Prize appeared to some as a potential resource for improving and internationalizing Swedish science.

In transforming Nobel's brief testament into a set of statutes regulating the prize institution, the committee left some features loosely defined, such as the meanings of "physics" and "chemistry" and "benefit to mankind." Crawford has shown in great detail how the statutes that finally were accepted arose from long deliberations, entailing compromise among the representatives from several institutions. Consequently, sufficient maneuverability remained to allow the different Nobel committees to develop their own strategies for interpreting and implementing the statutes.[4] Traditions could be developed based on precedent; strategies for breaking with tradition also eventually arose. During the first fifteen years of prize-awarding, committee members and the Academy worked with the statutes, feeling their way toward practical means for implementing the statutes and for establishing a smoothly running institutional arrangement.

This early period should also be understood as one in which the prize itself was developed as a resource. Of course the extraordinary amount of money involved brought enormous international and local attention from the start. By being able to award the early prizes to a number of universally respected researchers, the committees and the Academy could endow the prize with value over and above the great fortune involved. The prestige of the winners gave prestige to the prize itself. Still, on occasion, members of both chemistry and physics committees as well as the Nobel Foundation expressed concern about the level of excellence required for a prize and fears that the international scientific community was losing interest in the prizes.[5] Perceived success in bringing acclaim to the prizes and to Sweden, in principle, could give committee members greater prestige and authority within the Academy and within the Swedish scientific community.

During the first two decades of awarding the prizes, committee members often sought to have prizes given to works that reflected their own interests and specialties: eg., Arrhenius worked to award his circle of "Ionists" in physical chemistry, Peter Klason argued for industrial chemical advances, and Widman for organic chemists. It was not unusual for members of the respective five-person committees, as well as members of the Academy, to organize campaigns of nomination letters or to request that colleagues send in nominations for a certain candidate, and even, at times to suggest a motivation or the wording to be used.[6] Absolute numbers of nominations rarely proved decisive in persuading the committee. Moreover, only rarely after the first six years did any candidate receive a clear mandate from the nominators; and even when such consensus did exist, the relevant committee rarely took action.[7] When deliberating on how to award prizes, committee members worked with a number of concerns that, of course, changed over time. One example of how the predilection to reward work in near relation to the disciplinary orientation of individual committee members could help promote a candidate can be seen in the case of A. A. Michelson and the 1907 physics prize.

A.A. MICHELSON: HASSELBERG'S IDEAL PHYSICIST

Michelson's metrological investigations evoked considerable admiration from some members of the Nobel committee for physics.[8] Three of the five members of the committee during its first years belonged to an Uppsala tradition of experimental physics that stressed precision measurement, especially spectral measurements: Bernhard Hasselberg, Robert Thalén, and Knut Ångström. After Thalén's death, Gustaf Granqvist, who largely shared these experimentalist ideals for physics, joined the committee in 1904. A fourth member, the Uppsala professor of meteorology H. H. Hildebrandsson, was a product of the same physics milieu. Svante Arrhenius, professor of physics at the Stockholm Högskola, held theory in somewhat higher regard and tended to consider the pursuit of ever greater precision measurement for its own sake as a relatively sterile form of research.

Clearly when A. A. Michelson was first proposed for a prize in 1904 solely by the Harvard astronomer E. C. Pickering, his candidacy received a sympathetic reception. Hasselberg recognized in the American's spectral studies, work closely associated to his own. Hasselberg, who held the Academy of Sciences professorship in physics, was attracted to the application Michelson had made of his instrumental innovation -- the interferometer -- to spectroscopy and metrology. As Sweden's representative on the International Committee for Weights and Measures and one who used spectroscopic measurements himself for metrological investigation, Hasselberg prized Michelson's use of a natural constant for determining experimentally the length of the international meter. Similarly, Michelson's application of the interferometer to re-calculate wavelength tables for spectroscopy appealed to Hasselberg, who had conducted comparable investigations. These works of Michelson were praised in the committee's 1904 evaluative report for a "rigor and originality" that placed them at the forefront of contemporary physical research.[9] In claiming Michelson to be among those who should be considered for a prize, the committee underscored how Michelson's use of wavelengths of light for units in determining the length of the meter, had rendered the international standard of length "for all time to come, indestructible."[10] The committee concluded that Michelson, along with P. Lenard, E. Abbe, the combination J. Dewar and K. Olszewski, and Lord Rayleigh, should be considered for the 1904 prize and, in a very frank manner, continued by noting that the choice "to a large degree depends upon everyone's individual notion of the one or the other discovery's greater or lesser scientific or practical significance."[11] Lord Rayleigh received the prize.

When, in 1907, Michelson was next proposed -- which marked a return after a two-year absence of nominations from America --Hasselberg was prepared "to do all in my power to procure the prize for him."[12] In his confessions to G. E. Hale, who had nominated Michelson, Hasselberg noted that his own high regard of Michelson's work is "in some way an opinion of sympathy for an area closely connected with my own specialty . . . I cannot but prefer works of *high precision*."[13] Hasselberg prepared a detailed special report for the committee on the merits of Michelson for a prize. That

Hasselberg was determined to have Michelson receive a prize is also indicated by the fact that Michelson was proposed by only two other persons in 1907 (K. Prytz and H. Ebert) and only once before then.[14] Moreover, Michelson's work did not contain any specific discovery. Hasselberg's letters show that he was aware of the problem: he asked Hale for advice as to which of Michelson's works might constitute a discovery.[15] In short, Hasselberg's desire to reward precision measurement outweighed these otherwise negative considerations.

In a special report, Hasselberg emphasized that Michelson had perfected optical interference-measuring methods to a hitherto inconceivable level of precision. He described in considerable detail Michelson's metrological investigations with the interferometer. In principle, Michelson's work in perfecting the interferometer could be considered for a prize if it resulted in a discovery of great significance. Although no such discovery could be claimed, Hasselberg underlined that the extraordinarily sharp precision attained by Michelson constituted a precondition for discovery: "The history of the exact sciences shows that almost all major discoveries . . . are ultimately due to increased precision in measurement and that every advance in this direction in itself *can* contain the seed of new discoveries."[16]

In contrast to the metrological investigations, the now famous ether-drag experiment received scarcely a mention. Indeed, Hasselberg considered the experiment to be virtually irrelevant to the important contributions for which Michelson was being evaluated. This experiment, along with Michelson's efforts to measure the diameters of Jupiter's moons, is noted in passing at the close of the report. "Even though [the ether-drag experiment] only led to a negative result, it still possesses a significant historical interest in that Michelson first constructed the interferometer for precisely this purpose and with that made possible all the varied precision investigations about which we have attempted here to provide at least a preliminary introduction."[17]

Support to award Michelson undoubtedly came easily in the committee. Knut Ångström, the committee's chairman, shared Hasselberg's belief in the primacy -- if not actually exclusivity -- of precision measurement for the advance of physics. Perhaps the only opposition could have come from Arrhenius, both because of his and Hasselberg's long-standing antagonism and because of his distaste for defining physics by precision measurement. Hasselberg was aware that Arrhenius could well try to oppose the proposal and was jubilant when the full Academy approved the recommendation without incident.[18] The committee, followed by the Academy's physics section, and finally by the entire Academy, accepted in conclusion that Michelson's investigations can rightly be considered of "fundamental and epoch-making significance for all of precision physics." The Michelson-Morley experiment did not figure in the decision to award Michelson the 1907 Nobel physics prize for "his optical precision instruments and the spectroscopic and metrological investigations carried out with their aid."[19]

Hasselberg was interested in awarding a Nobel Prize to a person whose work perfectly exemplified his own vision for professional physics. At the award ceremony, Hasselberg extolled the virtues of precision measurement as the highest ideal of physics. Here was an opportunity to legitimize further a goal for physics that he -- and other members of the committee -- had endeavored to institutionalize.

When Hasselberg and Knut Ångström, for example, had served in 1898 as independent expert referees [*sakkunniga*] to judge and rank the candidates for the professorship in physics at Lund University, they opposed Janne Rydberg because he did not execute his own spectral analyses and instead used others' results for his studies that included "theoretical speculations" concerning the spectra of elements.[20] They claimed that a professor of physics must be a master of precision measurement. In contrast, the third "expert," the physics professor at Copenhagen University, C. Christiansen, ranked Rydberg first and expressed surprise when he learned that the two Swedish experts placed

Rydberg in third place behind Granqvist, who was then an associate professor, and C. A. Mebius, an assistant professor. Although letters from foreign physicists supported Rydberg, Hasselberg and Ångström maintained that Rydberg had shown little evidence of being able to conduct experiments. They expressed concern for the need for experimental performance: no matter how insightful the conclusions drawn from others' experiments might be, such work must be considered less worthy than a study in which conclusions are drawn from self-conducted experiments. The rigor of precision should be the measure of competence for a professorship in physics.

In awarding a Nobel Prize for such work, they could once again endeavor to reinforce and further to legitimize this opinion. At the awards ceremony Hasselberg claimed that precision measurement "is the very root, the essential condition, of our penetration deeper into the laws of physics --our only way to new discoveries."[21] Of course, not all physicists ascribed to this view. In an apparent reference to the Hasselberg and Uppsala school's perspective, Vilhelm Carlheim-Gyllensköld, the new acting professor of physics at the Stockholm Högskola, commented during a visit at this time to J. J. Thomson's laboratory that he was happy to see a laboratory where scientific work was conducted according to "an orderly plan and not for the sake of the instrument -- as in certain places."[22] Attempts to reward achievements in mathematical and theoretical physics at this time generally did not receive comparable favorable treatment.[23] In this respect the prizes could play some disciplinary role by legitimizing and bringing prestige to specific research orientations, specialties, and methods.

HALE: THE CHANGING ROLE OF ASTROPHYSICS IN SWEDISH PHYSICS

By the early 1920s a number of issues promoted a heightened understanding by many committee members that the prize institution was one of many resources that could be used in shaping and directing disciplinary growth. Not only did the committee's composition change after the deaths in 1922 of Hasselberg and Granqvist, but the importance of the prize for Swedish disciplinary considerations also changed. In part the greater desire to involve the prize-awarding with Swedish concerns arose in connection with a changed perception of how to organize research. A new generation of researchers, such as Hans von Euler-Chelpin, Carl Wilhelm Oseen, Manne Siegbahn, and The Svedberg, recognized that modern research was no longer the work of an individual professor in relative isolation from international trends. They sought to establish research schools in which several assistants could work on pieces of a larger problem in well-financed institutions. They and others wanted to introduce and institutionalize new specialties such as theoretical physics, biochemistry, physical-colloidal chemistry, experimental atomic physics, and -- in the 1930s -- nuclear physics and nuclear chemistry. Funding was meager; competition for funds from the state, from the few private foundations, and from the Royal Swedish Academy of Sciences was considerable. The prizes could figure in the political economy of disciplinary development directly and indirectly. Awarding or withholding prizes in certain research areas could conceivably assist in promoting and hindering the advance of these same areas in a Swedish context. I am not claiming that the process of awarding can be equated with or reduced to disciplinary politics, merely that at certain time periods disciplinary considerations, to various degrees and in different forms, figured in decision-making.

George Ellery Hale's candidacy illustrates some of the processes by which Nobel Prize deliberations became involved with local disciplinary concerns in the interwar period. The change in attitude toward rewarding an achievement in astrophysics arose from shifting preferences in disciplinary priorities and from the recognition of the need to limit the number of specialties drawing on the modest resources available to physics. First, comments on the formal "official" evaluation of Hale will provide a sense of the shift in attitude, and then a discussion of the Swedish disciplinary context will help illuminate what was at stake in striking Hale from the list of those who were considered certain to receive a Nobel Prize.

Hale was first nominated in 1909 by a group of American scientists for his invention of the spectroheliograph and the use of this instrument for discovering magnetic fields associated with sunspots. These and related studies by Hale in solar physics were praised in the committee's report. The committee also noted, however, that their recency and their surprising results required that more time be spent to see whether or not they would have a "thoroughgoing significance" in solar physics.[24]

When, in 1913, Hale was next nominated for his solar physics work, by Chicago professor O. H. Basquin, the committee linked his candidacy with that of Henri Deslandres, who had been nominated for his independent endeavor in creating and using a spectroheliograph.[25] In the report these two lines of investigation were hailed as the best in solar physics and their further progress as deserving to be "followed with the greatest attention, and which -- sooner or later and rightly so -- ought to come to be regarded as deserving of the Nobel physics prize."[26] The prize for 1913 went to H. Kamerlingh Onnes, who received seven nominations. Others who received strong support from nominators that year included: E. Amagat (6), K. Olszewski (4), M. Planck (5), and A. Righi (5). The strong standing Hale and Deslandres received in the report reflects the favorable disposition of some committee members rather than a mandate from the nominators. Hasselberg was primarily engaged with astrophysical spectroscopy; Arrhenius and Carlheim-Gyllensköld researched in what they called "cosmical physics" (a rubric then used for a physics of the macrophenomena of earth, seas, atmosphere, and space, especially before these areas became yet further professionally segregated), and Granqvist was continuing the recently deceased Knut Ångström's measurements of solar radiation.

This favorable disposition by committee members was again clear in 1914, when Hale was nominated by C. R. van Hise and T. W. Richards. The committee declared that Hale, along with M. von Laue and Planck, was among those who were most deserving for a prize. But, because Deslandres was not nominated that year, the committee declined to consider Hale for the 1914 prize: the two had to be evaluated and considered together. At the request of the committee, Arrhenius proposed the two together in 1915. Carlheim-Gyllensköld, however, proposed -- for reasons to be discussed below -- the Norwegian northern-lights researchers, Kristian Birkeland and Carl Størmer. In the report the committee expressed its agreement with those who had claimed that Hale and Deslandres deserved a Nobel Prize, but preferred to put them after W. H. and W. L. Bragg for that year's prize. The committee seemed especially anxious to reward the Braggs at this time; its proposal that the 1914 prize be awarded to von Laue was not acted on by the full Academy because of war. Lacking any statutory right to withhold prizes because of *force majeure*, the Nobel Foundation had petitioned and received permission from the government to postpone the prize decisions until 1915. Not knowing whether the war would end or whether the government would again allow deferring the prizes in 1915, the committee seems to have recognized that to be able to award Nobel Prizes for research in the same area to investigators from both sides of the warring nations was an "especially fortunate" circumstance.[27] Moreover, once Carlheim-Gyllensköld decided to advance his own candidate in cosmical physics, the possibility of gathering a majority of the committee for Hale and Deslandres became problematic.

When Hasselberg nominated Hale and Deslandres the following year, he could also point to nominations from van Hise, J. D. van der Waals, and P. Zeeman for Hale when preparing the evaluative report for 1916. In the draft that he prepared -- together with Allvar Gullstrand -- Hale and Deslandres were claimed to be worthy of being rewarded with a Nobel Prize. Nevertheless, they were again placed behind another candidate, this time Johannes Stark, who was proposed at first to receive that year's prize. When the full committee met to discuss the matter, the text was altered. The phrase concerning the astrophysicists being worthy of a prize was crossed out and the recommendation that the 1916 Nobel Prize in physics be awarded to Stark was crossed out as well. Having decided that the prize after all ought not be awarded during wartime, but not having legal authority to withhold prizes because of war, the committee agreed simply to invoke the statutory rule that allowed reserving a prize when no candidate's work meets the criteria of excellence for receiving a Nobel Prize. The evaluative report had to be revised to remove any claim that candidates existed who were worthy of being rewarded.[28]

In 1923 Hale again emerged as a leading contender for a Nobel Prize, at least in the opinion of some committee members. Although he had been nominated by six Americans in 1917, the committee was clearly determined that year not to award any prize. The 1916 prize money had been placed in a special fund and the 1917 prize had been reserved to the following year. But in 1923, Hale appeared to be a serious contender for a prize. During the spring, when the committee's chairman, Gullstrand, assessed the candidates for that year, he understood that based on the committee's prior criteria for awarding prizes and on the earlier evaluations, Hale -- together with Deslandres -- was the strongest candidate; that is, if the committee still wanted to consider astrophysics as part of physics.[29] Hasselberg, who had lead the support for astrophysics and who had had considerable authority within the Academy, had died. Carlheim-Gyllensköld alone was advocating dividing the prize between Hale and Deslandres. Oseen was inclined not to make such an award; he doubted that their contributions could be considered to belong to physics. He agreed, however, to yield to Arrhenius's authority as to whether the work should be considered part of physics.[30] Oseen was also prepared to follow the opinion of Siegbahn, who had recently joined the committee and who apparently had not himself decided as yet whether the astrophysical investigations should be included in the domain of physics.[31] Gullstrand urged Carlheim-Gyllensköld to prepare as quickly as possible a very preliminary draft of the special report on Hale and Deslandres, so that the committee could begin deliberating the matter before the summer. Although Carlheim-Gyllensköld wanted to follow the usual procedure of using the summer to prepare evaluations of candidates, he accepted Gullstrand's request, which had underscored the importance for members of the committee to form a clear consensus around a candidate: if they did not, the full Academy would surely support The Svedberg's nomination of Rutherford for the physics prize.[32]

Carlheim-Gyllensköld was prepared to press for awarding the prize to the astrophysicists. In his special report he extolled the significance of Hale and Delandres's work. He also planned to hold a lecture at a meeting of the Academy on cosmical physics research in conjunction with the committee's June meeting. His efforts were, however, of little consequence. After discussion among the remaining committee members, Arrhenius wrote a strongly worded statement in which he concluded that astrophysics had advanced so rapidly during recent years that it now encompassed all of astronomy. Astrophysics, therefore, should be identified as astronomy rather than as part of physics.[33] He and the others who were in opposition to an astrophysics prize followed Siegbahn's recommendation to award Robert Millikan in spite of the fact that the latter had received but one nomination that year. Carlheim-Gyllensköld protested. He objected strongly to Arrhenius's odd report stating that astrophysics no longer could be considered to be part of physics. Rejecting Arrhenius's claim that a prize to Hale and Delandres would precipitate a flood of nominations for astronomical work, Carlheim-Gyllensköld pointed out that physicists, after all, had nominated Hale.[34] What was at stake entailed more than merely who should receive the 1923 Nobel Prize for physics.

During the inter-war period leaders of the Nobel physics committee seem to have tried to restrict the definition of physics -- to pull in the boundaries to a central core of atomic and eventually also nuclear physics. Physics had been left broadly defined with regard to the Nobel Prizes and in the Academy. In response to an inquiry in 1900, Hasselberg wrote that the prize could be awarded not only to achievements in "pure Physics properly so-called," but also in "the sciences most closely connected with Physics and for the cultivation of which physical methods are employed." He included physical chemistry and astrophysics: "This I suppose will certainly answer to the proper meaning of the testator. Whereas thus . . . pure Astronomy or celestial dynamics . . . will scarcely be included, any work in Astrophysics of great importance deserves certainly a thorough consideration.[35] When the physics section of the Academy was allowed to expand to 10 members in 1904, it received the name "Physics and meteorology"; areas such as astrophysics, geophysics, meteorology, technical physics, and physical chemistry could be more or less accommodated under the label of physics. In part, this broad spectrum represented the research interests of the various Swedish physicists in the Academy and on the committee, but it also reflected the diffuse boundaries of professional physics around 1900. After the war several issues promoted a desire among most committee members to relegate astrophysics to a position outside the accepted boundaries of professional physics in Sweden.

By 1915, after the Royal Swedish Academy of Sciences relocated to a new building just north of Stockholm, it became clear that the Physics Institution of the Academy would soon be homeless. The new building did not have room and the Academy did not have funds to provide any other new facility. The Physics Institution consisted largely of a major collection of instruments dating to the eighteenth century, but also contained newer spectroscopic instruments, useful for contemporary research. To resolve the problem, the Academy's secretary, Christofer Aurivillius -- an entomologist -- suggested that perhaps the Nobel Committee for Physics might consider housing this collection in a special department of the planned Nobel Institute, which had been called for in the statutes for testing nominated discoveries and furthering research.[36] Although the only department of the Nobel Institute established at the time was Arrhenius's, for physical chemistry, the other departments belonging to the Academy -- for physics and for chemistry in general -- were expected to be erected by 1922 as interest on the funds for such centers continued to accrue.[37] Hasselberg, who led the Academy's Physics Institution, seems to have discussed the plan with the Academy's administrative committee and seems to have approved -- if not actually suggested -- the plan. In a letter to W. W. Campbell he discussed the problem: "As however the Nobel [Foundation] . . . will in some time erect a great laboratory for general physics it is perhaps possible, that the question can be solved in such a way, that my laboratory, in the form of an astrophysical department of it, can become a special pavilion in connection with the proposed new [Nobel] physical institution."[38] Such a strategy could even have strengthened astrophysical spectroscopy in Sweden as well as endowing greater prestige on Hasselberg and this field of research. Hasselberg's own eagerness to support Hale and Deslandres's candidacy for the Nobel Prize could well have been linked to his desire to campaign for a department of astrophysics within the Nobel Institute. Such a prize would of course legitimize astrophysics as an integral part of physics and would provide an occasion to draw attention to and to praise astrophysical precision measurement research.

Hasselberg was not alone in desiring to lead a division of the planned Nobel Institute. Carlheim-Gyllensköld also sought this goal and also sought to preside over the Academy's historically significant instrument collection. In addition to being involved with cosmical physics research, Carlheim-Gyllensköld possessed strong cultural interests, including the history of science; these interests prompted him eventually to try establishing a museum for the exact sciences.[39] To try promoting the need for a cosmical-- rather than purely astrophysical -- physics department Carlheim-Gyllensköld proposed Birkeland and Størmer for the 1915 Nobel Prize. In his special report on their northern-lights studies, Carlheim-Gyllensköld pointed out that unfortunately because no Nobel Institute for Cosmical Physics existed, a thorough examination of these experimental and theoretical investigations would not be possible.[40]   It is unclear whether Carlheim-Gyllensköld was trying to compete with Hasselberg or whether he sought to expand the scope from astrophysics to cosmical physics in the hope that following Hasselberg's retirement he might be able to claim the right to head such a department.   Regardless, by advancing his own candidates in this quirky maneuver, Carlheim-Gyllensköld hindered the awarding of Hale and Deslandres in 1915.

Whether to call the proposed department astrophysical or cosmical physics proved not to be the major issue. Hasselberg accepted "cosmical physics," which certainly would attract greater support from within the committee and Academy. He pointed out that he and Arrhenius both pursued research in cosmical physics. Moreover, he underlined cosmical physics' strong traditions in Sweden even though it actually was not formally represented in any university.[41] More significant, the Academy's decision not to provide housing for the instrument collection was seen by most committee members as a blow to the prestige of physics and its role within the Academy. When it became clearer that the Academy also had no desire to fill the professorship of physics after Hasselberg's eventual retirement, most committee members understood that Swedish physics was about to lose both a research facility and one of its very few professorships, one that was, by definition, devoted exclusively to research.

The Nobel physics committee repeatedly tabled motions in 1916 to make a formal reply to the Academy.[42]   In private, Arrhenius criticized the plans, noting that little desire existed for using the Nobel Institute for this purpose.[43] He expressed scorn for the two main supporters of the plan: Hasselberg, who managed his job so poorly that the Academy could regard the physics professorship as being "completely unnecessary" and Carlheim-Gyllensköld, a rather isolated and inactive researcher, who "is speculating about getting an appointment there."[44] Sensing Hasselberg's efforts to design the department for his personal professional interests, Granqvist expressed the need for planning a facility with more than spectroscopy in mind. Perhaps recognizing an opportunity for himself as well, he stressed the importance of not limiting the choice of leader by defining the position as one for a spectroscopist.[45]

Letters and reports in 1917 from the Nobel physics committee to the Academy urged that all possible solutions be considered that might permit the Academy to retain its physics institution. Only after the complete impossibility of retaining the institution had been demonstrated would the committee feel the need to consider the less satisfactory option of housing the Academy's collection in the Nobel Institute. The committee felt that the latter option would make it easier in the future for the Academy to leave the professorship unfilled, which would hardly be satisfactory given the tiny number of professorships in physics in Sweden. And, as the wartime economic situation worsened, the committee members understood increasingly that the Nobel Foundation's funds for the projected Nobel Institutes were becoming inadequate as inflation soared.   Still, plans were drawn up for alternative cost options for a cosmical physics department of the Nobel Institute.

By the end of the war, all of these plans were dropped: the economic situation prohibited erecting any new buildings. Even after it became clear that the money from some of the unawarded Nobel Prizes could be placed in the committees' own special funds, which then could be applied toward creating Nobel Institutes, the financial prospects were dim for establishing and maintaining any new Nobel Institutes in the immediate future. Most of the instruments belonging to the Academy's Physics Institution collection were placed in storage.   Still, as long as some committee members maintained a strong interest in astro- or cosmical physics, some pressure could be exerted to institutionalize these fields in a Nobel Institute once planning resumed.

Hale's re-emergence as major candidate in 1923 brought some of these issues into focus once again, but now Hasselberg and Granqvist were dead. Oseen and Siegbahn had replaced them on the committee; for their part, astrophysics only diverted resources away from what they considered more significant specialties. Oseen had long hoped to improve the status of Swedish physics, both at home and abroad.[46] When he and Siegbahn were elected to the Nobel physics committee, Oseen felt that "a new era for physics ought to begin, an era in which the institution's [Nobel physics committee's] resources for the first time will be totally utilized.[47] First and foremost, Oseen hoped to institutionalize theoretical physics in Sweden. In 1923 he conceded that toward this goal he would endeavor to have established a Nobel Institute for Theoretical Physics, where he might be able to escape university teaching and administrative burdens and devote himself to research and writing.[48] For his part, Siegbahn hoped to resume his research program in X-ray spectroscopy in Uppsala, where he had just replaced Granqvist. Experience at Lund University, where he began this work, showed that assistants and financial resources would be necessary to create a thriving experimental physics milieu. He also had learned that Swedish physics scarcely had adequate funds for improving itself. Even publishing physics articles through the Academy proved difficult: the Academy's poor economy and --according to some observers -- lack of appreciation of physics' needs resulted in serious delays in publishing, including one of Siegbahn's important articles on X-ray spectroscopy.[49]

One strategy that seemed apparent to Oseen and Siegbahn for raising the standard and esteem of Swedish physics was to concentrate resources on a limited number of research specialties, specialties that were also of considerable interest in international physics milieux. For them atomic physics held the best opportunities for Swedish physics to achieve better standards.[50] Siegbahn's research program in X-ray spectroscopy showed that significant work could be achieved under difficult conditions. An astrophysics department in the Nobel Institute would of course divert scarce funds from departments of theoretical physics and experimental atomic physics. Furthermore, as long as astrophysics and cosmical physics were considered legitimate parts of physics within the Academy, the few funds available would have to be shared with these fields. The issue of committee members' using the special fund was already apparent in 1923: the newly available research money obtainable from the committee's special fund had so far gone to Carlheim-Gyllensköld's study of secular variations in the earth's magnetic field[51] and Granqvist's solar radiation study. Although Granqvist had been working on measuring solar radiation, he used as a motivation for the grant the need to evaluate the nomination of the American C. G. Abbott for work on the solar constant.[52] Abbott was certainly not a major candidate, having been proposed for a divided prize by C. D. Walcott. Although the size of these grants was small, they pointed to a problem for the future: a committee member could more easily request a grant from the special fund and could better justify the need for a department of the forthcoming Nobel Institute, if nominations were at hand that could be said to require further evaluative investigation.

Awarding a Nobel Prize to Hale and Deslandres meant that astrophysics would be recognized as a legitimate part of physics, especially within the Academy. Oseen and Siegbahn had not been on the committee when Hale and Deslandres previously had been declared worthy of a Nobel Prize; they could with little difficulty try to reverse the evaluative record. Arrhenius had previously supported astro- and cosmical physics. His lack of enthusiasm for a relevant department in the planned Nobel Institute can be attributed both to his skepticism whether such a department under the leadership of Hasselberg or Carlheim-Gyllensköld would be a vital research institution and to his concern about money. During the war Arrhenius became increasingly obsessed with how the Academy and the Nobel Foundation's funds were used. In part he feared for his Nobel Institute for Physical Chemistry, and in part he worried about the ability of

Swedish researchers to partake in scientific-cultural activities. During the postwar economic crisis his institute came to a virtual standstill because of lack of funds. Arrhenius regarded the Academy's economic condition to be "quite terrible," which in turn prompted him to declare: "The prospects for the Academy's future are therefore quite hopeless."[53] As a means to improve local research capabilities, he often sought in the early and mid-1920s to reserve the prizes in order eventually to divert the money to the special fund. He, like Oseen and Siegbahn, understood that to use the few resources available to them for furthering their research areas and institutes, they had to try to restrict the number of specialties competing for access to these resources.

Having been associated with cosmical physics, Arrhenius could use his authority to declare as part of the committee's record that astrophysics had advanced so rapidly during recent years that it now encompassed astronomy and must be equated with astronomy, a science that Alfred Nobel did not choose to be among those open for a prize.[54] By removing astrophysics from among these potential users of the Academy and the Nobel physics committee's funds and future institutional development, they were trying to strengthen what they considered more significant areas of research.

Oseen's "new era" had begun. Astrophysics was not the only field to be banished from the scope of physics. At the same time the committee's majority also began trying to reduce the influence and number of meteorologists within the Academy's physics section.[55] As long as a majority of the committee could agree on priorities, the veracity of the argument need not be perfect. In general -- but not always -- the authority of the committee's majority could sway the Academy's Physics Section and the full Academy. Prior to 1923 astrophysics had sufficient support within the committee that any attempt to define this field outside of physics would have met with opposition. Indeed, when W. S. Adams was nominated, in 1920, he was not considered a major candidate. When the first draft of the committee report was presented to the committee, some disagreement arose on how to word the reason for not considering Adams worthy of a prize. In the draft his work was referred to as being astronomy and therefore not part of physics; in the final draft the work was left undefined as to whether it belonged to physics and simply dismissed as not having the significance for physics required for a prize.[56] Hasselberg and Carlheim-Gyllensköld probably requested this minor change. In 1923, however, Carlheim-Gyllensköld stood alone.

CONCLUSION AND FURTHER THOUGHTS

Whether Hale and Michelson deserved their respective fates with regard to the Nobel Prize is not the issue here. When a list of major American -- or other national-- candidates is examined, the question of why some were invited to Stockholm and others were not cannot be imputed by using the subsequent development of physics and chemistry to rationalize the choices that were made. One can agree or disagree, as one chooses, one can debate about so-called mistakes as seen in retrospect, but such armchair judgments should not be conflated with the actual thinking and intentions of the committees. Why some Americans received a Nobel Prize and others -- such as G. N. Lewis and R. W. Wood -- did not, cannot be answered simply by appealing to some timeless notion of excellence. Committee members' own changing understanding of physics and chemistry and disciplinary agendas played critical roles in making awards. The internal dynamics of the prize institution, including relations between the committees and the Academy in general, must also be comprehended.[57] This is especially important to recall because in most years no clear mandate was provided by the nominators.

When committee members had definite disciplinary concerns and interests, it was these that generally proved most significant for making decisions and not some impersonal statistical combination of how many nominations, by what category of nominator, or how many years a candidate must wait to be rewarded. Committee members clearly played roles as active actors in the prize-awarding. Hasselberg's role as a committee member proved crucial in ensuring that Michelson and Hale were regarded as serious candidates. Other examples abound. The Svedberg, who had a long standing interest in radiochemistry and nuclear research, took the initiative to nominate H. C. Urey and E. W. Washburn in 1934 for the discovery of heavy hydrogen. He sent the sole nomination for them, evaluated them, and convinced the committee to accept his recommendation for a prize to Urey.[58] At the same time Svedberg sought to award a *chemistry* prize to F. Joliot and I. Joliot-Curie for their dramatic work on artificially induced radioactivity, soliciting nominations on their behalf and sending in a nomination himself (virtually all the nominations for them were for a *physics* prize). I will make the argument elsewhere that Svedberg here was trying to claim nuclear studies as a legitimate part of chemistry, and not exclusively the domain of physics. He soon began considering initiating a Swedish nuclear chemical research program and institution. Both Svedberg and Siegbahn, anxious to acquire nuclear research facilities, courted Berkeley and Rockefeller Foundation connections, which certainly did not diminish the chances for a growing number of American candidates that they might receive prizes. Indeed, during the interwar years as America increasingly became a significant center for scientific research -- and the wealthiest -- Swedish researchers endeavored to improve communication and cooperation with American colleagues. The complex interrelations between these two national communities require further study in this connection. It should be clear, however, that the awarding of the Nobel Prizes, at least during the pre-World War II period, entailed more than a committee's somehow choosing the most deserving physicists and chemists according to fixed, timeless criteria.

NOTES

1. Elisabeth Crawford, *The Beginnings of the Nobel Institution: The Science Prizes, 1901-1915* (Cambridge: Cambridge Univ. Press; Paris: Maison des sciences de l'homme, 1984); Crawford and Robert Marc Friedman, "The Prizes in Physics and Chemistry in the Context of Swedish Science," in *Science, Technology and Society in the Time of Alfred Nobel*, ed. Carl Gustaf Bernhard, Elisabeth Crawford, and Per Sörbom (Oxford: Pergamon, 1982), 311-331; Friedman, "Nobel Physics Prize in Perspective," *Nature*, 1981, *292*: 793-798; Friedman, "Text, Context, and Quicksand: Method and Understanding in the Study of the Nobel Science Prizes," submitted to *Historical Studies in the Physical and Biological Sciences*.

2. Oscar Widman to Eduard Hjelt, 9 Jan. 1897, Hjelt Papers, Riksarkivet, Helsinki.

3. S. Arrhenius to Hjelt, 4 Jan. 1897, 26 Dec. 1898, 18 Jan. 1900, and 8 Jan. 1903, Hjelt Papers; Vilhelm Bjerknes to Fridtjof Nansen, 18 Jan. 1902, Nansen Papers, Universitetsbiblioteket, Oslo.

4. Crawford, *The Beginnings of the Nobel Institution*, Chap. 3. Friedman, "Nobel Physics Prize in Perspective," raises the issue of changing traditions for interpreting the statutes.

5. See, for example, Widman to Hjelt, 25 Nov. 1909, Hjelt Papers; C. W. Oseen to Bjerknes, 14 Dec. 1920 and 6 June 1923, Bjerknes Papers, Universitetsbiblioteket, Oslo; Nobelstiftelsen to KVA, 25 Jan. 1921, Appendix to Gemensamma sammanträde (joint-committee meeting), 1921.

6. Crawford has discussed Gösta Mittag-Leffler's campaign for Henri Poincaré; more modest efforts include Widman for Emile Fischer and Arrhenius, Carl Benedick's efforts for C. T. R. Wilson and H. Le Chatelier, Hans von Euler-Chelpin for several biochemists, and The Svedberg for F. Joliot and I. Joliot-Curie as well as for E. M. McMillian and G. T. Seaborg (See Friedman, "Text, Context, and Quicksand," forthcoming).

7. e.g., Max Planck (1914), Einstein for relativity theory (1921 and 1922), Walther Nernst (1910, 1920) and Otto Stern (1934).

8. Preliminary accounts of the Michelson prize can be found in Crawford, *The Beginnings of the Nobel Insitution*, Crawford and Friedman, "The Prizes in Physics and Chemistry," and Friedman, "Nobel Prize in Perspective." Gerald Holton made perceptive comments related to experimentalist philosophy of science and Michelson's Nobel Prize before the archives were open in *Thematic Origins of Scientific Thought: Kepler to Einstein* (Cambridge: Harvard Univ. Press, 1973), pp. 275-277.

9. KU, fysik, 1904 (Kommittéutlåtande, Nobelkommittén för fysik = committee report, Nobel committee for physics).

10. Ibid.

11. Ibid.

12. B. Hasselberg to G. E. Hale, 5 July 1907, Hale Papers, Niels Bohr Center for History of Physics, American Institute of Physics, New York.

13. Hasselberg to Hale, 29 Dec. 1907, Hale Papers.

14. In comparison E. Rutherford and G. Lippmann received seven nominations each and J. D. van der Waals, six nominations. Rutherford's nominations came from several nations, those for the other two were from their respective home nations.

15. Hasselberg to Hale, 5 July 1907.

16. KU, fysik, 1907.

17. Ibid.

18. Hasselberg to G. Mittag-Leffler, 24 Jan. 1910, Mittag-Leffler Correspondence, M-L Institute, Djursholm.

19. KU, fysik, 1907.

20. My discussion is based on Arvid Leide's detailed study based on archival documents, "Janne Rydberg och hans kamp för professuren," *Kosmos. Fysiska uppsatser*, 1954, *32*: 15-32.

21. *Les Prix Nobel en 1907* (Stockholm, 1909), 13.

22. V. Carlheim-Gyllensköld to Arrhenius, 7 June 1907, Arrhenius Papers, Stockholms universitetsbibliotek.

23. Friedman, "Nobel Physics Prize in Perspective," pp. 794-795.

24. Hale was nominated by W. Hallock, A. A. Michelson, E. F. Nichols, H. C. Parker, M. I. Pupin, F. Tufts, and M. Wolf. The prize went to F. Braun and G. Marconi, based on a nomination by committee member Granqvist; see Crawford, *The Beginnings of the Nobel Institution*, pp. 142-143, on background to the 1909 prize decision.

25. In all likelihood Hasselberg urged this linking. His own close relations with the French metrological and astronomical/astrophysical communities and the lack of any international nominations calling for such a divided prize support this assumption. In 1913, Deslandres was nominated by A. Pérot as a second choice after E. Amagat; all other nominations for Deslandres came from Nobel committee members who proposed a split with Hale.

26. KU, fysik, 1913.

27. Without mentioning the war directly, the committee pointed out the particularly fortuitous opportunity to make such a joint awarding of the 1914 and 1915 prizes.

28. Utkast [draft], KU, Protokoll, Nobelkommitté, fysik, 13 Sept. 1916.

29. A. Gullstrand to Arrhenius, 13 May 1923; Gullstrand to Carlheim-Gyllensköld, 17 May 1923, copies, Gullstrand Papers, Universitetsbibliotek, Uppsala.

30. Gullstrand to Arrhenius, 13 May 1923.

31. Gullstrand to Carlheim-Gyllensköld, 17 May 1923.

32. Ibid. Rutherford had received the Nobel Prize in chemistry in 1908; Svedberg and later Bohr, among others, wanted Rutherford to receive also a prize in physics.

33. Protokoll, NK, fysik, 5 Sept. 1923, Appendix F.

34. Protokoll, NK, fysik, 31 Jan. 1924, Appendix A.

35. Hasselberg to Simon Newcomb, 26 October 1900, Newcomb Papers, Library of Congress, Washington, D.C.

36. Protokoll, KVAs förvaltningsutskott, 8 Oct. 1915, KVA Archive; Protokoll, NK, fysik, 29 Jan. 1916, Appendix A.

37. The Nobel Foundation, ed., *Nobel: The Man and His Prizes*, 2nd edition (Amsterdam, London, New York, 1962), pp. 334-338, 521-523.

38. Hasselberg to Campbell, 24 Dec. 1916.

39. Wilhelm Odelberg, "Vetenskapsakademiens lärdomshistoriska samlingar," *Svenska museer*, 1962, *2*: 4-14.

40. KU, fysik, 1915, Appendix 3. Cosmical physics was not a clearly defined specialty; definitions varied.

41. Hasselberg, Arrhenius, and R. Åkerman to KVAs förvaltningsutskott, January 1916, copy in Protokoll, NK, fysik, 1916.

42. Protokoll, NK, fysik, 18 May and 21 Sept. 1916; Arrhenius to Bjerknes, 12 May 1916, Bjerknes Papers.

43. Arrhenius to Bjerknes, 12 May 1916.

44. Ibid.

45. Protokoll, NK, fysik, 24 March 1917, Appendix B., Granqvist, as chairman, writing on behalf of the Nobel Committee for Physics to the Academy's Administrative Committee.

46. Oseen to Bjerknes, 12 Nov. 1920, Bjerknes Papers; Oseen to Mittag-Leffler, 17 Oct. and 15 Nov. 1918, Mittag-Leffler Correspondence.

47. Oseen to Bjerknes, 3 Feb. 1923, Bjerknes Papers.

48. Ibid. Oseen to Bjerknes, 3 Feb. 1923, Bjerknes Papers.

49. M. Siegbahn to C. Aurivillius, 9 Dec. 1918, KVA Collection, Stockholm universitetsbibliotek; on difficulties in publishing, see also Arrhenius to A. Rindell, 22 Oct. 1922, Rindell Papers, Åbo akademi, Turku; Oseen to Benedicks, 25 March 1914, Benedicks Papers, Kungl. Biblioteket, Stockholm.

50. See Friedman, "Nobel Physics Prize in Perspective," pp. 794-796; and "Text, Context, and Quicksand" (forthcoming).

51. Protokoll, NK, fysik, 6 Dec. 1922.

52. Protokoll, NK, fysik, 6 March 1922.

53. Arrhenius to A. Rindell, 1 April 1923, Rindell Papers; see also Arrhenius to O. Aschan, 27 Oct. 1922, Aschan Papers, Åbo akademi; Arrhenius to Hjelt, 7 March 1916, Hjelt Papers.

54. Protokoll, NK, fysik, 5 Sept. 23, Appendix F.

55. Friedman, "Nobel Physics Prize in Perspective," pp. 796-798.

56. Utkast, KU, fysik, 8 Sept. and 22 Sept. 1920.

57. To analyze the committees' actions it is not sufficient to read the text of the evaluative reports; these have been written to justify the committee choices to the Academy. Processes by which consensus was formed need not at all be apparent from the text. An understanding of the rhetorical traditions used in the evaluative reports within a

historical context is necessary to reconstruct a wide range of meanings and intentions.

58. Washburn had died in the meantime. Svedberg considered proposing dividing the prize with G. N. Lewis, but felt that Lewis's work with heavy hydrogen was considerably less significant than Urey's and being eclipsed in importance by Hugh Taylor's Princeton group. I will be analyzing the evaluations of G. N. Lewis's different contributions to chemistry in a subsequent article.

PHYSICS AND ENGINEERING IN THE UNITED STATES, 1945-1965, A STUDY OF PRIDE AND PREJUDICE[*]

Nathan Reingold

This paper will not deal with aspects of a so-called "Michelson Age in American Science," simply because I do not believe in such an epoch. The term probably arises from the parochialism of at least some historians of physics who apparently identify their field with all of science. Although Michelson was a great experimental physicist, his work constituted neither the highpoint nor the model of scientific practice in the United States from 1870 to 1930. Opinions may differ, but the highpoints in my view are the theoretical contributions of J. Willard Gibbs, the classical genetics of Thomas Hunt Morgan and his school, and the investigations of Henry Norris Russell, the astrophysicist.

My intention is to use Michelson as a jumping off point for a consideration of the attitudes, contrasting and related, of U.S. physicists and engineers in the period roughly from World War II to the early stages of the Vietnam adventure. Specifically, this essay gives selected incidents illustrating how the two professional groups viewed their relations to one another, their attitudes to basic or fundamental research, and how these matters worked out in certain institutional policies. A few cases do not necessarily constitute proof. Nevertheless, what will be discussed are important, influential instances; many more of a similar nature can be cited. This is a sketch on how professional groups define themselves. Events at Stanford University will loom large in what follows.

Although recognized in America as a great experimental physicist even before receiving the Nobel Prize, during his lifetime Michelson did not dominate as the exemplar of pure research in public discourse. The most visible public push for pure research, associated with George Ellery Hale and Robert Millikan, invoked other names, notably Michael Faraday.[1] A cynic might conclude that a public exemplar of pure research had to meet two conditions: first, a dead exemplar was preferable, one who could not expound views at variance with a stated ideology of purity of research in the sense of total absorption with extending abstract knowledge. Throughout much of his career, Faraday, a notable contributor to theory, had intense involvements with applications of physics and chemistry. Second, the dead exemplar had to have an asserted linkage, no matter how remote, to contemporary favored technological activities. Electrical technology was loosely credited as resting upon Faraday's experiments. Similarly, while alive, Einstein achieved great public visibility but not as the cited exemplar of pure research. He did, after all, patent a refrigeration system with Leo Szilard. Einstein's findings had no obvious utility until Hiroshima. After his death, when no longer able to speak uncomfortable truths on public policy, Einstein emerged as a paragon, a model for emulation, a proponent of an ideology of pure research. By that time, the long dead Michelson had acquired a place in the pantheon of the same ideology.

During his lifetime Michelson's influence in this regard was quite limited. Even his role as a president of the National Academy of Sciences did not enable him to halt Millikan's move to elect Thomas Alva Edison to membership in 1927.[2] Much earlier, Michelson contributed a striking phrase to the literature when urging his student Frank Baldwin Jewett not to join what later became the Bell Telephone Laboratories -- "prostituted physics." Disregarding Michelson, Jewett had a spectacularly successful career as an administrator of industrial research. But Jewett disclosed his teacher's

*The support of the Lounsbery Foundation and the Smithsonian's Scholarly Studies Program is gratefully acknowledged.

advice during a New Deal investigation.[3] The charge obviously rankled. Michelson's words implied a belief both in the nature of physics and in the role of the unprostituted physicist.

During World War II and immediately afterwards, Jewett served as president of the National Academy of Sciences. Although he was the only scientist to testify against increasing federal support of research and development in the immediate postwar science policy hearings, Jewett is clearly one of the architects of the new dispensation emerging out of the wartime experiences.[4] On two occasions, in 1944 and in 1947, he reflected on the lessons of a lifetime. In the first, during a National Research Council conference on restoring international scientific cooperation, Jewett remembered that more than forty years previously "Mr. Michelson told me in no uncertain terms that I was a renegade [i.e. for going to industry] and some fifteen years afterwards he made generous amends for that." Speaking to the meeting as a representative of industrial science, Jewett remembered that years before, men from industry were not welcomed at scientific societies and had little part in their activities. "Today," he continued, "we all participate," and he cited his presidency of the academy as proof of U.S. uniqueness in this respect. Men in industry had "the same aspirations and the same willingness to serve as men who were devoting their time to fundamental science." Although much of Bell Telephone Laboratory's work was engineering, Jewett made no distinction between the scientists and the engineers in industry.[5] As Reich also recently noted, in the Bell Telephone Laboratory it was often difficult to distinguish science from technology.[6]

In 1947 Jewett gave a farewell address to the academy dwelling on its intrinsic nature and its future needs. He emphasized the statutory obligation to serve the nation. For this, he urged a wider membership, presumably including more scientists and engineers from industry. Anticipating charges of biased background, Jewett made an avowal of faith:[7]

> In justification of my point of view I can only say three things, viz.:
>
> 1. While most of my creative life has been devoted to technology and applied science, my main interest has been in fundamental science and its advancement.
>
> 2. Although I think I have made contributions to both fundamental and applied science and might have made more, the conditions under which I have worked made it impossible to attach my name to anything specific. To have sought to do so would have destroyed my ability to achieve an objective which involved many men.
>
> 3. My point of view is an honest one.

Michelson's student disclosed a closeness between fundamental science and applied research and development at variance with the moments of tension, public and private, upon which so much attention is lavished in the literature. The moments of tension are important but best understood within particular contexts -- institutional, intellectual, ideological, and, sometimes, the personal. From such contexts richer and more convincing general viewpoints will emerge.

Certain contexts involving relations of engineering with basic science advances and with basic scientists raise important questions needing further exploration. Three such contexts are the large-scale research and development programs in defense and space; the attempts of engineering fields to develop a presence in the spectrum of basic research;

and what was happening in engineering education in the decade after World War II.

Consider, for example, the attitudes of Rear Admiral Julius Augustus Furer, the Navy's Coordinator of Research and Development in World War II. Disturbed by the views of James Forrestal and Rear Admiral Harold Bowen, slated to succeed him in 1945, Furer noted in his diary:

> . . . they [Forrestal and Bowen] do not realize that they can't make a scientist think by giving him an order. They don't realize that scientists are essentially creative artists, and that the creative thinker never produces anything unless he gets interested . . . Forrestal's attack of the whole problem has been that you can accomplish anything by giving orders and that the profit motive is as potent with scientists as it is with businessmen . . . [8]

Furer was an EDO (Engineering Duty Only) officer who got on well with Vannevar Bush and resented Admiral Ernest King's exclusion of EDO's from his staff as chief of Naval Operations. Furer worked for the Navy but, like other technical officers in the armed services, had attitudes reflecting his background that influenced policies and programs after 1945.

The space program was and is overwhelmingly dominated by engineering and by engineers but has had a science component from its inception. The relationship of the science to the engineering remains an important and contentious issue in space policy. NASA's first head, T. Keith Glennon, an engineer, noted of scientists in his diary on 5 May 1960 that "these are good people and their voices must be heard," after observing that "the scientific community, as such, is a bunch of spoiled individuals -- the higher they rise in the hierarchy, the more spoiled they become." By 8 July, he could write that "we can repair this dis-affection by supporting the individual members of the so-called community in the manner to which they would like to become accustomed. Unfortunately, we don't have enough money for this."[9] And his successor, the attorney James Webb, would reflect the high status of science in the era by averring that loss of the race to the moon could be partially offset by scientific triumphs.[10] Presumably, this was a reason why scientific voices had to be heard: The status of science, however defined, reflected popular and administrative perceptions of the lessons of World War II derived from the experiences of the Office of Scientific Research and Development and the Manhattan Project.

One legacy of the OSRD experience, the National Science Foundation, early provided a bittersweet experience for engineers. During the fashioning of *Science, the Endless Frontier*, Vannevar Bush had objected to the omission of engineering research from the contemplated agency by one of his advisory committees.[11] Bush was a notable electrical engineer and engineering educator; his influence made engineering one of the areas covered by the new agency, originally lodged in a unit designated "MPE" -- Mathematics, Physical and Engineering Sciences. Very soon after NSF began operating, professional engineers became conscious of their poor showing in competition for grants against the mathematicians, the physicists, and the chemists, particularly the first. Working against engineering applicants, among other factors, was the sense that NSF was the preserve of basic science.[12] By the standards of some, engineering simply did not have a significant distinctive basic component.

In 1955 Frederick Terman, an electrical engineer and provost of Stanford, joined the MPE Advisory Committee. Terman had done his dissertation under Bush at MIT. Immediately, Terman and others had to consider just what was basic engineering science. Ordinary engineering, they concluded, is "an application where cost considerations are involved and which meets a need." Basic engineering, in contrast is "of general applicability (as distinct from one process or leads to better understanding...) Thus basic is something that does not change with time."[13] Such ruminations did little to change the situation.

Basic scientists, not only physicists, had great hostility to the presence of engineering in NSF. In 1957, H. J. Muller, the Nobel Prize geneticist, responded with alarm to the perceived "threat to the concept the NSF emphasizes basic science" of the appointment of a geophysicist from industry to head MPE. Later that year Muller stated his ideological position on the relation of theory and practice, a position by no means unique to him:[14]

> the applied branches . . . although they should be closely coordinated with the basic science Department, cannot effectively be combined under it without swamping of developments in basic science itself, that feeds them best when it forgets them.

Still another blow to professional engineering occurred about the same time Muller wrote. Eisenhower's first science advisor, J. R. Killian, Jr., discovered with alarm the existence of an appreciable drop in undergraduate engineering enrollment. Right after World War II, Bush and leaders of professional engineering education moved to increase the amount of science and mathematics in the undergraduate curriculum, as well as to increase graduate work in the various engineering sciences. They believed that the World War II experience demonstrated the applicability of abstruse parts of mathematics and the physical sciences, and that such applications had increasingly fallen into the hands of scientists, especially the physicists, an outcome that disturbed them. For a greater role in R and D and because various industrial sectors had incorporated formerly abstruse knowledge, engineers also had to get suitable graduate training. Resulting curricular reforms actually continued trends antedating World War II.

Why the drop in enrollment? In Killian's office various explanations appeared: (1) the boredom of routine engineering (i.e. being "chained" to the drafting table); (2) upgrading the curriculum, which scared away marginal students; (3) the lure of business schools and the M.B.A. degree; (4) the glamorous aura of science; (5) the trend to increased graduate work. Noted only in passing was the incongruity of calling for more students while defense cuts resulted in engineers being laid off by industry.[15]

What underlay all these events before the Space Age and heavier defense budgets came to engineering's rescue? What occurred were the visible signs of a long-term shift in engineering. Earlier it had been a mass profession largely based upon training to the bachelor's level, members of which typically manned posts of a restricted professional scope (or even outside of professional practice to some extent). Although still a numerous profession, it now required more formal education of an abstruse nature aimed at a different vocational division (e.g., more R and D and less technical sales).

Engineering as a profession continued an old struggle to encompass all of technology. Formerly, engineering had to deal with inventors, empirics, craftsmen. Now, engineering had to adjust to an upsurge of practitioners of technology deriving from the physical sciences and mathematics. During the same period, engineers experienced an apparent squeezing out of the higher reaches of the corporate world by lawyers and financial experts. Layoffs in industry raised the specter of unionization with implications of a loss of professional status.

When some in Congress reacted to Sputnik by proposing the establishment of a Department of Science and Technology, the official representatives of the engineering community objected only to the use of "Technology." They urged the substitution of "Engineering." The former term, they insisted, related only to aspects of industry. (Jewett, of course, would consider that both proper and adequate.) When the senators politely pointed out that engineering did not include applications in agriculture, medicine, and forestry, the engineers persisted. Engineering had to have a basic component and also primacy in meeting human material needs.[16]

During the same period, the devotees of basic research had a different set of concerns. The case of the Stanford Physics Department is instructive. From a relatively modest position at the start of World War II, physics in Stanford after 1945 expanded in terms of both quality and quantity. Prior to the war, the department chairman, David Locke Webster, had hired Felix Bloch, a theoretician of Swiss citizenship, whose Jewish extraction had ended his career in the German university world. The other outstanding member of the department was W. W. Hansen. Before the outbreak of the conflict, Hansen had collaborated with the Varian brothers on improving what later became known as the klystron, a device for producing very high frequency currents of electrons. Stanford University acquired a share of the patent rights. During the war Hansen worked for Sperry. Bloch did research for the Manhattan Project in its early stages but left after a brief period at Los Alamos. He went to Cambridge to join Terman at OSRD's Radio Research Laboratory, which investigated radar countermeasures. Here he developed knowledge and skills ultimately contributing to his Nobel Prize work.

In 1944 Hansen, supported by Bloch and by Terman, successfully proposed the establishment by Stanford of a microwave laboratory, now bearing his name. The move was pushed in the absence of Webster. His opposition to an increased emphasis on applied work, added to the objections of others, might have impeded the move. Hansen died in 1949 at a relatively young age. By then the work he had launched began to yield results, ideas, and ambitions leading to the construction in the late 1950s of the Stanford Linear Accelerator (SLAC), one of the big machines of American particle physics.[17]

Assuming the chairmanship of the department in 1948 was Leonard Schiff, a notable quantum theorist. In effect Schiff was the heir to the earlier initiatives of Hansen, Webster, Bloch, and Terman. Although the department clearly had attained a high status, Schiff had problems, some regarded as quite serious by him and his colleagues at the time. In 1949, for example, a conference at Stanford on "high powered klystrons and related power devices" was sponsored by the Microwave Laboratory and the Office of Naval Research's (ONR) Panel on Electron Tubes. The department had no official connection, although three Stanford physicists attended. The department used the occasion to affirm its opposition to any restricted research.[18]

In 1955 Schiff complained that the Stanford Research Institute (SRI) had a program cutting across all fields. Schiff objected to any physics at Stanford not being conducted within the department, although he graciously conceded that similar arrangements were all right for chemistry.[19] Founded in 1946 by the university, SRI did classified and applied research supposedly unsuitable for the regular academic departments. Schiff and Bloch clearly faced in the university the situation implicit in Jewett's words: the blurring of distinctions between basic and applied work.

By 1959 Bloch and Schiff had concerns about the relationship of departmental appointments with those for SLAC and for applied physics research, specifically joint appointments at the assistant professor level:

> It will be considered by the administration as a breach in our wall
> of principles through which Terman can hope to see his fond dreams
> of unprincipled opportunism come closer to fulfillment.[20]

To continue with these selected incidents: In 1962 Terman was perturbed that Arthur Schawlow was not listed by the department, to minimize the association with applied physics.[21] (Schawlow, a future Nobel laureate in physics, succeeded Schiff as chair in 1966.) In the same year, 1962, Robert Hofstadter, another Stanford physics Nobelist, in a public address, attacked the growth of research organizations in universities but supported by outside organizations with narrower purposes than academic departments:

> Though bigness is the essence of the research organization, the thoughts they produce fall into various small corners of the universities' range of knowledge and interest.

Hofstadter preferred stressing individual thinkers even in the presence of massive devices and large administrative organizations.[22] The theme recurs in correspondence in the Schiff papers. To sum up, in this highly visible physics department serious concern existed about the trend to applied research; the danger of neglecting the full range of physics as a discipline; the impact of secrecy; and the increasing role of non-departmental organizations in research.

On the national level, many leading scientists and such bodies as the National Academy of Sciences maintained a state of alert concern for basic research. Unlike physicists in Stanford and elsewhere who did exult in increased funding and big machines, unlike those engineers who were so impressed with the new status of basic research after World War II, many leaders of the scientific community were apprehensive, if not pessimistic, about the national situation. In 1958, when only 6 percent of federal funds went to the support of basic research, George Kistiakowsky, soon to be Eisenhower's second science advisor, protested allowing the reclassification of industrial defense contracts from applied to basic. In his view, little basic research occurred in industry. Since real basic research was in danger, the reclassification could produce a reaction against all such investigations.[23] In 1965 the federal share for basic research had reached 10 percent. To Harvey Brooks, that figure was swollen by recent DOD reclassifications of its expenditures and other bookkeeping devices.[24] Federal support of basic research constituted a minute portion of the R and D budget at best. Knowledgeable scientists believed the published levels were inflated They had little sympathy with engineers and others who assumed the hegemony of basic or pure research in national science policy.

A more fundamental problem concerned at least some adherents of basic research. Bookkeeping classifications are open to revision. An anxiety prevailed that much else passing as basic research did not really merit that appellation, particularly in the universities. In 1959, for example, the Internal Revenue Service, with the tacit blessing of the National Science Foundation, considered removing the tax exemption from the nonbasic research and development activities of universities and other nonprofit institutions like SRI and the Mellon Institute. In assuming managerial roles for facilities for Defense and other agencies and in their increasing involvement with applied research and development, the universities, some argued, were abusing tax exemptions predicated upon education and the advancement of knowledge defined as basic research.[25] But if this doubt of the university's dedication to basic research was, as I suspect, a minority view, the minority included influential individuals whose views resonated with those of other individuals concerned over the low level of federal funding for basic research and the heavy pressure for applied research and development.

One such influential individual was Warren Weaver, a mathematician who headed the Natural Sciences Division in the Rockefeller Foundation. For many years Weaver chaired the ONR advisory committee. Using funds from the Sloan Foundation (to which he moved on 1 August 1959), Weaver chaired the arrangements committee and coordinated the program for a Symposium on Basic Research held in New York City on 14-16 May 1959.[26] It was an event meant to make a splash, to influence policy makers and key administrators. The president of the United States himself addressed the dinner session with warm words for basic research. Except for one talk by the physicist Merle A. Tuve of the Carnegie institution of Washington,[27] the papers are bland (and sometimes banal) recitals of belief in virtue (basic research) and assertions of the need and intention to provide for more.

Weaver authored a preamble to the invitations asserting the inadequacy of the nation's support of basic research. He stressed the large support of applied research and the "massive present support of development," and the "lip service" to basic research of industry. As to the universities, Weaver wrote:[28]

> Are not universities so deeply invaded by the demands for solving immediate problems and by the temptation of income for so doing, that there are all too few cases of competent scholars pondering about problems simply because it interests them to do so? . . .

Accompanying his draft of the preamble in the planning stage was a general description of the program in which he asserted that the best men in the "Bell Labs . . . are almost wholly free (in various ways more effectively free than professors are) to do just what they want to do."[29] (Were he alive, Jewett would have applauded.)

Weaver opened up the symposium by defining basic research as "the never-ending search for better understanding of man himself and of the total world, animate and inanimate, in which he lives ... stimulate[d] [by] insatiable curiosity ... and the intellectual and aesthetic satisfaction that comes from understanding." Among other issues raised by Weaver, but not in the printed proceedings, was, "Are we, as a nation, mature enough to give general support to a philosophy of favoring the elite few?"[30]

Never one to pull punches, Tuve declared that most of the research supported as basic was actually efforts to generate data for technology. To his mind, that also included much of the work funded by the National Institutes of Health. Real basic research, to Tuve, consisted of individuals dealing with ideas. Tuve favored research professorships and research institutions like the Carnegie. Tuve, Weaver, and others exalted the motivation of the research and the freedom to investigate as one saw fit.[31] As one attendee later wrote, "it is a question of how one does the work rather than what one does."[32] Tuve struck a responsive chord in an era when many did not like being told that this was the age of "big science," and, later, that most basic research was "normal science." The heavy role of national security in funding went largely unmentioned by the speakers. Nor was there any real consideration of the proliferation of research organizations (now known as Federally Funded Research and Development Centers) administered by industry, universities, and other nonprofit entities. The microbiologist René Dubos was in the audience and later wrote Tuve of his emotional reaction: "it became painfully clear that I was slowly drifting into a type of para-scientific activity often of a trivial nature . . . I merely wish to thank you for having helped me to save my soul."[33]

Although universities certainly figured in the program, one could argue that the emphasis overall was on increasing the participation of industry and of foundations in the support of basic research. The Sloan Foundation's efforts were cited as exemplary. Tuve and Weaver did not expect much more from the federal patron, so obviously prone to the applied and development end of the spectrum. But, as late as 1965, in contrast to the calls for more basic research, such diverse individuals as Harvey Brooks, Arthur Kantrowitz, and Edward Teller would assert that the dubious 10 percent for basic or pure research so dominated that applied science and engineering suffered from inadequate leadership.[34] That may only have continued the situation Jewett complained of earlier in the century, a second-class status of "industrial science"[35] that a historian of technology recently traced to the "elitist" views of George Ellery Hale and Robert A. Millikan, both associates of Albert Michelson.

One very vigorous and effective leader of engineering, Vannevar Bush's student F. E. Terman, did not attend the symposium. Nor did Bush, who had read an advance copy of Tuve's remarks. Bush had reservations about the "kind of discussion" slated for the symposium.[36] Terman often acknowledged his indebtedness to Bush. He even consulted Bush on Stanford affairs during the postwar years. In 1964, after reading a Bush statement in *Science*, Terman was moved to write to his teacher a statement of faith, an affirmation of his fidelity to Bush's views.[37]

In the *Science* article, taken from his remarks at a Congressional hearing, Bush argued against big crash programs because "great scientific steps forward originate in the minds of gifted scientists, not in the minds of promoters." He criticized the universities for assuming the management of secret defense programs:

> It should never be forgotten that the main task of the universities is to educate men .... Every research program placed in a university should be so ordered that its product is not only new knowledge but skilled educated men.

Terman averred he had turned down the opportunity to manage a large secret program; that the School of Engineering at Stanford now had 1,300 graduate students and produced 90 doctorates annually; and that the engineering research program consisted of "many grants reflecting the interest of individual faculty members." His avowal placed Terman fairly close to Tuve and Weaver. And in spite of clashes and differences, Terman's record at Stanford shows much in common with the views of physicists like Schiff and Bloch.

Like Schiff, Terman disliked SRI's increasing involvement with basic research from 1955.[38] SRI competed with Stanford while not contributing to graduate education. By 1964-65, the relations between the university and its offspring were badly strained. In 1966 the effects of the Vietnam War became acute. In that year Terman noted that since 1948 Stanford had granted doctorates for 30 classified theses.[39] Eventually Stanford transferred all its classified research to SRI, which became independent of the university in 1970.

Not that Terman was against defense contracts. Stanford had the same problems, on a smaller scale, that had afflicted MIT in 1966. J. A. Stratton observed in a conversation with Terman that Vietnam illustrated a poor use of science. And Stratton pointedly stated that the secret labs in MIT resulted from faculty desires.[40] Yet, Terman made clear where he stood. There was no apparent incompatibility with his role as a promoter of what we now call Silicon Valley. There was no anguishing over whether or not technology was applied science. Terman stood, as it were, with the Michelsons of the world, the university investigator seeking truth, "something that does not change with time."

Despite differences, Terman was not a polar opposite to physicists such as Leonard Schiff. Both exalted the university. Both placed great faith in the individual investigator. Both subscribed, albeit in differing ways, to an ideology stressing the essentiality of basic research. One can speculate that both were enthralled in that period by a nationalistic hubris acting as a means of evading reality. Both were reacting, as it were, to Jewett's perception of the erosion of the distinction between industrial and university scientists. Perhaps Schiff had an easier time dealing with Terman than our stereotypes might indicate. During the thirties, Schiff received his undergraduate degree from Ohio State, a B.S. in engineering physics. Nineteen thirty-three was a very good year for engineering physics at Ohio State. Schiff's classmate William A. Fowler would later receive the Nobel Prize in physics. Despite harsh words in moments of tension, contrary to the beliefs of some historians about the relations of technology with science, in post-World War II America, there were engineers and physicists who perceived their fields as occupying differing positions on common spectrums, intellectual and institutional.

NOTES

1. Discussed in N. Reingold, "The Case of the Disappearing Laboratory," *American Quarterly* 1977, *29*: 79-101.

2. R. C. Cochrane, *The National Academy of Sciences* . . . (Washington, 1978), pp. 283-284.

3. The incident is discussed in S. Weart, "The Rise of Prostituted Physics," *Nature* 1976, *262*: 13-17.

4. For Jewett's role, see N. Reingold, "Vannevar Bush's New Deal for Research ...," *HSPS* 1987, *17*: 299-344.

5. National Research Council, "Transcript of Conference on Problems of Restoring International Scientific and Scholarly Cooperation, December 13, 1944," 53f. Copy in D. W. Bronk Papers, Rockefeller Archive Center.

6. L. Reich, *The Making of American Industrial Research* . . . (Cambridge, UK: 1985), p. 2O6.

7. Jewett's speech was read before the Academy on 17 November 1947. Subsequently, it appeared in the *Proceedings of the National Academy of Sciences*, 1962, *48*: 481-90. I have used the text in the Archives of NAS. The quotation is from p. 6.

8. Entry of 29 May 1945, Furer Papers, Library of Congress. For Admiral King, see the entry of 7 Aug. 1942.

9. I have used the copy of the Glennon diary in the Eisenhower Library, Abilene, Kansas.

10. Webb to V. Kennedy, 16 May 1966; to Lyndon Johnson, 23 May 1960; and to E. F. Buryan, 18 July 1961, Webb Papers, Truman Library, Independence, Missouri.

11. Memo, V. Bush to Carroll L. Wilson, 15 January 1945, Series 2, Box 3, OSRD records, RG 227.

12. E. Hille to T. K. Sherwood, 29 Oct. 1955, and Sherwood to Hille, 8 Nov. 1955, box 2, John von Neumann papers, Library of Congress. I am indebted to M. McMahan's paper, "Engineering Education as 'Best Practice': Historical Reflections on the Crisis," delivered at the National Research Council's Conference on Engineering Interactions with Society, 19-21 July 1983.

13. Based on materials in box 10, series VI, Terman Papers, Stanford University Archives, especially note of 25 Aug. 1955 conversation with G. H. Hickox.

14. G. Beadle memo to Muller, 3 July 1957, and Muller to V. Cohen, 18 Oct. 1957, box 21, Muller Papers, Lilly Library, Indiana University. J. M. England, *A Patron for Pure Science* (Washington, 1982), pp. 214, 275.

15. Based on materials in box 8 in the Education Panel 4, 1957-60, Records of the Office of the Special Assistant for Science, Eisenhower Library, Abilene, Kansas. See especially, R. M. Biber memo and attachments of 20 Feb. 1959 and M. P. O'Brian to Killian, 30 Jan. 1959.

16. Senate Committee on Government Operations. *Hearings . . . to Create a Department of Science and Technology*, 16 and 17 April 1959 (Washington, 1959), pp. 8-16.

17. This is a very compressed account largely based on materials in Hansen's papers in the Stanford University Archives. I am also indebted to S. W. Leslie's paper "Redefining Post-War Science and Engineering at Stanford," presented at a conference at Johns Hopkins, 17-18 April 1986.

18. Memo of 21 Sept. 1949, Schiff Papers, Stanford University Archives. The copy does not indicate the recipient of the original.

19. Schiff memo to Terman, 26 May 1955, box 54, series III, Terman Papers, Stanford University Archives.

20. F. Bloch to Schiff, 21 Oct. 1959. Schiff Papers, Stanford Archives.

21. Terman memo to A. Bowker, 23 Oct. 1962, in Series III, box 3, folder 6. Terman favored joint appointments to prevent the physicists' insulating themselves from applied physics and to enable the applied physicists to have a voice in the Physics Department. In 1963-64, Applied Physics became a division in the Physics Department. In 1965-66 a separate Department of Applied Physics came into existence.

22. The text of this commencement address at Hofstadter's alma mater, CCNY, "The Free Spirit in the Free University" is in folder 4 of box 31, series III, Terman Papers, Stanford Archives.

23. Kistiakowsky memo to Killian, 10 Oct. 1958, box 6. Special Assistant for Science records, Eisenhower Library.

24. *Basic Research and National Goals* (Washington, 1965), pp. 79, 308-9. The latest volume of *Science Indicators* for 1985, pp. 36-37 and tables 2-6 and 2-3, has a 12.5 percent for 1985 share for basic research in current dollars. In constant 1972 dollars the figure is 5 percent. In relative terms the situation has not appreciably changed since 1958.

25. R. Adams to W. Weaver, 1 April 1959, Symposium on Basic Research file in NAS-NRC Central Policy Files 1957-1961, Archives, National Academy of Sciences.

26. Dael Wolfle, ed., *Symposium on Basic Research* (Washington, 1959).

27. Tuve's paper appears on pp. 169-184 of the published proceedings. Tuve's files on the symposium, very helpful indeed, are in boxes 185 and 187 of his papers in the Library of Congress. The latter has interesting reactions to the talk in print and in correspondence.

28. Weaver's preface to the printed proceedings, xi-xv, contains the text of the Preamble.

29. Enclosed in Weaver to Program Committee, 30 Dec. 1958, Symposium on Basic Research file, NAS archives.

30. "Basic Research," in box 2, folder 1 of A. Hollaender Papers. (MS 1105), in the Archives of Radiation Biology, in the University of Tennessee, Knoxville, Tenn. Hollaender was head of biology at the Oak Ridge National Laboratory. The comment on "the elite few" is from the description cited in fn 29, p. 4.

31. There is a distinction here between the modifiers "pure" and "basic" for research. For a discussion of this distinction, see N. Reingold, "American Indifference to Basic Research: a Reappraisal," in G. H. Daniels, ed., *Nineteenth-Century American Science: A Reappraisal* (Evanston, 1972), pp. 38-62. Tuve was challenging the tendency to equate "pure" and "basic," in this century. A basic point of this essay is that *all* such terms used to categorize aspects of R and D simply do not match the institutional and intellectual realities.

32. A. Hollaender to Weaver, 25 May 1959, Hollaender Papers.

33. René J. Dubos to Weaver, 28 May 1959, box 187, Tuve Papers.

34. *Basic Research and National Goals*, 15, 105.

35. George Wise, *Willis R. Whitney, General Electric, and its Origins of U.S. Industrial Research* (New York: 1985), p. 169.

36. Bush to Tuve, 7 April 1959, box 187, Tuve Papers.

37. "Vannevar Bush Speaks," *Science* n.s. 142 (27 Dec. 1963) 1623. Terman to Bush, 3 Jan. 1964, folder 1, box 1, series XIII, Terman Papers, Stanford Archives. The quotes that follow are from these sources.

38. See boxes 52-56 of series III, Terman Papers, for the many documents bearing on this.

39. Based upon folder 8, box 10, series III of Terman Papers, which deals with classified research 1966-70, particularly 1966.

40. Notes of talks with Stratton [1966] in *idem*. See Dorothy Nelkin, *The University and Military Research: Moral Politics at M.I.T.* (Ithaca, 1972).

LIST OF CONTRIBUTORS

James E. Beichler is a candidate for the Ph.D. in the Program in the History of Science and Technology at the University of Maryland. He is currently teaching in the University of Maryland's European Division.

Jed Z. Buchwald is Associate Professor at the Institute for History and Philosophy of Science and Technology at the University of Toronto. His latest book is entitled *The Rise of the Wave Theory of Light.*

Elisabeth Crawford is Senior Research Fellow with the Centre National de la Recherche Scientifique in Paris. She has published several works on the early history of the Nobel Prizes and is currently doing research on international science, 1880-1939.

Chris J. Evans is Visiting Professor at the University of Wisconsin and a guest researcher at the Precision Engineering Division at the National Bureau of Standards in Gaithersburg, Maryland.

Robert Marc Friedman is temporarily living and working in Scandinavia while writing a book on the Nobel Prizes in physics and chemistry, 1915-1939. Cornell University Press will be publishing his *Appropriating the Weather: Vilhelm Bjerknes and the Construction of Modern Meteorology.*

Stanley Goldberg is Visiting Professor of History at the University of Maryland, Baltimore County, and consultant to the Department of the History of Science and Technology at the Smithsonian Institution's National Museum of American History.

Barbara Haubold is at the Central Institute for Astrophysics of the Academy of Sciences of the German Democratic Republic at Potsdam.

Hans J. Haubold is Professor at the Central Institute for Astrophysics of the Academy of Sciences of the German Democratic Republic at Potsdam.

Gerald Holton is Malinckrodt Professor of Physics and Professor of the History of Science at Harvard University.

Daniel J. Kevles is the J. O. and Juliette Koepfli Professor of Humanities at the California Institute of Technology.

William A. Koelsch is Professor of History and Geography and the University Historian at Clark University.

Ole Knudsen is Associate Professor in the History of Science Department at the University of Aarhus, in Aarhus, Denmark.

Edwin T. Layton is Professor of the History of Science and Technology in the Department of Mechanical Engineering at the University of Minnesota in Minneapolis.

Horst Melcher is Professor in the Mathematics and Physics Department of the Teaching College in Erfurt in the German Democratic Republic.

John Michel is a candidate for the Ph.D. in the Department of the History of Science at the University of Wisconsin at Madison.

Albert E. Moyer is Associate Professor of the History of Science at Virginia Polytechnic Institute and State University, in Blacksburg, Virginia. His special interest is the development of American physics.

Nancy J. Nersessian is Assistant Professor in the Program in the History of Science and the Department of History at Princeton University. She is the author of *Faraday to Einstein: Constructing Meaning in Scientific Theories* and is editor of *Selected Works of H. A. Lorentz*.

Kathryn Olesko is Assistant Professor of History at Georgetown University. Her book *Physics as a Calling: Discipline and Practice in the Königsberg Seminar for Physics* will be published in 1989 by Cornell University Press.

Lewis Pyenson is Chair Professor in the Department of History at the University of Montreal. He is writing volume 3 of his *Science and Imperialism*.

Nathan Reingold is Senior Historian in the Department of the History of Science and Technology at the Smithsonian Institution's National Museum of American History.

Darwin Stapleton is the Director of the Rockefeller Archive Center in North Tarrytown, New York. He is writing a history of University Circle in Cleveland.

Roger Stuewer is a Professor in the School of Physics and Astronomy at the University of Minnesota in Minneapolis.

Loyd Swenson is Professor of History at the University of Houston in Houston, Texas.

Deborah Jean Warner is Curator of Physical Science in the Department of the History of Science and Technology at the Smithsonian Institution's National Museum of American History.

AIP Conference Proceedings

		L.C. Number	ISBN
No. 1	Feedback and Dynamic Control of Plasmas – 1970	70-141596	0-88318-100-2
No. 2	Particles and Fields – 1971 (Rochester)	71-184662	0-88318-101-0
No. 3	Thermal Expansion – 1971 (Corning)	72-76970	0-88318-102-9
No. 4	Superconductivity in d- and f-Band Metals (Rochester, 1971)	74-18879	0-88318-103-7
No. 5	Magnetism and Magnetic Materials – 1971 (2 parts) (Chicago)	59-2468	0-88318-104-5
No. 6	Particle Physics (Irvine, 1971)	72-81239	0-88318-105-3
No. 7	Exploring the History of Nuclear Physics – 1972	72-81883	0-88318-106-1
No. 8	Experimental Meson Spectroscopy –1972	72-88226	0-88318-107-X
No. 9	Cyclotrons – 1972 (Vancouver)	72-92798	0-88318-108-8
No. 10	Magnetism and Magnetic Materials – 1972	72-623469	0-88318-109-6
No. 11	Transport Phenomena – 1973 (Brown University Conference)	73-80682	0-88318-110-X
No. 12	Experiments on High Energy Particle Collisions – 1973 (Vanderbilt Conference)	73-81705	0-88318-111–8
No. 13	π-π Scattering – 1973 (Tallahassee Conference)	73-81704	0-88318-112-6
No. 14	Particles and Fields – 1973 (APS/DPF Berkeley)	73-91923	0-88318-113-4
No. 15	High Energy Collisions – 1973 (Stony Brook)	73-92324	0-88318-114-2
No. 16	Causality and Physical Theories (Wayne State University, 1973)	73-93420	0-88318-115-0
No. 17	Thermal Expansion – 1973 (Lake of the Ozarks)	73-94415	0-88318-116-9
No. 18	Magnetism and Magnetic Materials – 1973 (2 parts) (Boston)	59-2468	0-88318-117-7
No. 19	Physics and the Energy Problem – 1974 (APS Chicago)	73-94416	0-88318-118-5
No. 20	Tetrahedrally Bonded Amorphous Semiconductors (Yorktown Heights, 1974)	74-80145	0-88318-119-3
No. 21	Experimental Meson Spectroscopy – 1974 (Boston)	74-82628	0-88318-120-7
No. 22	Neutrinos – 1974 (Philadelphia)	74-82413	0-88318-121-5
No. 23	Particles and Fields – 1974 (APS/DPF Williamsburg)	74-27575	0-88318-122-3
No. 24	Magnetism and Magnetic Materials – 1974 (20th Annual Conference, San Francisco)	75-2647	0-88318-123-1
No. 25	Efficient Use of Energy (The APS Studies on the Technical Aspects of the More Efficient Use of Energy)	75-18227	0-88318-124-X